T0245206

Utilisation of grazed grass in temperate animal systems

Utilisation of grazed grass in temperate animal systems

Proceedings of a satellite workshop of the XXth International Grassland Congress, July 2005, Cork, Ireland

edited by:
J.J. Murphy

Subject headings:
Pasture systems
Animal production
Decision support systems for grazing

ISBN 9076998760

First published, 2005

Wageningen Academic Publishers
The Netherlands, 2005

The Cork Satellite Workshop of the XX International Grassland Congress held in association with The European Grassland Federation took place in Cork from 3 July to 6 July 2005.

Local Organising Committee

Mr. D. Cliffe, Chairperson
Dr. F. Buckley
Dr. P. Dillon
Dr. C. Ferris
Dr. P French
Dr. B. Horan
Dr. J. Humphreys
Dr. J. J. Murphy
Dr. B. O' Brien
Dr. E. O' Callaghan
Dr. M. A. O' Donovan
Dr. G. Stakelum

Reviewers

Dr. L. Delaby
Dr. F. Buckley
Dr. P. Dillon
Dr. C. P. Ferris
Dr. P. French
Dr. J. Humphreys
Dr. J. F. Mee
Dr. J. J. Murphy
Dr. M. A. O' Donovan
Dr. J. L. Peyraud

Chairpersons

Mr. Jim Flanagan	Pasture based animal production in Ireland
Dr. Trevor Gilliland	Appropriate plants for grazing
Dr. Roel Veerkamp	Appropriate animals for grazing
Dr. Myles Rath	Developments in modelling of herbage production and intake by grazing ruminants
Dr. Conrad Ferris	Decision support systems for grazing
Dr. David Leaver	Constraints and opportunities for animal production from temperate pastures
Dr. Sinclair Mayne	Open debate: what should grassland/grazing research focus on now?

Sponsors Cork Satellite Workshop of the XX International Grassland Congress

Allied Irish banks
Dairygold Co-Op
Dairymaster Ltd.
FBD Trust
Glanbia
Gold Crop
Kerry-Agri

Foreword

This book contains a compilation of invited and offered papers presented at the Cork satellite Workshop of the XX International Grassland Congress held association with The European Grassland Federation in Morans Silver Springs Hotel, Cork from 3 July to 6 July 2005.

The Satellite Workshop brought together grassland and ruminant production scientists, mathematical modellers working on grazing systems, extension workers, students of agriculture and animal production and progressive livestock farmers.

The Satellite Workshop was divided into a number of thematic areas, appropriate plants for grazing, appropriate animals for grazing, developments in modelling of herbage production and intake by grazing animals, decision support systems for grazing, constraints and opportunities for animal production from temperate pastures and what should grassland/grazing research focus on now. The first paper is designed to provide an overview of Agriculture in Ireland especially for visitors from other countries.

Over 120 delegates from 19 countries attended the Satellite Workshop. The 9 invited and over 90 offered papers in this book came through a thorough review and editorial process. They cover areas such as the nutritive value of pasture, plant characteristics conducive to high animal intake and performance, modelling of grass growth intake and animal production in grazing systems, optimising financial returns from grazing, decision support systems, optimal animal breeds and traits for grazing systems and the challenges and opportunities for animal production in the immediate future. One paper will present a comprehensive overview of animal production from pasture in Ireland. Four Discussion Openers on the topic of grassland/grazing research for the future have also synthesised their thoughts into one-page contributions at the end of this book.

John Murphy

Table of contents

Foreword 7

Opening of conference address 15
J. Flanagan

Keynote presentations 17

Overview of animal production from pastures in Ireland 19
M.J. Drennan, A.F. Carson and S. Crosse

Plant and sward characteristics to achieve high intake in ruminants 37
W.J. Wales, C.R. Stockdale and P.T. Doyle

New insights into the nutritional value of grass 49
D.H. Rearte

Genetic characteristics required in dairy and beef cattle for temperate grazing systems 61
F. Buckley, C. Holmes and M.G. Keane

Grass growth modelling: to increase understanding and aid decision making on-farm 79
P.D. Barrett and A.S. Laidlaw

Modelling of herbage intake and milk production by grazing dairy cows 89
R. Delagarde and M. O'Donovan

Decision support for temperate grasslands: challenges and pitfalls 105
J.R. Donnelly, L. Salmon, R.D.H. Cohen, ZL. Liu and XP. Xin

Challenges and opportunities for animal production from temperate pastures 119
D.A. Clark

Optimising financial return from grazing in temperate pastures 131
P. Dillon, J.R. Roche, L. Shalloo and B. Horan

Section 1: Appropriate plants for grazing 149

Intake characteristics of diploid and tetraploid perennial ryegrass varieties when grazed by Simmental x Holstein yearling heifers under rotational stocking management 151
R.J. Orr, J.E. Cook, K.L. Young, R.A. Champion and A.J. Rook

The effect of early and delayed spring grazing on the milk production, grazing management and grass intake of dairy cows 152
E. Kennedy, M. O'Donovan, J.P. Murphy, L. Delaby and F.P. O'Mara

Performance of meat goats grazing winter annual grasses in the Piedmont of the southeastern USA 153
J-M. Luginbuhl and J.P. Mueller

The importance of patch size in estimating steady-state bite rate in grazing cattle 154
E.D. Ungar, N. Ravid, T. Zada, E. Ben-Moshe, R. Yonatan, S. Brenner, H. Baram and A. Genizi

Spring calving suckler beef systems: influence of grassland management system on herbage availability, utilisation, quality and cow and calf performance to weaning 155
M.J. Drennan, M. McGee, S. Kyne and B. O'Neill

Production and plant density of Sulla grazed by sheep at three growth stages 156
H. Krishna and P.D. Kemp

Management of pasture quality for sheep on New Zealand hill country 157
D.I. Gray, J.I. Reid, P.D. Kemp, I.M. Brookes, D. Horne, P.R. Kenyon, C. Matthew, S.T. Morris and I. Valentine

Perennial ryegrass variety differences in nutritive value characteristics 158
T.J. Gilliland, R.E. Agnew, A.M. Fearon and F.E.A. Wilson

A survey of European regional adaptation in Italian ryegrass varieties 159
T.J. Gilliland and A.J.P. van Wijk

Effect of perennial ryegrass cultivars on the fatty acid composition in milk of stall-fed cows 160
A. Elgersma, H.J. Smit, G. Ellen and S. Tamminga

Survey of tetraploid and diploid perennial pastures in the Waikato for number of spores produced by the fungus *Pithomyces chartarum* 161
J.P.J. Eerens, W.W. Nichol, J. Waller, J.M. Mellsop, M.R. Trolove and M.G. Norriss *161*

Diverse forage mixtures effect on herbage yield, sward composition, and dairy cattle performance 162
M.A. Sanderson, K. Soder, N. Brzezinski, S. Goslee, H. Skinner, M. Wachendorf, F. Taube and L. Muller

Potential yield of cocksfoot (*Dactylis glomerata*) monocultures in response to irrigation and nitrogen 163
A. Mills, D.J. Moot, R.L. Lucas, P.D. Jamieson and B.A. McKenzie

Intake and milk production of lactating dairy cows grazing diverse forage mixtures over two grazing seasons 164
K.J. Soder, M.A. Sanderson, J.L. Stack, L.D. Muller

A comparison of perennial ryegrass cultivars differing in heading date and grass ploidy for grazing dairy cows at two different stocking rates 165
M. O' Donovan, G. Hurley, L. Delaby and G. Stakelum

In situ rumen degradability of perennial ryegrass cultivars differing in ploidy and heading date in Ireland 166
V. Olsson, J.J. Murphy, F.P. O'Mara, M. O'Donovan and F.J. Mulligan

Caucasian clover is more productive than white clover in temperate pastures 167
A.D. Black, R.J. Lucas and D.J. Moot

Grazing behaviour of beef steers grazing Kentucky 31 endophyte infected tall fescue, Q4508-AR542 novel endophyte tall fescue, and Lakota prairie grass 168
H.T. Boland, G. Scaglia, J.P. Fontenot, A.O. Abaye and R. Smith

Yield components in a Signal grass-Clitoria mixture grazed at different herbage allowance 169
R. Jiménez-Guillen, S. Rojas-Hernández, J. Olivares-Pérez, A. Martínez-Hernández and J. Pérez-Pérez

Response of warm-season grass pasture to grazing period and recovery period lengths 170
B.E. Anderson and W.H. Schacht

Renovation-year forage quality of grass pastures sod-drilled with Kura clover 171
P.R. Peterson, P. Seguin, G. Laberge and C.C. Sheaffer

Forage yield and quality of Signal grass-Clitoria mixture grazed at different frequencies 172
R. Jiménez-Guillen, J. Olivares-Pérez, S. Rojas-Hernández and A. Martínez-Hernández

Section 2: Appropriate animals for grazing **173**

Farm performance from Holstein-Friesian cows of three genetic strains on grazed pasture 175
K.A. Macdonald, B.S. Thorrold, C.B. Glassey, J.A.S. Lancaster, G.A. Verkerk, J.E. Pryce and C.W. Holmes

Genetic alternatives for dairy producers who practise grazing 176
H.D. Norman, J.R. Wright and R.L. Powell

Suitability of small and large size dairy cows in a pasture-based production system 177
M. Steiger Burgos, R. Petermann, P. Hofstetter, P. Thomet, S. Kohler, A. Munger, J.W. Blum and P. Kunz

The effect of stocking rate and lamb grazing system on sward performance of *Trifolium repens* and *Lotus corniculatus* in Uruguay 178
F. Montossi, R. San Julián, M. Nolla, M. Camesasca and F. Preve

Effect of stocking rate and grazing system on fine and superfine Merino wool production and quality on native swards of Uruguay 179
I. De Barbieri, F. Montossi, E.J. Berretta, A. Dighiero and A. Mederos

Cattle and sheep mixed grazing 1: species equivalence 180
R.D. Améndola-Massiotti, S.J.C. González-Montagna and P.A. Martínez-Hernández

Cattle and sheep mixed grazing: 2: competition 181
S.J.C. González-Montagna, P.A. Martínez-Hernández and R.D. Améndola-Massiotti

Effect of strain of Holstein-Friesian cow and feed system on reproductive performance in seasonal-calving milk production systems over four years 182
B. Horan, J.F. Mee, M. Rath, P. O'Connor and P. Dillon

Pasture intake and milksolids production of different strains of Holstein-Friesian dairy cows 183
J.L. Rossi, K.A. Macdonald, B.S. Thorrold and C.W. Holmes

Does the feeding behaviour of dairy cows differ when fed ryegrass indoors vs. grazing? 184
A.V. Chaves, A. Boudon, J.L. Peyraud and R. Delagarde

Variation between individuals in voluntary intake and herbage intake of grazing dairy cows 185
H.M.N. Ribeiro Filho, R. Delagarde, L. Delaby and J.L. Peyraud

Relationships between traits other than production and longevity in New Zealand dairy cows 186
D.P. Berry, B.L. Harris, A.M. Winkelman and W. Montgomerie

Section 3: Developments in modelling of herbage production and intake by grazing ruminants 187

Modelling the effect of breakeven date in spring rotation planner on production and profit of a pasture-based dairy system 189
P.C. Beukes, B.S. Thorrold, M.E. Wastney, C.C. Palliser, G. Levy and X. Chardon

Development of a model simulating the impact of management strategies on production from beef cattle farming systems based on permanent pasture 190
M. Jouven and R. Baumont

Intake by lactating goats browsing on Mediterranean shrubland 191
M. Decandia, G. Pinna, A. Cabiddu and G. Molle

The impact of concentrate price on the utilization of grazed and conserved grass 192
P. Crosson, P. O'Kiely, F.P. O'Mara, M.J. Drennan and M. Wallace

Adapting the CROPGRO model to predict growth and perennial nature of bahiagrass 193
S.J. Rymph, K.J. Boote and J.W. Jones

Modelling urine nitrogen production and leaching losses for pasture-based dairying systems 194
I.M. Brookes and D.J. Horne

A model to evaluate buying and selling policies for growing lambs on pasture 195
P.C.H. Morel, B. Wildbore, I.M. Brookes, P.R. Kenyon, R.W. Purchas and S. Ramaswami

Sensitivity analysis of a growth simulation for finishing lambs 196
P.C.H. Morel, B. Wildbore, I.M. Brookes, P.R. Kenyon, R.W. Purchas and S. Ramaswami

Modelling winter grass growth and senescence 197
D. Hennessy, S. Laidlaw, M. O'Donovan and P. French

The meal criterion estimated in grazing dairy cattle: evaluation of different methods 198
P.A. Abrahamse, D. Reynaud, J. Dijkstra and S. Tamminga

Effect of nitrogen on the radiation use efficiency for modelling grass growth 199
R. Lambert and A. Peeters

Radiation use efficiency of ryegrass: determination with non cumulative data 200
R. Lambert and A. Peeters

Modelling the digestibility decrease of three grass species during spring growth according to the age of the grass, the thermal age and the yield 201
M.E. Salamanca, R. Lambert, M. Gomez and A. Peeters

Visual Modelling of Alfalfa Growth and Persistence under Grazing 202
S.R. Smith, Jr., L. Muendermann and A. Singh

A new agro-meteorological simulation model for predicting daily grass growth rates across Ireland 203
R.P.O. Schulte

Section 4: Decision support systems for grazing 205

Pâtur'IN: a user-friendly software tool to assist dairy cow grazing management 207
L. Delaby, J.L. Peyraud and P. Faverdin

A farmer-based decision support system for managing pasture quality on hill country 208
I.M. Brookes and D.I. Gray

Understanding Livestock Grazing Impacts: a decision support tool to develop goal-oriented grazing management strategies 209
S.J. Barry, K. Guenther, G. Hayes, R. Larson, G. Nader and M. Doran

Enhancing grasslands education with decision support tools 210
H.G. Daily, J.M Scott and J.M. Reid

Simulation of pasture phase options for mixed livestock and cropping enterprises 211
L. Salmon, A.D. Moore and J.F. Angus

A farmer friendly feed budget calculator for grazing management decisions in winter and spring 212
M. Curnow and M. Hyder

GrassCheck: monitoring and predicting grass production in Northern Ireland 213
P.D. Barrett and A.S. Laidlaw

Grass growth profiles in Brittany 214
P. Defrance, J.M. Seuret and L. Delaby

Effect of strategy of forage supplementation and of turnout date in a medium stocking rate system on the main characteristics of dairy cows grazing 215
P. Defrance, L. Delaby, J.M. Seuret and M. O'Donovan

Using the GrassGro decision support tool to evaluate the response in grazing systems to pasture legume or a grass cultivar with improved nutritive value 216
H. Dove and J.R. Donnelly

Pasture land management system decision support software 217
G.E. Groover, S.R. Smith, N.D. Stone, J.J Venuto, Jr and J.M. Galbraith

Forecast of herbage production under continuous grazing 218
K. Søegaard, J. Berntsen, K.A. Nielsen and I. Thysen

A decision support tool for seed mixture calculations 219
B.P. Berg and G. Hutton

Section 5: Constraints and opportunities for animal production from temperate pastures 221

Environmental clustering of New Zealand dairy herds 223
J.R. Bryant, N. López-Villalobos, J.E. Pryce and C.W. Holmes

Risk-efficiency assessment of haying 224
A.J. Romera, J. Hodgson, S.T. Morris, S.J.R. Woodward and W.D. Stirling

Irish dairy farming: effects of introducing a Maize component on grassland management over the next 50 years 225
A.J. Brereton and N.M. Holden

Lucerne crown and taproot biomass affected early-spring canopy expansion 226
E.I. Teixeira, D.J. Moot, A.L. Fletcher

Autumn root reserves of lucerne affected shoot yields during the following spring 227
D.J. Moot, E.I. Teixeira

Milk production performance based on grazed grassland in Switzerland 228
P. Thomet and H. Menzi

Extending the grazing season with turnips 229
P. Thomet and S. Kohler

The effect of inclusion of a range of supplementary feeds on herbage intake, total dry matter intake and substitution rate in grazing dairy cows 230
S.J. Morrison, D.C. Patterson, S. Dawson and C.P. Ferris

White clover soil fatigue: an establishment problem on large and intensive dairy farms 231
K. Søegaard and K. Møller

Effect of farm grass cover at turnout on the grazing management of spring calving dairy cows 232
M. O'Donovan, L. Delaby and P. Defrance

What supplementation type for spring calving dairy cows at grass in autumn? 233
M. O'Donovan, E. Kennedy, T. Guinee and J.J. Murphy

Manipulation of grass growth through strategic distribution of nitrogen fertilisation 234
M. Stettler and P. Thomet

The effect of closing date and type of utilisation in autumn on grass yield in spring 235
E. Mosimann, M. Lobsiger, C. Hofer, B. Jeangros and A. Lüscher

A comparison of three systems of milk production with different land use strategies 236
L. Shalloo, P. Dillon and J.J. Murphy

Project Opti-Milk: optimisation and comparison of high yield and low input milk production strategies on pilot farms in the lowlands of Switzerland 237
H. Menzi, T. Blaettler, P. Thomet, B. Durgiai, S. Kohler, R. Staehli, R. Mueller and P. Kunz

Assessment of grass production and efficiency of utilisation on three Northern Ireland dairy farms 238
A.J. Dale, P.D. Barrett and C.S. Mayne

The effect of grassland management on bovine nitrogen efficiency 239
N.J. Hoekstra, R.P.O. Schulte, E.A. Lantinga and P.C. Struik

The effect of stocking rate and initial grass height on herbage production and utilization, and milk production per unit area under set stocking by lactating dairy cows 240
H. Nakatsuji, T. Endo, S. Bawm, T. Mitani, M. Takahashi, K. Ueda and S. Kondo

Section 6: Other aspects of grassland/grazing 241

Better dairy farm management increases the economic return from phosphorus 243
J.D. Morton

The effect of two magnesium fertilisers, kieserite and MgO, on herbage Mg content 244
M.B.O'Connor, A.H.C.Roberts and R.Haerdter

Effect of different phosphorous sources and levels on the productive behaviour of a *Lotus pedunculatus* cv. Grasslands Maku oversown pasture 245
R.E. Bermúdez and W. Ayala

Field testing of a turnip growing protocol on New Zealand dairy farms 246
J.P.J. Eerens and P.M.S. Lane

Labour input associated with grassland management on Irish dairy farms 247
B. O'Brien, K. O'Donovan, J. Kinsella, D. Ruane and D. Gleeson

Factors affecting Italian ryegrass (*Lolium multiflorum* L.) seed distribution 248
R.D. Williams and P.W. Bartholomew

Practical application of a one-parameter approach to assess the accuracy of two different estimates of diet composition in sheep 249
C. Elwert and M. Rodehutscord

Use of alkanes to estimate dry matter intake of beef steers grazing high quality pastures 250
G. Scaglia, H.T. Boland, I. Lopez-Guerrero, R.K. Shanklin and J.P. Fontenot

Effects of rumen fill on intake and milk production in dairy cows fed perennial ryegrass 251
A.V. Chaves and A. Boudon

Both grass development stage and grazing management influence milk terpene content 252
G. Tornambé, A. Cornu, N. Kondjoyan, P. Pradel, M. Petit and B. Martin

Supplementation under intensive grazing, silage- or grain-based diets for beef production on steer performance and meat fatty acid composition 253
J. Martínez Ferrer, E. Ustarroz, C.G. Ferrayoli, A.R. Castillo and D. Alomar

Section 7: Open debate - What should grassland/grazing research focus on now? 255

Research into the types of cows and systems required to utilise grazed pastures sustainably in 100 years from now 257
C. Holmes

Farming for fun & profit 259
M. Murphy

Grassland research: goals for the future 260
J.R. Roche

What research is required for economically and environmentally sustainable farming? 261
W. Taylor

Keyword index 263

Author index 267

Opening of conference address

J. Flanagan
Director of Teagasc

Summary

I am very pleased to welcome delegates from many places in the world to the Cork Satellite Workshop that is being held as part of the XX International Grassland Congress. Ireland and the United Kingdom are honoured to host this Congress and its five associated satellite meetings. This Cork Satellite is being held in association with the European Grassland Federation and I wish to thank the Federation for its help and support.

The theme of the meeting is "The utilisation of grazed grass in temperate animal systems". Grass is the most important crop produced in Ireland. It is the base of our animal production which dominates farming. Of the total area of the country of about 7 million ha, about 4.4 million ha is productive in farming. Less than 400,000 ha is cultivated for crops. The rest is grassland which is the main feed for our 7 million cattle and 7 million sheep.

The agri-food sector continues to be a very important part of the Irish economy accounting for 8.8% of GDP, 9.1% of employment and 8.4% of exports. Given the relatively small import content in agri-food exports, the sector accounts for about 20% of net exports. Exports of meat and dairy products were valued at €3,760 million in 2003.

Thus, grass is truly important to farming in Ireland and not just to farming but also to our national wellbeing. Our temperate climate and long growing season makes grass an ideal crop which can be grazed for most months of the year and gives Ireland a competitive advantage in cost of production of milk, beef and sheepmeat. Grassland research is an important part of our agricultural research programme and you will get a good appreciation of this work during this meeting. We very much look forward to all the presentations. There is a very good spread of presentations covering the most important temperate grassland areas of the world.

Finally, I very much appreciate that the meeting includes a visit to the Teagasc Research Centre at Moorepark where you will see grassland research at first hand. We very much look forward to your visit.

Keynote presentations

Overview of animal production from pastures in Ireland

M.J. Drennan[1], A.F. Carson[2] and S. Crosse[3]
[1]*Teagasc, Grange Research Centre, Dunsany, Co. Meath, Ireland*
Email: mdrennan@grange.teagasc.ie
[2]*Agricultural Research Institute of Northern Ireland Hillsborough, Co. Down BT26 6DR*
[3]*Teagasc, Oak Park, Co. Carlow*

Key points

The importance of grassland to agriculture in Ireland is indicated by the fact that:

1. Sixty percent of agricultural output is from grassland as cattle, milk and sheep products.
2. Over 90% of the total farmed area is in grass.
3. Livestock are almost entirely dependent on grazed grass for 200 to 235 days of the year.
4. Grass conserved as silage is the main source of fodder in winter.
5. To improve competitiveness changes are continuously taking place, which include:
   - increased suckler herd size and a movement to late maturing continental cattle breeds;
   - movement in the dairy herd towards Holstein with increased production per animal;
   - increased importance, post CAP reform, of technical efficiency to maintain competitiveness in a more market-orientated era, and
   - greater influence on future livestock systems of agri-environmental support schemes and environmental legislation.

Keywords: grassland, animal production, systems

Introduction

Animal production from pasture accounted for over 60% of the total Agricultural output on the island of Ireland in 2003 (€4,975 million) (Table 1). Beef cattle, milk and dairy products and sheep accounted for 34%, 25% and 5% of output, respectively. There are over 130,000 farmers in the Republic of Ireland (ROI) and 34,000 in Northern Ireland (NI). Of this total in the ROI, 27% are involved in dairying, 51% mainly in beef, 17% in sheep and less than 6% mainly in tillage (Connolly *et al.*, 2004) (Table 2). Average family farm income varies from €7,337 for those mainly involved in cattle rearing to €30,138 for those exclusively in dairying. Corresponding Utilisable Agricultural Areas (UAA) for these two groups are 26.4 and 40.1 ha. The number of livestock units (LU) per holding varies from 27.7 on cattle rearing farms to 86.4 on farms with dairying and other enterprises. The overall stocking density was shown to be 1.53 livestock units per ha devoted to livestock but varies considerably between the different enterprises. Figures for NI show a similar trend.

Land use

Ireland has a total land area of 8.24 million ha with 6.89 million ha in ROI and 1.35 million ha in NI (Table 3). Over 90% of the land area is grassland in both ROI (91%) and NI (96%), while the figure for Great Britain (GB) is 71%.

Table 1 Output value (million €) in agriculture, 2003

	Gross output at producer prices		Direct* payments		Total		% of total	
	ROI	NI	ROI	NI	ROI	NI	ROI	NI
Milk	1,445	494	-	-	1,445	494	25	24
Cattle and Calves	1,23	291	850	276	2,079	567	36	27
Sheep and Lambs	193	58	109	29	303	87	5	4
Pigs and Poultry	434	283	-		434	283	8	14
Other	1,429	604	134	13	1,563	617	27	30
Total	4,731	1,229	1,093	245	5,824	2,048	100	100

*Various EU premiums; Source: CSO (2004); DARD (2004)

Table 2 National farm survey data by farming system - all farms, 2003

Farm system	Dairying	Dairying and other	Cattle rearing	Cattle other	Mainly sheep	Mainly tillage	All systems
% of Population	16.2	10.7	27.5	23.3	16.8	5.7	100.0
Mean FFI (€)	30,138	24,656	7,337	8,106	13		15,054
UAA (ha)	40.1	53.0	26.4	29.4	37.3	63.7	36.1
Total LU	74.3	86.4	27.7	40.3	48.4	40.8	48.6
LU/UAA	1.91	1.85	1.06	1.41	1.32	1.46	1.53

FFI = Family Farm Income; UAA = Utilisable Agricultural Area; LU = Livestock Units

Table 3 Land use in Ireland and GB

	Ireland		GB
	ROI	NI	
Total land area (m.ha)	6.89	1.35	NA
Forestry (m.ha)	0.60	0.08	NA
Total farmed (m.ha)	4.42	1.07	16.5
Total crops (%)	9	5	30
Grassland (%)	81	80	36
Rough grazing (%)	10	15	35

Source: DAF (2004), DARD (2003), Hopkins (2000)

Farm size

The average UAA per holding in the ROI and NI is 31.4 and 38.0 ha, respectively (Table 4). The average UAA for the EU15 is 18.7 ha per holding with the UK being at the top end of the scale (67.7 ha) and Italy at the lower end (6.1 ha). As an indicator of change, average farm size in ROI has increased gradually from 22.3 to 32.0 ha/holding between 1975 and 2002.

Table 4 Utilised agricultural area (UAA) per holding

	UAA (m.ha) 2002	No. of holding ('000) 2000	UAA per holding 2000
EU15	130.1	6,771	18.7
France	29.6	664	42.0
Ireland: (ROI)	4.4	142	31.4
(NI)	1.1	28	38.0
Italy	15.3	2,154	6.1
Netherlands	1.9	102	20.0
UK	15.7	233	67.7

Source: Eurostat (2002)

Climate

Rainfall, temperature and radiation are the most important climatic components affecting grass production. Ireland is suited to grassland farming. Located between 51°N and 55°N latitude it has a temperate, humid climate, influenced by the prevailing westerly winds and the proximity of the ocean and the gulf stream. Annual average rainfall in lowland areas (elevation less than 100 metres) varies from about 750 mm in parts of the east and northeast to greater than 1,200 mm in the west, northwest and south-west. While there are no well defined dry and wet seasons, the year may be divided into a relatively dry half, February to July, and a relatively wet half, August to January. There is considerable year-to-year variability in total annual rainfall.

The mean annual temperature over Ireland has a distinct north-northeast to south-southwest gradient. For example, at Hillsborough in the north-east, the mean annual temperature is 8.6°C, whilst in the south, at Moorepark, the mean temperature stands at 9.8°C (Table 5). Monthly mean temperature decreases by approximately 1°C for each 150 metres increase in altitude. Grass growth is considered to be continuous at temperatures over about 6°C in Irish conditions. Year-to-year fluctuations are comparatively small.

Table 5 Monthly mean temperature (°C) and rainfall (mm) at [1]Hillsborough (north-east) and [2]Moorepark (south) averaged over a 30 (1961-90) or 20 (1982-2001) year period

	J	F	M	A	M	J	J	A	S	O	N	D	Year
Temperature													
N	4.0	3.9	5.3	7.1	9.7	12.6	14.2	14.0	12.2	9.7	6.0	4.8	8.6
S	5.2	5.6	7.1	8.2	11.0	13.6	15.7	15.2	12.9	10.2	7.3	6.0	9.8
Rainfall													
N	87	60	70	57	62	64	57	83	85	94	82	84	885
S	109	92	81	66	61	68	54	92	78	114	101	109	1025

N = north-east; S = South

Source: [1]Cruickshank (1997); [2]Shalloo *et al.* (2004)

Grassland

Grassland in Ireland is composed predominantly of long-term permanent pastures with only about 3 percent reseeded yearly. With such a high proportion of land in grass, it is not surprising that cattle and sheep largely rely on grazed and conserved grass as sources of feed. A typical grass growth curve for Ireland shows that growth commences in March, reaches a peak of about 80 kg of dry matter per ha per day in late May, with a second lower peak of about 65 kg in early August followed by a rapid decline until growth almost ceases in November (Figure 1).

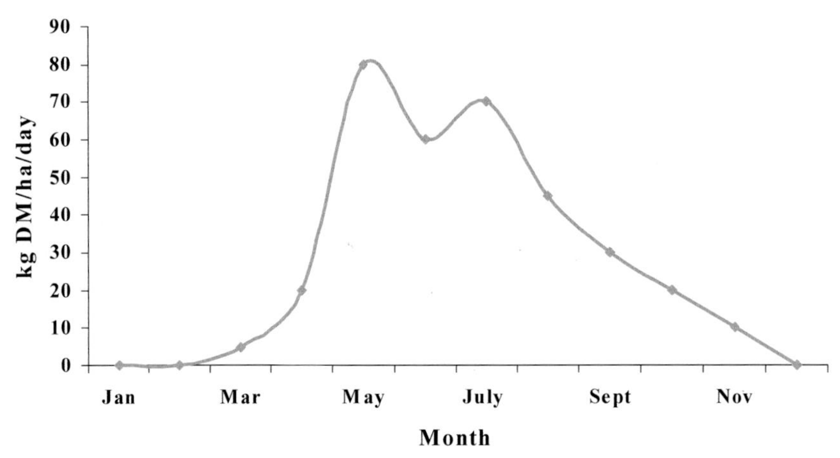

Figure 1 Typical grass growth curve for Ireland

Total annual grass dry matter production varies from about 15 t/ha in the southwest to 11 t/ ha in the northeast in an average year (Figure 2) (Brereton, 1995). The length of the grass growing season varies from about 8 months in the north-east of the island up to 11 months in the extreme south-west (Keane 1992). The estimated starting dates of the grazing season vary from March 25 in the southwest to April 20 in the northeast (Brereton, 1995) (Figure 3). Thus, the grazing season varies from about 235 days (mid March to early November) in the south and southwest to about 200 days (mid April to late October) in the midlands and north. Soil type has a major effect as poorly drained soils have a shorter grazing season due to utilisation problems and have a correspondingly longer winter feeding period. Moisture deficit is generally not a problem in relation to grass production in Ireland with only small losses in production potential which (<1.5 tonnes DM/ha) are confined to a narrow coastal strip in the east and southeast (Brereton and Keane, 1982).

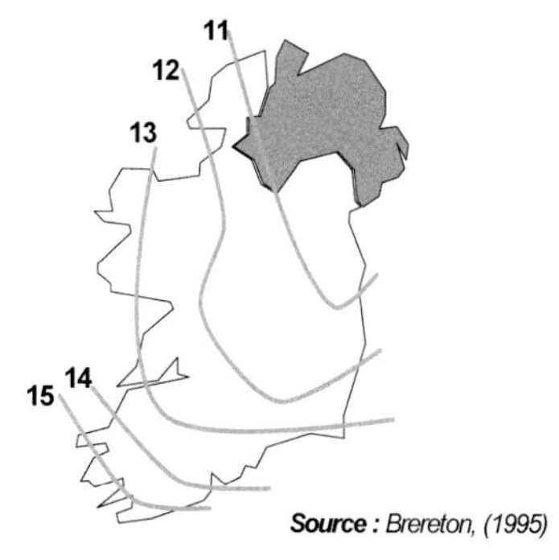

Figure 2 Model estimates of annual dry matter grass production (t ha^{-1})

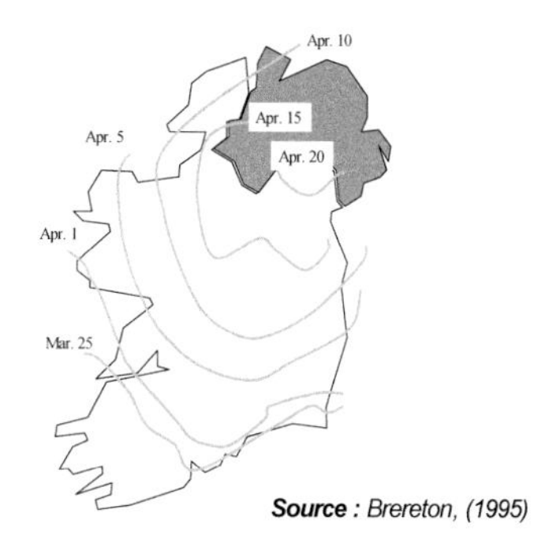

Figure 3 Estimated starting dates of the grazing season in Ireland

Provision of winter feed

Grass silage is the principal source of winter feed for livestock in Ireland. Indeed the proportion of farms that make silage continues to increase, now standing at 86% of all farms in the ROI (O'Kiely *et al.*, 2000) (Table 6). The total area harvested for grass silage in 1999 was 1.24 million ha providing 4.6 million t. of edible silage DM. First cut accounts for over 70% of the silage harvested with second harvests from the same area accounting for most of the remainder. It is estimated that baled silage accounts for 35% of the area harvested for silage. Virtually all baled silage, and almost 75% of conventional silage, is made without the application of additives. It is estimated that 0.2 million ha of grass is harvested for hay each year. Similar trends in the provision of winter feed are seen in NI with an estimated 0.32 million ha of grassland now harvested for silage yearly producing 1.2 million t of edible silage DM (DARD, 2004). Although increasing in recent years, the total quantity of maize silage harvested in the ROI amounted to only 19,600 ha in 2003 or about 3% of conserved forage DM.

Table 6 Trends in the percentage of farms that make silage within the main farming enterprises

	1991/92	1999
Dairying	91	99
Beef	52	86
Sheep	50	76
All systems	65	86

Source: O'Kiely *et al.* (2000)

Fertiliser use

Recent data from the ROI (Coulter *et al.*, 2002) show that fertiliser use is greater on grassland areas used for silage than for grazing or hay areas (Table 7). Application rates of fertiliser nitrogen (N) were shown to be 109, 133 and 53 kg/ha on grazing, silage and hay areas, respectively. Within farming systems, fertiliser use was shown to be greater for dairying than for beef cattle or sheep systems (Table 7).

Table 7 Estimated nitrogen, phosphorus and potassium fertilizer applied (kg/ha) to grassland for grazing, silage and hay and for different farming systems

	Nitrogen	Phosphorus	Potassium
Grazing	109	9	21
Silage	133	15	49
Hay	53	11	27
Dairying	176	12	26
Cattle	48	8	17
Sheep	48	6	13

Source: Coulter *et al.* (2002)

National cow herd

There have been substantial changes in the composition of the cow herd over the last twenty years. In the early 1980's total cow numbers in the ROI were just over 2 million, of which 80% were dairy cows and 20% suckler cows (Figure 4, CSO publications). The introduction by the European Union (EU) of milk quotas in 1984 and increased milk production per cow has resulted in a gradual decline in dairy cow numbers from 1.65 million in 1984 to 1.16 million in 2004. The corresponding change from 1984 to 2004 in suckler cow numbers was from 0.44 to 1.21 million. The average number of cows in dairy and suckler herds is 37 and 15 respectively (CSO, 2001). In NI, significant quantities of milk quota have been imported from GB, leading to an expansion in overall milk output (37% increase in milk quota held by NI producers over the last 10-years). Whilst overall numbers of dairy cows in NI have remained relatively unchanged (0.28 million), average milk yield has increased from 4,639 to 6,290 litres per cow over the period from 1984 to 2003, with average herd size now standing at 61 (DARD, 2004). The milk yield increase in the ROI was from 5,080 l. in 1985 to 6,166 l. in 2003 (ICBF 2003).

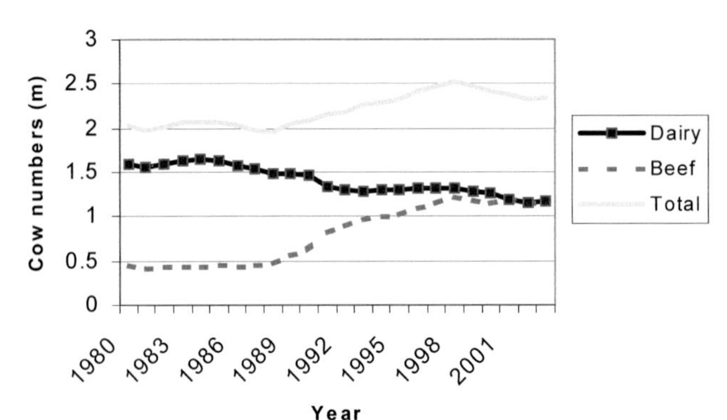

Figure 4 Trend in cow numbers (million) in the ROI

Breed composition of the cattle herd

The dairy cow herd in the ROI is predominantly Holstein-Friesian (98%) and has changed little over recent years (Drennan, 1999a). Fifty percent of dairy cows in the ROI are bred to Holstein-Friesian sires, about 28% to late maturing beef breeds (e.g. Charolais) and 22% to early maturing breed sires (Hereford and Aberdeen Angus).

There has been major changes in the composition of the beef herd, on both the dam and sire side in both the ROI and NI. Between 1992 and 2003 the proportion of beef cows comprised of late maturing breeds increased from 40 to 71%. Increasing usage of late maturing breeds is also evident on the sire side. Approximately 85% of suckler cows are now bred to continental breed sires, of which over 40% are bred to Charolais sires (Table 8).

Table 8 Breed composition (%) of the beef cow herd in 1992 and 2003 and breed of sire used (%) on beef dams in 2003

	Beef Cow Herd				Sire Breed	
	1992		2003		2003	
	ROI	NI	ROI	NI	ROI	NI
Aberdeen Angus	9	23	12	16	6	10
Hereford	35	12	20	6	5	5
Friesian	20	11	-	1	-	-
Simmental	9	23	17	18	8	7
Limousin	8	12	19	33	25	25
Belgian Blue	-	-	-	-	7	6
Charolais	7	7	21	12	44	40
Other breeds	12	2	11	11	5	8

Source: Drennan (1999a), ICBF (2003), DARD (2004), Kirkland *et al.* (2004)

Calving pattern

In the ROI, both the beef and dairy herds are predominately spring calving, which indicates the dependence on grazed grass. In the beef herd, 16% of calvings are in January-February, 43% in March-April, 22% in May-June with only 19% in the remaining 6 months of the year (Figure 5). The corresponding percentage figures for the dairy herd are 34, 41, 12 & 13. The marginally earlier calving in the dairy than in the beef herd reflects the fact that dairying systems are mainly in the southern part of the country, which as discussed previously, has earlier grass growth than northern areas.

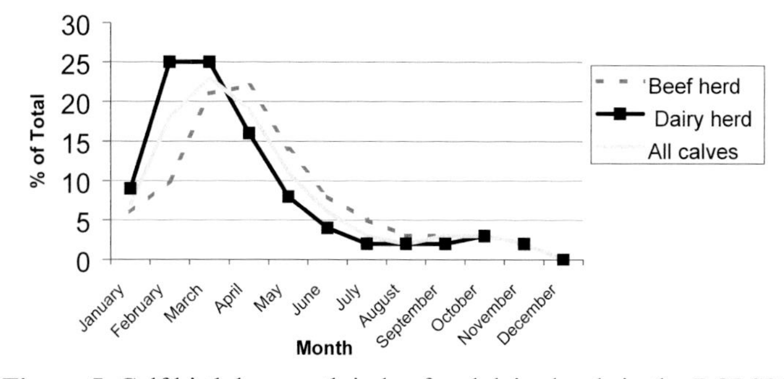

Figure 5 Calf birth by month in beef and dairy herds in the ROI 2003

In NI, the majority of dairy cows calve over the autumn-winter period with only 18% calving in the January to April period (CAFRE, 2003). This is mainly a result of the higher output systems practised in NI along with the shorter growing season and more difficult ground conditions early and late in the season.

Cattle slaughterings

For orderly marketing, an even supply of beef throughout the year is desirable. However, in contrast with most other EU countries, beef production in Ireland, because it is based on grazed grass, has tended in the past to have a pronounced seasonality in production (Figure 6). In 1990, 40% of prime cattle slaughterings in ROI were in the final quarter of the year. This has changed in recent years and the corresponding figures for 2003 were 28% for ROI and 26% for NI. Various EU schemes including the eligibility to meet premium payments have contributed to this change, which may again be altered following the decoupling of subsidy payments from production systems. Average carcass weights in 2003 for steers, young bulls, heifers and cows in ROI were 341, 327, 273 and 294 kg respectively. Corresponding percentages with carcass conformation classes of EUR were 59, 82, 67 and 12%. Similar carcass weights and carcass quality were recorded in the NI beef industry.

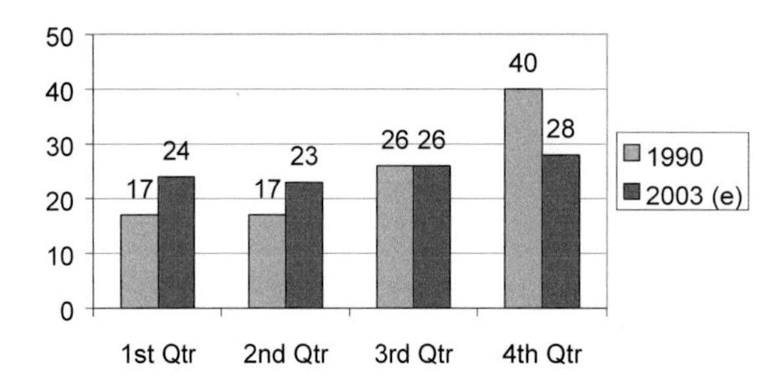

Figure 6 Seasonality of prime cattle supplies in the ROI (%)

Cattle and beef production and exports

In the ROI, total cattle disposals in 2003 were 2.08 million, of which 1.86 million were slaughtered and 0.22 million were exported live (Bord Bia 2004). Total beef availability was 583,000 t carcass weight equivalent (includes 20,000 t imported) of which 14% was used for the home market, with the remainder exported. The home market is supplied almost entirely by heifer beef. Live cattle exports have varied widely from year to year. Between 1995 and 2003 exports varied from 57,000 head in 1997 to peaks of over 400,000 head per year in 1999 and 2000 (*Appendix Table 1*). In the peak export years, three-quarters were to continental EU countries (Spain, followed by Italy and Holland being the main markets) with minimal numbers to non-EU markets. In contrast, non-EU markets accounted for 71 and 73% of total live exports in 1995 and 1996, respectively with Egypt and Libya accounting for practically all exports in these years. Between 1995 and 2003 beef carcass exports varied from 345,000 t in 2001 to 554,000 t in 1999 (*Appendix Table 2*). While non-EU markets accounted for 40 to over 60% of exports in the period up to 2000 most have been exported to EU countries in recent years. In 2003, 53, 30 and 17% of total meat exports were to the UK, continental EU and non-EU markets, respectively.

In NI, cattle disposals in 2003 were 408,000 head all of which were slaughtered (DARD 2004). BSE (Bovine Spongiform Encephalopathy) export restrictions on beef have changed markets considerably over the past decade. In 1995, 52% of prime beef production in NI was exported outside the UK, chiefly to continental Europe. In 2003, of the 0.41 million of prime beef cattle slaughtered, 80% were exported to GB with the remainder used largely for home consumption (LMC, 2004).

Beef production systems

The data outlined shows that the climate in Ireland is ideal to grow grass and thus a suitable feed source is available which is the major cost factor in animal production. Beef, dairy and sheep production systems were designed to make optimum use of grass. As indicated earlier grass growth is confined, on average, to 7 to 8 months of the year when grass can be grazed cheaply *in situ*. For the remainder of the year it is necessary to conserve the grass from the time of most rapid growth (spring) to use in the period when grass growth is negligible in winter. In the past, grass conservation was as hay, which because of our wet climate was not ideal. Conservation of grass as silage was first introduced in the 1950's and has since continued to increase. It has the advantage over hay of being less dependent on weather conditions, allowing somewhat greater scope for mechanisation and permitting harvest at an earlier stage of grass maturity thereby allowing the production of higher quality conserved feed. In general, the systems are based on spring calving/lambing thereby ensuring that animal feed requirements are lowest when feed costs are greatest. When feed requirements are greatest in lactation the animals are at pasture and the management systems are designed to provide high quality leafy pasture throughout the grazing season. Compensatory growth is also availed of in that growing cattle are fed for moderate rates of gain in winter and subsequently high growth rates are attained at pasture. The grassland management practice results in beef cows being in good body condition at the start of winter and studies have shown that these cows can tolerate substantial losses in body reserves over the winter period without ill effects on subsequent cow or calf performance.

There is practically no veal production in Ireland with young bull beef production accounting for only 6% of male slaughterings in the ROI (Bord Bia 2004) and 18% in NI (LMC, 2004). Although declining, slaughter age of steers is generally between 24 and 30 months of age, while heifers are slaughtered 4 to 6 months earlier than steers. Most animals are housed in winter. Animal housing includes slatted floor sheds, straw-bedded courts, and cubicle accommodation.

Although numerous production systems are operated at farm level, the majority in the ROI involve spring born calves. Target weights at different stages of growth for these systems, based on studies at Grange Research Centre, are presented in the following sections for calves from suckler (Drennan 1999b, 2004, Drennan and Keane 2001) and dairy (Keane 2001, Drennan and Keane, 2001) herds.

Suckling systems

In studies of beef systems at Grange Research Centre, Limousin x Friesian and Simmental x (Limousin x Friesian) cows are used. Mature cows are bred to Charolais sires of high beef merit or Simmental sires for breeding herd replacements. Heifers are managed to calve at 2 years of age and are bred to easy calving Limousin sires.

Average calving date is mid March with the cows and calves turned out to pasture in April. Calves are weaned in October-November, when all animals are housed. Weaned calves are offered grass silage plus 1 kg of concentrates daily, normally over a 5-month winter period, following which they are put to pasture for a second grazing season. Heifer progeny are slaughtered in November at 20 months of age having received 3 kg of concentrates daily with grass (or silage) for the final 3 months. Steers are housed in mid October and offered silage plus 4 to 5 kg concentrates daily until slaughter in early March at two years of age.

Both semi-intensive and more extensive systems have been examined (Table 9). In the semi-intensive system 0.81 ha of grassland is allowed per cow unit, (cow, progeny and 25% replacements) with 225 kg of fertiliser N applied per ha and two silage harvests taken yearly. Fifty-five percent of the area is harvested for silage in late May (good quality for progeny with a dry matter digestibility (DMD) of about 740 g/kg) with a further 30% harvested in July (for cows with a DMD of about 650 g/kg). The extensive system has a lower stocking rate (1.0 ha /cow unit), less than half the level of fertiliser N applied (100 kg/ha) with one silage harvest half of which is in May (high DMD for progeny) and the remainder in June (lower DMD for cows). As no second silage cut is planned in the extensive system, although some may be harvested to maintain grass quality, there is an opportunity to accumulate sufficient grass as autumn approaches to allow the heifers to be finished outdoors. In both systems, flexible paddock rotational grazing programmes are operated with the objective of providing adequate supplies of leafy pasture throughout the season. Similar animal performance levels were obtained on both systems and the mean weights achieved by steers and heifers at different stages are shown in Table 10. With the same concentrate inputs per animal similar high animal performance levels can be obtained from semi-intensive or extensive grassland management systems. Consequently, beef output per ha is greater on the semi-intensive system (510 versus 410 kg/ha) but, due to lower costs in the extensive system, margins per ha were shown to be similar for the two systems.

Table 9 Details and performance of semi-intensive and extensive sucking systems

	System	
	Semi-intensive	Extensive
Stocking rate: ha/cow unit*	0.81	1.0
Nitrogen: kg/ha	225	100
Number of silage cuts	2	1
Percent of area harvested	85	55
Silage tonnes/cow unit	14.5	13.5
Heifers finished with concentrates on	Silage	Grass
Concentrates/cow unit (kg)	700	700

*Cow plus progeny to slaughter plus 25% replacements

Table 10 Animal weights (kg) and age at slaughter (days)

	Steer	Heifer
Weaning weight	316	288
Yearling weight	404	373
Slaughter weight	700	565
Carcass weight	396	309
Age at slaughter	725	606

Dairy calf to beef systems

Calves are born in February-March and are at pasture from May to November. Calves are rotationally grazed ahead of the yearling cattle. At a fertiliser N application rate of 114 kg/ha, the grassland area required for late maturing beef breed x Friesian and Friesian steers taken to slaughter is 0.55 ha. To provide adequate winter feed (total of 1.5t of silage DM consumed per animal) 55% of the total area is harvested in late May. Total lifetime concentrate inputs are 1000 kg (130 kg in calf stage and end of first grazing season, 120 kg during the first winter and 750 kg (5 kg/day) during finishing). Because of their earlier slaughter, the concentrate requirements of early maturing breed crosses are about 300 kg less than for Friesians and continental crosses. Lifetime live weight targets for both Holstein-Friesian and continental x Holstein-Friesian steers slaughtered at 2 years of age are shown in Table 11. The weights for Friesians also apply to early maturing breed crosses but the latter would have a shorter finishing winter and a lighter slaughter weight (carcass weight 295 kg).

Table 11 Target weights and gains for Holstein/Friesian (FR) and Continental x Friesian (CT) steers slaughtered at 2 years of age

Date	System event	No. days	Weight (kg)		Age (weeks)
			FR	CT	
Mid March	Purchase		45	50	
Mid May	To pasture	58	80	90	8
Mid November	To house	189	230	240	35
Late March	To pasture	126	300	320	53
Mid October	To house	210	490	510	83
Mid March	Slaughter	147	620	650	104
Overall		730	620	650	
Kill-out (g/kg)			520	538	
Carcass wt (kg)			320	350	

Dairying

In 2003 the volume of milk sold off farms in Ireland totalled 6,972 million litres (Table 12). Close to 90% of this milk output was manufactured into butter, cheese, cream, and whole milk powder, with 10% produced for the liquid milk market. In the ROI, 57% of the milk used in manufacture was for butter and 20% for cheese. In NI, the main milk products are milk powder and cheese, using an estimated 20 and 50% of milk produced, respectively.

Table 12 Milk output and disposal[1], 2003 (m.l whole milk only unless otherwise stated)

Manner of disposal	ROI	NI
Milk sold off farms	5,200	1772
Milk used in farm households[2]	45	
Imported milk Intake	349	9
Total milk available	5,594	1782
Of which		
Used for liquid consumption	505	213
Used in the manufacture of:		
Butter	3,216	34 cream
Cheese	1,106	200 + 64 cream
Cream[3]	220	587 + 270 cream
Whole Milk Powder	247	587 +270 skim milk
Chocolate Crumb	129	
Miscellaneous Products	717	

[1]Milk output and disposal will not reconcile due to the existence of different production processes in the production of milk based products
[2]Including milk used for the production of farm butter, cream and cheese and milk given as payment in kind to agricultural employees
[3]Includes milk used for the manufacture of cream by creameries and pasteuries
Source: DAF (2004), DARD (2004)

Spring-calving systems

Milk production in the ROI is predominantly based on spring-calving systems. Thus grazed grass, makes a major contribution to the feed budget of dairy cows. The Blueprint (Dillon and Stakelum 1999) for efficient dairying based on calving at the start of grazing in spring sets a target of 6,000 litres of milk per cow with an average fat content of 3.9% and a protein content of 3.4%. This level of performance is achievable at a stocking rate of 2.5 cows/ha, a N input of 325 kg/ha and a mean calving date in mid February-early March. The inputs per cow include 500 kg of concentrates, 3.6 t (DM) of grazed grass and 1.4 t (DM) of silage. The blueprint is applicable for dry land in the south and it will change to reflect differences in soil type and location within Ireland.

The objective over the main grazing season (May to September) is to achieve high cow performance from an almost complete grass diet. This is achieved by allocating an adequate daily supply of high quality grass. The provision of adequate silage for the winter is also important over this period. The aim of autumn grazing management (September to November) is to maximise the amount of grass utilised while at the same time finish the grazing season with the desired farm grass cover so as to set up the farm for early spring grass. The timing of autumn supplementation depends on grass growing conditions, stocking rate, calving pattern and milk yield.

It is recommended that, on dry land, all of the farm should be grazed initially, starting in early-March if grass supply and weather conditions permit. This may not be possible in all years. Early grazing is facilitated by early applications of N fertiliser and the correct timing of final autumn defoliation. However, due to the low growth rate in early spring, grass supply will not be adequate to meet the dairy cow's demand when first turned out to grass. With compact spring-calving and stocking rates of 2.5 cows/ha, daily grass growth will not be adequate to meet the cows demand until mid to late April. Therefore, up to that date and depending on turnout date, supplementary concentrates and silage will be provided with grass. It is important that the first rotation should not finish before mid to late April. At a stocking rate of 2.5 cows per hectare, 45 to 50% of the total area can be closed for silage on the first week of April.

During the early part of the grazing season (late April to June), tight grazing (residual sward height of 6 cm) is critical. First cut silage is taken during mid to late May with a second silage crop (35% of farm closed) cut 7 to 8 weeks later (mid to late July). The two cuts will provide a total of 7 t of silage (20% DM) per cow. From mid to late August onwards, the total farm is available for grazing. During this period (July to September), grazing pressure may be relaxed to allow a post-grazing sward surface height of 7-8 cm in order to increase milk yield per cow without resulting in deterioration in sward quality afterwards.

Autumn calving systems

Research at the Agricultural Research Institute of Northern Ireland has examined systems to allow high nutrient intake to support milk production from high genetic merit Holstein-Friesian cows which are widespread in the NI dairy industry (Ferris *et al.*, 2004). Systems examined incorporated the following broad approaches to increasing nutrient intakes:

- improving the feed value of the silage offered or increasing concentrate feed level during the winter.
- offering a high allowance of high quality pasture without supplementation or tighter grazing regimes combined with concentrate supplementation during the summer.

Although total milk outputs were similar with each of the four systems (7,900 litres/cow), milk protein contents were higher with systems involving high concentrate inputs during the winter. System had only minor effects on the degree of tissue loss/gain during lactation and on the fertility of the cows involved. However, the land requirement associated with each of the systems was very different ranging from 2.3 to 3.3 cows/ha. Consequently gross margin per ha increased with increasing stocking rate, while gross margin per cow and gross margin per litre of milk produced were relatively unaffected by system. Thus the profitability of different grassland production systems for autumn calving cows is to a large extent influenced by the fixed costs which arise on the individual farm,

Sheep production

The ewe population is 3.9 million in ROI and 1.1 million in NI. Corresponding average flock sizes are 113 and 126 ewes (CSO, 2004; DARD, 2004). Over the past 10 years, total ewe numbers have fallen by approximately 25% in both the ROI and NI. In the ROI, most of the decline in ewe numbers has been in hill areas, whereas in NI hill ewe numbers have remained relatively unchanged with the decline being more evident in the lowland sector.

Sheep production in 2003 amounted to around 4 million lambs in the ROI and 0.8 million lambs in NI. In the ROI, 70% of lambs were exported, mainly to France (70% of exports). In NI, 39% of lambs were exported to the ROI for processing. Of the lambs slaughtered in NI the majority were exported to GB (63%), with 22% to continental Europe (mainly France) and 15% marketed for home consumption (LMC, 2004).

Hill sheep systems are predominately based on Scottish Blackface and Cheviot ewes either bred pure or crossed with prolific breeds (e.g. Belclare, Blue-Faced Liecester) to produce crossbred female replacements for the lowland sector. Typical levels of performance in hill sheep systems are presented in Table 13. In most hill sheep systems lambs are mainly sold at weaning for finishing in the lowland sector or housed and finished off on concentrate diets.

Table 13 Output from Scottish Blackface and Cheviot ewes on hill farms across Northern Ireland (Carson *et al.*, 2001)

	Scottish Blackface	Wicklow Cheviot
No. lambs born per ewe mated	1.29	1.29
No. lambs weaned per ewe mated	1.14	1.20
Lamb live weight at weaning (kg)	30.5	31.5
Age at slaughter (months)	9.7	8.5
Carcass weight (kg)	17.8	18.3

The dominant system of lowland sheep production is grass-based. The great majority of ewes lamb in spring and are managed in an integrated grazing/silage/housing system, often mix-stocked with cattle or in association with tillage enterprises.

Developments in recent years have seen the emergence of a significant core of specialist lowland sheep producers who have invested in relatively large flocks ranging from 400 to 800 ewes for economy of scale and labour efficiency. Research in sheep production has been focussed in particular on two major determinants of production efficiency, namely, the number of lambs reared per ewe joined (Hanrahan, 1997; Dawson and Carson, 2002) and the number of ewes stocked per ha of pasture. The set target for ewe productivity is 1.7 lambs reared per ewe mated (Flanagan 2003, 2004).

The significance of ewe productivity and stocking rate, was evident in the comparative performance of flocks managed in intensive and extensive systems at Knockbeg (Flanagan, 2003) (Table 14). Lamb output per ha in the intensive and extensive systems were 450 and 342 kg of carcass per ha, respectively.

Table 14 Flock performance and output at Knockbeg Sheep Unit, Carlow: Pooled results for 1999 and 2000

System	Grazing/silage/housing	Extended grazing
No. ewes per ha	14	10
Ewes lambing (%)	95	96
No. lambs reared/ewe joined	1.76	1.78
Carcass wt. (kg)	18.8	19.3
Age at slaughter (days)	160	146
Lamb output: kg/ewe	33.3	34.1
kg/ha	450	342

Environmental issues

In NI over 11,000 farmers participate in voluntary agri-environmental schemes, covering over a quarter of the farmland area. The schemes, which have been developed under the EU Rural Development Regulation (EC 99/1257), focus on maintaining and improving biodiversity through the positive management of wildlife habitats, improving water quality of rivers and lakes by nutrient management planning and the adoption of the Codes of Good Agricultural Practice; and the maintenance of landscape and heritage features by integration of their management into farming system.

European Union legislation including the Water Framework Directive (EC 2000/60) and the Nitrates Directive (EC 91/676) will have implications for grassland production systems in Ireland. NI has adopted a 'total territory' approach within the Nitrates Directive, and will produce a legally binding action programme during 2005. This will impose restrictions on the spreading periods for both organic and inorganic manures, define a minimum storage period for organic manure, and set maximum limits on phosphorus balances on individual farms.

The Council of Ministers of the European Union has recognised that farmers in receipt of direct agricultural support have important responsibilities towards the protection of the environment, animal health and welfare, and public health. Farmers will therefore be required to observe certain conditions in these areas in return for receipt of direct agricultural support, which post-reform of the Common Agricultural Policy (CAP), is now decoupled from production.

In the ROI, the Rural Environment Protection Scheme (REPS) was introduced in 1994 and now almost 44,000 farmers participate in the scheme. In this scheme inputs of both organic and inorganic N on grassland are limited, a nutrient management plan developed and pollution avoidance is critical. In the third version of the REPS scheme, which has been introduced in the last year, there is greater emphasis on broader environmental objectives with farmers expected to be managers of the natural heritage. The action programme with regard to the Nitrates Directive is presently being finalised with the ROI.

Appendix tables

Appendix Table 1 Irish live cattle exports from the ROI, 1992-2003 ('000 head)

	1995	1996	1997	1998	1999	2000	2001	2002	2003
Total	370	190	57	171	416	401	101	147	220
of which to:									
International markets	263	139	7	29	74	65	11	32	37
Continental EU	89	41	23	137	324	311	40	73	143
United Kingdom	18	10	27	5	18	27	50	42	40

Source: Bord Bia

Appendix Table 2 Destination of beef exports from the ROI, 1995-2003 ('000 tonnes cwe)

	1995	1996	1997	1998	1999	2000	2001	2002	2003
Total	440	425	450	510	554	495	345	460	500
of which to:									
UK	100	60	95	85	95	110	220	255	265
Continental EU	158	100	90	130	150	135	72	116	150
International markets	183	265	265	295	309	250	50	89	85

Source: Bord Bia

Acknowledgements

The authors thank Miss D. Hennessy for assistance in data assembly and Ms. M. Weldon and Miss S. Caffrey for typing the manuscript.

References

Brereton, A.J. & T. Keane (1982). The effect of water on grassland productivity in Ireland. *Irish Journal of Agricultural Research,* 21, 227-248.

Brereton, A. J. (1995). Regional and year to year variation in production. In: Jeffrey, D.W., Jones, M.B. & J.H McAdam (eds.) 1995 Irish grasslands – their biology and management: *Dublin Royal Irish Academy,* 12-22.

Bord Bia (2004). Irish Food Board. *Meat and Livestock Review and Outlook 2003/04*, 80pp.

Carson, A.F., B.W. Moss, L.E.R. Dawson & D.J. Kilpatrick (2001). Effects of genotype and dietary forage to concentrate ratio during the finishing period on carcass characteristics and meat quality of lambs from the hill sector. *Journal of Agricultural Science, Cambridge,* 137, 205-220.

Central Statistics Office (CSO) (2001) Publications.

Central Statistics Office (CSO) (2004) Publications.

College of Agriculture, Food and Ruaral Enterprise (CAFRE) (2003). *Greenmount Dairy Benchmarking Report*

Connolly, L., A. Kinsella & G. Quinlan (2004). *National Farm Survey 2003,* 16pp.

Coulter, B.S., W.E. Murphy, N. Culleton, E. Finnerty & L. Connolly (2002). A survey of fertiliser use in 2000 for grassland and arable crops. *Teagasc, End of project report,* 34pp.

Cruickshank, J.G. (1997). Soil and Environment: Northern Ireland. Published by the Agricultural and Environmental Sciences Division, DANI and the Environmental Science Department, The Queen's University of Belfast.

Dawson, L.E.R & A.F. Carson (2002). Options for genetic improvement within the NorthernIreland sheep industry. In: *75th Annual Report of the Agricultural Research Instituteof Northern Ireland,* 58-68.

Department of Agriculture and Food (DAF) (2004). Annual Review and Outlook for Agriculture and Food.

Department of Agriculture and Rural Development for Northern Ireland (DARD) (2003). The Agricultural Census in Northern Ireland – Results for 2003. *A National Statistical Publication.*

Department of Agriculture and Rural Development for Northern Ireland (DARD) (2004). Statistical Review of Northern Ireland Agriculture 2003. *A National Statistical Publication.*

Dillon, P. & G. Stakelum (1999). Spring Milk Production. In: J. Murphy (ed.) A new agenda for dairying, 3-7.

Drennan, M.J. (1999a). Breed composition of the Irish Cattle Herd. *Teagasc, Beef Production, Series,* No. 22, 26pp.

Drennan, M.J. (1999b). Development of a competitive suckler beef production system *Teagasc, Beef Production Series* No. 18, 32pp.

Drennan, M.J. (2004). Quality suckler beef from low and high input grassland management systems. *Teagasc, Beef Production Series* No. 48, 15pp.

Drennan, M.J. & M.G. Keane (2001). Producing cattle in the current policy and market environment. R and H Hall *Technical Bulletin,* 12pp.

Eurostat (2002). Farm Structure 1999/2000 Survey.

Ferris, C.P., D.C. Patterson & J. Murphy (2002). Grassland-based systems of milk production for autumn calving dairy cows: a three-year comparison. In: 75th Annual Report of the Agricultural Research Institute of Northern Ireland, 44-57.

Flanagan, S. (2003). Extended grazing by sheep as a substitute for silage and housing. In: Proceedings of Agricultural Research Forum, Tullamore, March, p109.
Flanagan, S. (2004). Feed demand and grass supply profiles for mid-season lamb production. In: Proceedings of Agricultural Research Forum, Tullamore, March, p26.
Hanrahan, J.P. (1997). Exploiting genetic resources to enhance the competitiveness of lowland sheep systems. *Irish Grassland and Animal Production Association Journal*, 31, 106-115.
Hopkins. A. (2000). In: A. Hopkins (ed.) Grass. Its production and utilization, 1-12.
Irish Cattle Breeding Federation (ICBF) (2003). In: Andrew Cromie (ed.) Irish Cattle Breeding Statistics, 24pp.
Keane, T. (1992). Chapter 3: Temperature In: T. Keane (ed.) Irish Farming, Weather and Environment. AGMET Group Publication, 21-27.
Keane, M.G. (2001). Development of an intensive dairy calf to beef system and associated grassland management. *Teagasc, Beef Production Series,* No. 41, 34pp.
Kirkland, R.M., T.W. J. Keady, P.A. Ingram, R.W.J. Steen, J. Comerford, D.C. Patterson & C.S. Mayne (2004). In: Proceedings of the British Society of Animal Science, p38.
Livestock and Meat Commission (2004). *Livestock and Meat Commission Yearbook 2004.*
O'Kiely, P., K. McNamara, D. Forristal & J.J. Lenehan (2000). Grass silage in Ireland. *Farm and Food* 10, 33-38.
Shalloo, L., P. Dillion, J. O'Loughlin, M. Rath & M. Wallace (2004). Comparison of a pasture-based system of milk production on a high rainfall, heavy-clay soil with that on a lower rainfall, free-draining soil. *Grass and Forage Science,* 59, 157-168.

Plant and sward characteristics to achieve high intake in ruminants

W.J. Wales, C.R. Stockdale and P.T. Doyle
Primary Industries Research Victoria (PIRVic), Department of Primary Industries, Kyabram Centre, 120 Cooma Road, Kyabram, Victoria, 3620, Australia
Email:bill.wales@dpi.vic.gov.au

Key points

1. Intake is affected by complex interactions between signals from the digestive tract, intermediary metabolism and energy supply, and behavioural signals associated with learned behaviours or sensory signals.
2. The ideal sward needs to have characteristics that are similar to total mixed rations to achieve high intake and animal performance.
3. Genetic manipulation of plants may offer an accelerated rate of plant improvement, but benefits need to be demonstrated in a systems context.

Keywords: intake, grazing ruminants, herbage mass, herbage allowance, sward height

Introduction

The primary limit to performance of grazing livestock is energy available to the tissues for productive purposes. This occurs because intake at grazing is less than potential intake. Secondary limits to performance appear to occur because the balance and synchrony between energy and amino acid availability to rumen organisms, or to the tissues, leads to inefficiencies in nutrient utilisation. This lack of synchrony between energy and amino acid availability is probably greater in lactating cows fed energy supplements. However, experimental quantification of the implications of poor synchrony in grazing cows is rare. Tertiary limits may be imposed by inadequacies in the supply of other essential nutrients at the rumen or tissue level. The ideal sward would contain plants that enable high rates of intake, long meal duration and optimum supplies and synchrony of energy yielding substrates and essential nutrients for rumen organisms and tissues.

Livestock systems relying on grazed herbage face constraints to animal production from seasonal variations in pasture growth and nutritive characteristics. These constraints limit carrying capacity and individual animal performance. In this review, we have not considered these seasonal constraints in detail, but have focused on the characteristics of vegetative plants and swards that are associated with high intake. For example, the ideal plant for dairy cows would need to provide sufficient metabolisable energy, metabolisable protein, fibre, and other essential nutrients to sustain high yields of milk solids. Currently, total mixed rations (TMR) can be formulated to allow dairy cows to approach their genetic potential for intake and milk production. Cows in early lactation consuming TMR have produced 2.7 kg protein + fat/day; (44.1 kg milk/day; Kolver & Muller 1998), levels of production that cannot be achieved by grazing pasture alone where maximum daily milk yields are only about 30 kg/cow (Doyle *et al.*, 2001).

This raises questions as to what limits intake and nutrient supply when animals with the genetic potential for high rates of growth or milk production graze high quality herbage, and are there options to overcome these constraints. To examine these questions, we consider the characteristics of vegetative swards and plants that influence nutrient intake and production by grazing ruminants in relation to theories of intake regulation.

Intake regulation

Herbage intake by grazing animals is affected by the characteristics of the sward, animal factors (physiological state and species), the environment, and interactions between these (Doyle *et al.,* 2000). In all grazing systems, characteristics of swards, such as pasture mass, the spatial distribution of plants and their nutritive characteristics, affect intake. Under strip or small paddock rotation systems that are common in dairy farming, herbage allowance is also a key determinant of herbage intake.

One theory on the regulation of intake has been the concept of limits to intake due to rumen fill, with differences between the capacity of forages to fill the rumen responsible for the observed differences in intake (Mertens, 1987). This concept suggests a key role of the structural and slowly digestible components of plants, generally measured as neutral detergent fibre (NDF), in limiting intake. This concept has been useful when applied to TMR fed to appetite. However, the regulation of intake is more complex and includes interactions between limitations imposed by, or signals from, the digestive tract and intermediary metabolism, involving signals generated by supply of energy and essential nutrients (Weston, 1982). Rumen fill appears to have a key role in intake regulation when digestibility of the diet is less than 75% (Dove, 1996), but vegetative herbages often have digestibility values above this. Ketelaars & Tolkamp (1992) proposed that the intake of highly digestible feeds is physiologically determined, and physical restrictions to intake are less important.

When animals graze vegetative herbage, simple relationships between intake, digestibility and rumen fill do not exist. For example, in cows consuming the same amount of Persian clover (*Trifolium resupinatum)* or perennial ryegrass (*Lolium perenne* L.) with similar digestibility, rumen fill was lower on clover (Williams *et al.,* 2005). In a subsequent study with grazing cows consuming different amounts of Persian clover, rumen fill varied throughout the day with eating bouts, but there were no differences in average rumen fill, DM or NDF loads across the range of intakes (5.5 to 20.4 kg DM/day) (Williams 2003). In contrast to this, rumen fill appeared to have an unexpected role in intake regulation with highly digestible ryegrass. With sheep fed low digestibility forage diets, Doyle *et al.* (1987) indicated rumen fill was not always the major factor limiting intake and that nutrient imbalances were important. However, they suggested that for these types of forage, the setting at which signals associated with fill of the rumen influenced intake might change with type of diet and nutrient supply. It would also seem that with highly digestible herbage or TMR, no single factor such as rumen fill, nutrient supply or deficits to the tissues alone will regulate intake.

At grazing, factors such as the time available for grazing and rumination (Rook, 2000), dietary preferences (Provenza, 1995) and sensory factors such as palatability (Weston, 1985), may also play roles in intake regulation. In mixed swards, animals spend time searching for preferred components, and there are upper limits to the time animals will spend each day ingesting and ruminating feeds. Buckmaster *et al.* (1997) suggested that intake is reduced when grazing time is less than 8 hours per day.

These complexities mean that predicting intake from simple relationships based on single factors, such as digestibility or NDF concentration, will not be universally applicable across the extremes of grazing and TMR feeding systems. Hence, in considering sward and plant characteristics conducive to high intake, an understanding of the complex interactions between signals from the digestive tract and digestive processes, intermediary metabolism and sufficiency of energy and essential nutrients, and behavioural signals associated with learned behaviours or sensory signals is important.

Characteristics of grazed herbage compared with TMR

Conceptually, an ideal sward would have characteristics, nutrient profile and physical attributes similar to a TMR formulated to provide nutrients in relation to requirements while having the physical characteristics necessary to stimulate rumen function and rumination. Because a TMR offers control over the nutritive characteristics of the diet, when offered in sufficient quantities, it allows animals to approach their potential intake and provide the nutrient requirements for high animal performance. Kolver and Muller (1998) compared the nutritive characteristics of a pasture diet based on a mixed grass/clover sward and a TMR consumed by dairy cows in early lactation. Despite both diets having similar digestibilities, there are obvious differences in the concentrations of essential nutrients, as well as pasture having lower DM and non-structural carbohydrate concentrations and higher NDF concentrations. This comparison also does not illustrate the differences in physical characteristics of the diets where clearly the particle sizes in a TMR are conducive to rapid rates of removal from the rumen.

In comparison with TMR, cows consuming grazed pasture even when supplemented with grain, had lower DM intake, milk production, milk protein and fat concentrations, lost more body condition and had lower liveweight (Kolver & Muller, 1998; Bargo *et al.*, 2002). In high producing dairy cows, DM intake would need to approach 5% of the cows' liveweight at grazing to achieve similar intakes to that observed by cows consuming TMR. Stockdale (1993) reported that cows grazing Persian clover consumed up to 4.5% of their liveweight, representing one of the highest reported intakes of herbage in the literature. However, Kolver & Muller (1998) concluded that current pasture species are unlikely to provide the nutrients in sufficient quantities to achieve similar milk yields to TMR.

Importantly, at very high water contents, intake of herbage is reduced. For dairy cows, the critical water content was estimated to be about 82%, with a depression of 0.34 kg DM intake for each percentage increase in water content above this level (Verite & Journet, 1970). When water was added to the rumen *per fistulum*, there were no detrimental effects on the intake of forages by sheep (Lloyd Davies, 1962), indicating the effects of water content on herbage intake may be associated with palatability or the large volumes of fresh herbage that need to be processed during ingestion. In cattle, Cabrera Estrada *et al.* (2004) showed that intake and eating rate was restricted by internal water of grass, but not by external water.

Achieving high intake from grazed herbage

To maximise intake, animals need to consume plants that have characteristics that allow rapid consumption and lead to fast rates of passage through the rumen. Intake of herbage has been defined as the product of the rate of eating (R) and the time spent eating (T) (Allden & Whittaker, 1970).

$$\text{Daily intake} = R \times T \quad (1)$$

This relation has been further refined by Rook (2000), who described rate of eating as the product of bite mass and bite rate, and time spent grazing as the meal duration and number of meals per day.

$$\text{Daily intake} = (\text{bite mass} \times \text{bite rate}) \times (\text{meal duration} \times \text{number of meals}) \quad (2)$$

Grazing ruminants vary bite dimensions, bite rate and grazing time in response to changes in sward conditions (Hodgson, 1981; Milne *et al.*, 1982; Penning *et al.*, 1991a; Gibb *et al.*, 1997). The animal's mouth size (Taylor *et al.*, 1987; Illius, 1989; Laca *et al.*, 1992), the proximity of the bite to the ground (Hughes *et al.*, 1991; Mitchell, 1995) and the effort required to break the pasture (Hughes *et al.*, 1991) influence bite dimensions and, hence, bite mass.

The complexity of the interactions between factors can be illustrated by examining bite mass. Increases in sward height and bulk density have been shown to increase bite mass (from 0.25 to 4 g DM) for cattle offered micro-swards of lucerne (*Medicago sativa* L.) and paspalum (*Paspalum dilatatum* Poir.) (Ungar, 1996). However, although increasing sward height increases bite depth in sheep (Edwards *et al.*, 1995) and cattle (Laca *et al.*, 1992), increases in bulk density leads to decreases in bite depth, particularly with longer swards (Laca *et al.*, 1992). Bite mass is, therefore, influenced by the height of the sward, its bulk density and the effect of the density on reducing bite depth.

Hughes *et al.* (1991) suggested that the structural strength of accessible pasture components would determine the bite dimensions and bite mass, and that the upper limit to the force ruminants are prepared to exert to sever a bite may be important. Further to this idea, Illius *et al.* (1995) demonstrated that the number of tillers constrains bite mass and is determined by the force required to sever a mouthful. In support, Tharmaraj *et al.* (2003) showed that the bite fracture force, a measure of the resistance to breaking, increased down the sward profile.

Bite rate is related to ease of prehension, herbage shear force, and bite mass, as smaller mouthfuls often lead to an increase in bite rate. It has proven difficult to isolate the individual effects of sward characteristics, such as height, mass, leaf area and nutritive characteristics on ingestive behaviour, since these factors are linked, and experiments that have attempted to vary sward height, for example, have generally varied mass as well. These interrelationships between characteristics of swards lead to confounding when trying to isolate the importance of a particular characteristic.

In short term studies with fasted cows grazing perennial ryegrass, sward height and sward density were shown to have marked effects on hourly intake rate (and presumably potential intake rate) due to differences in bite mass (Mayne *et al.*, 1997). Intake rates were maximised at 3.5 to 4 kg DM/hour when sward heights, measured using a sward stick, were greater than 18 cm. However, intake rates were still very high (3 kg DM/hour) when sward height was 15 cm. With shorter swards, bulk density becomes important, with intake rates varying from 1 to 2.5 kg DM/hour for swards varying in bulk density from 1.7 to 3.1 t DM/ha. Thus, high intake rates by dairy cows may be achieved by grazing tall swards (greater than 15 cm) or by grazing denser short swards (less than 15 cm). However, these high short-term intake rates measured on experimental swards are difficult to translate into grazing systems as swards change during grazing, and utilisation of available herbage is an important consideration.

Where intake rates have been estimated over 24 hours, similar principles apply. For example, Wales *et al.* (1999) reported intake rates of 2.7 kg DM/hour for cows grazing dense, tall swards (12.6 cm using a rising plate meter), but only 1.9 kg DM/hour at the same pasture allowance, but with short (5.6 cm) swards. The grazing time on both swards was not different at 8.3 hours/day.

In many instances, management decisions, such as herbage allowance in strip grazing systems and/or pre-grazing pasture mass, have marked effects on intake (Wales *et al.*, 1998; Wales *et al.*, 1999). Curvilinear relationships exist between herbage allowance and intake for grazing dairy cows, with intake increasing with increasing allowance (Stakelum, 1986a; Stakelum, 1986b; Stakelum, 1986c; Holmes, 1987; Stockdale, 2000). In contrast to strip grazing, intake by sheep and cattle in continuous grazing systems is maximised at a green pasture mass between 1.5 and 2.5 t DM/ha, although sheep in particular tend to patch graze when pasture mass exceeds 1.5 t DM/ha (Doyle *et al.*, 1993). Thus, herbage mass, which is influenced by the height of the sward and its bulk density, is a key determinant of diet selection and intake by grazing animals (Kenney & Black, 1984; Black & Kenney, 1984; Laca *et al.*, 1992; Edwards *et al.*, 1995; Gibb *et al.*, 1997; Concha & Nicol, 2000; Pulido & Leaver, 2001). A number of studies have attempted to quantify the effect of herbage mass on intake, with increases in intake by dairy cows under strip grazing of between 1.1 and 2.3 kg DM for each additional t DM/ha (Stockdale, 1985) and between 1.1 and 2.6 kg OM for each additional t OM/ha (Stakelum, 1986b; Stakelum, 1986c).

Meal duration is not only influenced by sward characteristics, but potentially by the capacity of the rumen-reticulum, the need for rumination to breakdown ingested material and the rate of passage of digesta from the rumen. The nutritive characteristics of different plant species are important and studies of the kinetics of digestion at grazing are needed to explore the importance of these characteristics further. For example, Williams *et al.* (2005) found that at the same intakes, perennial ryegrass resulted in higher rumen fill than Persian clover. Cows grazing perennial ryegrass at a high allowance spent less time eating and more time ruminating than those grazing the clover. However, there were no differences in average DM in the rumen for cows grazing Persian clover, with intakes between 5.5 and 20.4 kg DM/day. Although DM in the rumen varied throughout the day as meals were consumed, little time was spent ruminating and it appeared the primary effect of increasing intake was increased passage from the rumen on this herbage type (Williams, 2003).

Relative advantages of legumes and grasses

In general, legumes have characteristics that lead to higher animal performance compared with grasses. An early study in southern Australia (Rogers *et al.*, 1986) highlighted the advantage of white clover (*Trifolium repens*) compared with perennial ryegrass for milk production when intake was not restricted by pasture allowance. Cows consuming the white clover pasture produced more milk (5750 vs. 4740 L) and milk fat (236 vs. 194 kg) and gained more liveweight (85 vs. 80 kg) due to a 30% higher intake (Rogers *et al.*, 1982). More recently, Harris *et al.* (1997) showed that milk yield was increased by 20% when dairy cows consumed a diet with 55 – 65% DM clover, with the balance as perennial ryegrass, compared to a diet with only 20% clover. No further advantage in animal performance was achieved by offering diets with 80% clover. Sheep (Gibb & Treacher, 1983; Penning *et al.*, 1991b) and cattle (Thomson, 1984; Beever *et al.*, 1986b) also eat more and grow faster when consuming diets of pure clover, because of the superior nutritive characteristics of white clover (Beever *et al.*, 2000).

Clover frequently comprises a minor component of mixed white clover/perennial ryegrass swards, particularly in dairy production systems using strip grazing, and given the preference for the legume (Newman *et al.*, 1992), there are likely to be energy costs and restrictions imposed on DM intake as animals search for and select clover. This preference for the clover reduces its presence (Parsons *et al.*, 1994b), which may partly explain why white clover rarely comprises more than 20% of the available herbage for grazing dairy cows (Doyle *et al.*, 2000).

Clovers contain less structural carbohydrate leading to more rapid rates of breakdown of OM, nitrogen (N) and cell walls (Beever & Siddons, 1986; Aitchison *et al.*, 1986; Beever *et al.*, 1986a) and the retention time is less compared with ryegrass (Ulyatt, 1981). The faster rate of passage of legume compared with grass has been ascribed to differences in particle shape (Troelsen & Campbell, 1968; Moseley & Jones, 1984). The breakdown of grass produces long, thin threadlike structures, while the clover produces more blockish, irregular shapes. Closer examination has shown that grass particles consist of strands of vascular tissue and epidermal sheets, where the fracture lines run longitudinally along the length of the leaves and stems. Clover particles consist largely of epidermal tissue with little evidence of vascular structure, while the fracture lines occur in all planes with equal frequency, giving rise to irregular shaped particles with no dominant axis.

Despite the clear advantages in intake of legumes over grasses, there are other issues that need to be considered. Firstly, the cost of increased prevalence of bloat and the additional costs of maintaining swards need to be addressed. Secondly, while this more rapid breakdown and fermentation is an advantage, it also has some negative consequences. Time spent ruminating by cows consuming clover is low compared with grass (Williams *et al.*, 2000), and rumen fluid pH can be below 6.0 for considerable periods on these legume diets (Williams, 2003; Williams *et al.*, 2005). Pure clover swards may not provide sufficient NDF for efficient rumen function, with potential negative consequences for milk fat production in dairy cows. Low rumen pH predisposes animals to acidosis when concentrate supplements are fed. Thirdly, associated with the high crude protein concentrations in clovers, rumen ammonia concentrations (up to 500 mg/L) (Stockdale, 1993) can be much higher than microbial requirements. This has energy costs in converting absorbed N into urea for excretion, and has been estimated to be as high as the equivalent amount of energy required to produce 2 kg milk/day for cows grazing irrigated clovers (Cohen, 2001). One possible strategy to improve the utilisation of the excess N is to feed high-energy supplements, but an unintended consequence of this approach is the increased prevalence of sub clinical and clinical acidosis.

An interesting observation is that sheep may show a preference for white clover in the morning, but this preference diminishes through the day in favour of a preference for grass (Parsons *et al.,* 1994a). In theory, this type of preference should lead to increases in DM intake, as rate of intake of clover is faster than grass, and could involve post-ingestive feedback with propionate (Francis, 2002).

Despite the clear advantages of clovers over grass in intake and animal performance, clovers are not an ideal plant from the perspective of rumen function or synchrony of supply of N and energy to the rumen organisms. An alternative approach is to present choices to grazing livestock. This has prompted research into systems where choice is offered to grazing ruminants. When given a choice between grass and clover monocultures, sheep consumed 50 - 70% white clover and 30 - 50% ryegrass (Newman *et al.,* 1992; Parsons *et al.,* 1994b). Cows also prefer 70% of their diet DM as white clover (Cosgrove *et al.,* 1999), and providing them with free choice between perennial ryegrass and white clover has increased milk yield by 10 - 30% compared with a conventional interspersed mixed pasture (Marotti *et al.,* 2001). Offering free choice of pure swards may overcome management and competition issues associated with maintaining optimal amounts of plant species in a mixed sward. However, it also presents challenges in presenting each monoculture in an ideal state (height, density, and nutritive characteristics) for high intake. This may be achieved by offering the alternate pastures at different times of day or using other classes of less productive stock to utilise residual herbage. However, the additional costs and increased level of management complexity may limit the widespread adoption of this approach.

Modification of the nutritive characteristics of legumes and grasses

Plant breeding objectives have expanded from the traditional focus on improving yield and pest and disease resistance to those that have effects on animal health, fertility and characteristics of the animal product (Caradus *et al.*, 2000). For example, techniques to genetically modify the plant have enabled the development of plants with elevated concentrations of ruminal undegraded dietary protein and high-energy yielding compounds, such as starch or triacylglycerides (Spangenberg *et al.*, 2001; Roberts *et al.*, 2002). An alternative to increasing carbohydrates such as starch in leaves, which are readily transported to storage sites, is to introduce the ability to synthesise fructans, storage compounds based on sucrose. Roberts *et al.*, (2002) described research investigating the potential for modifying white clover through the introduction of genes that code for the ability to synthesise fructans from a bacterium and globe artichoke. Their research has indicated that elevated levels of fructans will improve the nutritive value of the herbage. Genetic modification of plants to introduce desirable attributes has the potential to accelerate the selection of plants with desirable animal production traits.

Increasing the digestibility of the herbage is an example of a well-established strategy for increasing intake in ruminants. Stehr & Kirchgessner (1976) demonstrated that herbage intake increased by 5.5 kg for every 10 unit increase in OMD from 64 - 80%. In lambs consuming legumes and grasses, Freer & Jones (1984) reported a linear intake response with OMD over the range 57 to 83%. Grazing animals have demonstrated a strong preference for herbage fractions with high soluble carbohydrate concentrations (Jones & Roberts, 1991; Dove *et al.*, 1992; Simpson & Dove, 1994; Ciavarella *et al.*, 2000). Dove & Milne (1994) reported that the efficiency of microbial protein synthesis in sheep grazing perennial ryegrass swards was halved in autumn when water soluble carbohydrate concentrations were lower than in that measured in summer, despite the digestibility of the swards being the same. In zero grazing studies, dairy cows offered pasture with high water soluble carbohydrate concentrations consumed more DM and produced more milk than cows fed grasses with lower concentrations (Miller *et al.*, 1999; Moorby *et al.*, 2001).

The benefits of many of these modified cultivars have yet to be demonstrated in animal systems.

Conclusions

Seasonality of pasture growth and of the nutritive characteristics of plants within swards will always present challenges when attempting to provide pastures to achieve intakes near the potential of high producing animals. With vegetative swards, legumes offer significant potential to increase intake of grazed pasture compared with grasses. However, they also present challenges in terms of rumen stability and disposal of excess N. While choice grazing systems (involving legumes and grasses) offer potential to achieve synergistic effects on intake and animal performance, management of such systems will bring complexities, and alternatives that include supplementation or involve partial mixed rations may be more realistic options. Genetic manipulation of plants may offer an accelerated rate of plant improvement, but benefits need to be demonstrated in systems.

Acknowledgments

We would like to acknowledge financial support from the International Grasslands Congress and the Department of Primary Industries, Victoria.

References

Aitchison, E.M., M. Gill, M.S. Dhanoa & D.F. Osbourn (1986). The effect of digestibility and forage species on the removal of digesta from the rumen and the voluntary intake of hay by sheep. *British Journal of Nutrition*, 56, 463-476.

Allden, W.G. & I.A.M. Whittaker (1970). The determinants of herbage intake by grazing sheep: the interrelationship of factors influencing herbage intake and availability. *Australian Journal of Agricultural Research*, 21, 755-766.

Bargo, F., L.D. Muller, J.E. Delahoy & T.W. Cassidy (2002). Performance of high producing dairy cows with three different feeding systems combining pasture and total mixed rations. *Journal of Dairy Science*, 85, 2948-2963.

Beever, D.E., M.S. Dhanoa, H.R. Losada, R.T. Evans, S.B. Cammell & J. France (1986a). The effect of forage species and stage of harvest on the processes of digestion occuring in the rumen of cattle. *British Journal of Nutrition*, 56, 439-454.

Beever, D.E., H.R. Losada, S.B. Cammell, R.T. Evans & M.J. Haines (1986b). Effect of forage species and season on nutrient digestion and supply in grazing cattle. *British Journal of Nutrition*, 56, 209-225.

Beever, D.E., N. Offer & E.M. Gill (2000). The feeding value of grass and grass products. In: A. Hopkins, (ed.) Grass Its production and utilization. Blackwell Science Ltd. Oxford, UK. Pages 140-195.

Beever, D.E. & R.C. Siddons (1986). Digestion And Metabolism In The Grazing Ruminant. In: L. P. Milligan, W. L. Grovum, and A. Dobson (eds.) Control of Digestion and Metabolism in Ruminants. Prentice Hall. Englewood Cliffs, New Jersey, 479-497.

Black, J.L. & P.A. Kenney (1984). Factors affecting diet selection by sheep. II. Height and density of pasture. *Australian Journal of Agricultural Research*, 35, 551-563.

Buckmaster, D.R., L.A. Holden, L.D. Muller & R.H. Mohtar (1997). Modelling intake of grazing cows fed complementary feeds. *Proceedings of the XVIII International Grassland Congress,* Vol. 1, 2-9-2-10.

Cabrera Estrada, J.L., R. Delagarde, P. Faverdin & J.L. Peyraud (2004). Dry matter intake and eating rate of grass by dairy cows is restricted by internal, but not external water. *Animal Feed Science and Technology*, 114, 59-74.

Caradus, J.R., D.R. Wood field and H.S. Easton (2000). Improved grazing value of pasture cultivars for temperate environments. In: G.M. Stone (ed.). Animal production for a consuming world, Vol B, 5-8.

Ciavarella, T.A., H. Dove, B.J. Leury & R.J. Simpson (2000). Diet selection by sheep grazing *Phalaris aquatica* L. pastures of differing water-soluble carbohydrate content. *Australian Journal of Agricultural Research*, 51, 757-764.

Cohen, D.C. (2001). Degradability of crude protein from clover herbages used in irrigated dairy production systems in northern Victoria. *Australian Journal Agricultural Research*, 52, 415-425.

Concha, M.A. & A.M. Nicol (2000). Selection by sheep and goats for perennial ryegrass and white clover offered over a range of sward height contrasts. *Grass and Forage Science*, 55, 47-58.

Cosgrove, G.P., G.C. Waghorn & A.J. Parsons (1999). Exploring the nutritional basis of preference and diet selection by sheep. *Proceedings of the New Zealand Grassland Association,* 61, 175-180.

Dove, H. (1996). Constraints to the modelling of diet selection and intake in the grazing ruminant. *Australian Journal of Agricultural Research*, 47, 257-275.

Dove, H. & J.A. Milne (1994). Digesta flow and rumen microbial protein production in ewes grazing perennial ryegrass. *Australian Journal of Agricultural Research*, 45, 1229-1245.

Dove, H., C. Siever-Kelly, B.J. Leury, K.L. Gatford & R.J. Simpson (1992). Using plant wax alkanes to quantify the intake of plant parts by grazing animals. *Proceedings of the Nutritional Society of Australia*, 17, 149.

Doyle, P.T., M. Grimm & A.N. Thompson (1993). Grazing for pasture and sheep management in the annual pasture zone. In: D. R. Kemp and D. L. Michalk, (eds.) Technology for the 21st century. CSIRO Australia. Melbourne, Australia.

Doyle, P.T., G.R. Pearce & A. Djajanegara (1987). Intake and digestion of cereal straws. *The 4th Animal Science Congress of the Asian-Australasian Association of Animal Production Societies,* 4, 59-62.

Doyle, P.T., C.R. Stockdale, A.R. Lawson & D.C. Cohen (2000). 'Pastures for Dairy Production in Victoria.' Department of Primary Industries, Victoria, Australia.

Doyle, P.T., C.R. Stockdale, W.J. Wales, G.P. Walker & J.W. Heard (2001). Limits to and optimising of milk production and composition from pastures. *Recent Advances in Animal Nutrition in Australia*, 13, 9-17.

Edwards, G.R., A.J. Parsons, P.D. Penning & J.A. Newman (1995). Relationships between vegetation state and bite dimensions of sheep grazing contrasting plant species and its implications for intake rate and diet selection. *Grass and Forage Science*, 50, 378-388.

Francis, S.A. (2002). Investigating the role of carbohydrates in the dietary choices of ruminants with an emphasis on dairy cows. PhD thesis, The University of Melbourne. Australia.

Freer, M. & D.B. Jones (1984). Feeding value of subterranean clover, lucerne, phalaris and Wimmera ryegrass for sheep. *Australian Journal of Experimental Agriculture and Animal Husbandry*, 24, 156-164.

Gibb, M.J., C.A. Huckle, R. Nuthall & A.J. Rook (1997). Effect of sward surface height on intake and grazing behaviour by lactating Holstein Friesian cows. *Grass and Forage Science*, 52, 309-321.

Gibb, M.J. & T.T. Treacher (1983). The performance of lactating ewes offered diets containing different proportions of fresh perennial ryegrass and white clover. *British Society of Animal Production*, 37, 433-440.

Harris, S.L., D.A. Clark & E.B.L. Jansen (1997). Optimum white clover content for milk production. *Proceedings of the New Zealand Society of Animal Production*, 57, 169-171.

Hodgson, J. (1981). Variation in the surface characteristics of the sward and the short-term rate of herbage intake by calves and lambs. *Grass and Forage Science*, 36, 49-57.

Holmes, C.W. (1987). Pastures for Dairy Cows. In: A. M. Nicol (ed.) Livestock Feeding on Pasture. New Zealand Society of Animal Production. 133-143

Hughes, T.P., A.R. Sykes, D.P. Poppi & J. Hodgson (1991). The influence of sward structure on peak bite force and bite weight in sheep. *Proceedings of the New Zealand Society of Animal Production,* 51, 153-158.

Illius, A.W. (1989). Allometry of food intake and grazing behaviour with body size in cattle. *Journal of Agricultural Science, Cambridge*, 113, 259-266.

Illius, A.W., I.F. Gordon, J.D. Milne & W. Wright (1995). Costs and benefits of foraging on grasses varying in canopy structure and resistance to defoliation. *Functional Ecology*, 9, 894-903.

Jones, E.L. & J.E. Roberts (1991). A note on the relationship between palatability and water-soluble carbohydrates content in perennial ryegrass. *Irish Journal of Agricultural Research*, 30, 163-167.

Kenney, P. A. & J.L. Black (1984). Factors affecting diet selection by sheep. I. Potential intake rate and acceptibility of feed. *Australian Journal of Agricultural Research*, 35, 551-563.

Ketelaars, J.J. M.H. & B.J. Tolkamp (1992). Toward a new theory of feed intake regulation in ruminants 1. Causes of differences in voluntary feed intake: critique of current views. *Livestock Production Science*, 30, 269-296.

Kolver, E.S. & L.D. Muller (1998). Performance and nutrient intake of high producing Holstein cows consuming pasture or a total mixed ration. *Journal of Dairy Science*, 81, 1403-1411.

Laca, E.A., E.D. Ungar, N. Seligman & M.W. Demment (1992). Effects of sward height and bulk density on bite dimensions of cattle grazing homogeneous swards. *Grass and Forage Science*, 47, 91-102.

Lloyd Davies, H. (1962). Intake studies in sheep involving high fluid intake Adelaide, Australia. *Proceedings of the Australian Society of Animal Production*, 4, 167-171

Marotti, D.M., G.P. Cosgrove, D.F. Chapman, A.J. Parsons, A.R. Egan & C.B. Anderson. (2001). Novel methods of forage presentation to boost nutrition and performance of grazing dairy cows. *Australian Journal of Dairy Technology*, 56, 159.

Mayne, C.S., D.A. McGilloway, A. Cushnahan & A.S. Laidlaw (1997). The effect of sward height and bulk density on herbage intake and grazing behaviour of dairy cows. *Proceedings of the XVIII International Grassland Congress*. Vol. 1, 2-15-2-16.

Mertens, D.R. (1987). Predicting intake and digestibility using mathematical models of ruminal function. *Journal of Animal Science*, 64, 1548-1558.

Miller, L.A., M.A. Neville, D.H. Baker, R.T. Evans, M.K. Theodorou, J.C. MacRae, M.O. Humphreys & J.M. Moorby (1999). Milk production from dairy cows offered perennial ryegrass selected for high water soluble carbohydrate concentrations compared to a control grass. *Proceedings of the British Society of Animal Science*, 208.

Milne, J.A., J. Hodgson, R. Thompson, W.G. Souter & G.T. Barthram (1982). The diet ingested by sheep grazing swards differing in white clover and perennial ryegrass content. *Grass and Forage Science*, 37, 209-218.

Mitchell, R.J. (1995). The effects of sward height, bulk density and tiller structure on the ingestive behaviour of Red deer and Romney sheep. PhD thesis, Massey University, New Zealand.

Moorby, J.M., L.A. Miller, R.T. Evans, N.D. Scollan, M.K. Theodorou & J.C. MacRae (2001). Milk production and N partitioning in early lactation dairy cows offered perennial ryegrass containing a high concentration of water soluble carbohydrates. *Proceedings of the British Society of Animal Science*, 6.

Moseley, G. & J.R. Jones. (1984). The physical digestion of perennial ryegrass *(Lolium perenne)* and white clover (*Trifolium repens*) in the foregut of sheep. *British Journal of Nutrition*, 52, 381-390.

Newman, J.A., A.J. Parsons & A. Harvey (1992). Not all sheep prefer clover: diet selection revisited. *Journal of Agricultural Science, Cambridge*, 119, 275-283.

Parsons, A.J., J.H. M. Thornley, J. Newman & P.D. Penning (1994a). A mechanistic model of some physical determinants of intake rate and diet selection in a two-species temperate grassland sward. *Functional Ecology*, 8, 187-204.

Parsons, A.J., J.A. Newman, P.D. Penning, A.Harvey & R.J. Orr (1994b). Diet preference of sheep, effects of recent diet, physiological state and spatial abundance. *Journal of Animal Ecology*, 63, 465-478.

Penning, P.D., A.J. Parsons, R.J. Orr & T.T. Treacher (1991a). Intake and behaviour responses by sheep to changes in sward characteristics under continuous stocking. *Grass and Forage Science*, 46, 15-28.

Penning, P.D., A.J. Rook & R.J. Orr (1991b). Patterns of ingestive behaviour of sheep continuously stocked on monocultures of ryegrass or white clover. *Applied Animal Behaviour Science*, 31, 237-250.

Provenza, F.D. (1995). Postingestive feedback as an elementary determinant of food preference and intake in ruminants. *Journal of Range Management*, 48, 2-17.

Pulido, R.G. and J.D. Leaver (2001). Quantifying the influence of sward height, concentrate level and initial milk yield on the milk production and grazing behaviour of continuously stocked dairy cows. *Grass and Forage Science*, 56, 57-67.

Roberts, N.J., K.R. Hancock & D.R. Woodfield (2002). Genetic modification to create novel high quality forages. *Proceedings of the New Zealand Society of Animal Production,* 62, 278-281.

Rogers, G.L., R.H.D. Porter & I. Robinson. (1982). Comparison of perennial rye-grass and white clover for milk production. Production. *Proceedings of the Conference on Dairy Production from Pasture, New Zealand and Australian Societies of Animal Science.* 213-214

Rogers, G.L., I.B. Robinson & P.J. Moate (1986). Milk production of cows grazing white clover and perennial ryegrass. *Proceedings of the Australian Society of Animal Production,* 16, 427.

Rook, A.J. (2000). Principles of foraging and grazing behaviour. In: A. Hopkins, ed. Grass Its production and utilization. Blackwell Science Ltd. Oxford, UK., 229-246

Simpson, R.J. and H. Dove. (1994). Plant non-structural carbohydrates, diet selection and intake. 1994. *Proceedings of the Australian Society of Animal Production*, 20, 59-61.

Spangenburg, G., R. Kalla, A. Lidgett, T. Sawbridge, E.K. Ong & U. John (2001). Breeding forage plants in the genome era. In: G. Spangenburg (ed.) Molecular breeding of forage crops. Kluwer Academic Publishers, The Netherlands. 1-39

Stakelum, G. (1986a). Herbage intake of grazing dairy cows 1. Effect of autumn supplementation with concentrates and herbage allowance on herbage intake. *Irish Journal of Agricultural Research*, 25, 31-40.

Stakelum, G. (1986b). Herbage intake of grazing dairy cows 2. Effect of herbage allowance, herbage mass and concentrate feeding on the intake of cows grazing primary spring grass. *Irish Journal of Agricultural Research*, 25, 41-51.

Stakelum, G. (1986c). Herbage intake of grazing dairy cows 3. Effects of herbage mass, herbage allowance and concentrate feeding on the herbage intake of dairy cows grazing on mid-summer pasture. *Irish Journal of Agricultural Research*, 25, 179-189.

Stehr, W. and M. Kirchgessner (1976). The relationship between the intake of herbage grazed by dairy cows and its digestibility. *Animal Feed Science and Technology*, 1, 53-60.

Stockdale, C.R. (1985). Influence of some sward characteristics on the consumption of irrigated pastures grazed by lactating dairy cattle. *Grass and Forage Science*, 40, 31-39.

Stockdale, C.R. (1993). The productivity of lactating dairy cows fed irrigated Persian clover *(Trifolium resupinatum). Australian Journal of Agricultural Research*, 44, 1591-1608.

Stockdale, C.R. (2000). Levels of pasture substitution when concentrates are fed to grazing dairy cows in northern Victoria. *Australian Journal of Experimental Agriculture*, 40, 913-921.

Taylor, S.S., J.I. Murray & A.W. Illius (1987). Relative growth of incisor arcade breath and eating rate in cattle and sheep. *Animal Production*, 45, 453-480.

Tharmaraj, J., W.J. Wales, D.F. Chapman & A.R. Egan (2003). Defoliation pattern, foraging behaviour and diet selection by lactating dairy cows in response to sward height and herbage allowance of a ryegrass-dominated pasture. *Grass and Forage Science*, 58, 225-238.

Thomson, D.J. (1984). The nutritive value of white clover. *Proceedings of a Symposium organised by the British Grassland Society*, 16, 78-92.

Troelsen, J.E. and J.B. Campbell (1968). Voluntary consumption of forages by sheep and its relation to the size and shape of particles in the digestive tract. *Animal Production*, 10, 289-296.

Ulyatt, M.J. (1981). The Feeding Value of Herbage. In: G. W. Butler and R. W. Bailey (eds.) Chemistry and Biochemistry of Herbage. Academic Press. London, New York, 131-178.

Ungar, E.D. (1996). Ingestive behaviour. In: J. Hodgson and A. W. Illius, (eds.) The ecology and management of grazing systems. CAB International. Wallingford,UK. 185-218

Van Soest, P.J. 1994. Nutritional ecology of the ruminant. Comstock Publishing Associates, Ithaca.

Verite, R. & M. Journet (1970). Influence de la teneur en eau et la deshydration de l'herbe sur la valeur alimentaire pour les vaches laitiers. *Annales de Zootechnie*, 19, 225-268.

Wales, W.J., P.T. Doyle & D.W. Dellow (1998). Dry matter intake and nutrient selection by lactating cows grazing irrigated pastures at different pasture allowances in summer and autumn. *Australian Journal of Experimental Agriculture*, 38, 451-460.

Wales, W.J., P.T. Doyle, C.R. Stockdale & D.W. Dellow (1999). Effects of variations in herbage mass, allowance, and level of supplement on nutrient intake and milk production of dairy cows in spring and summer. *Australian Journal of Experimental Agriculture*, 39, 119-130.

Weston, R.H. (1982). Animal factors affecting feed intake. In: J. B. Hacker (ed.) Nutritional Limits to Animal Production from Pastures. Commonwealth Agricultural Bureaux. Farnham Royal, UK. Pages 183-198.

Weston, R.H. (1985). The regulation of feed intake in herbage-fed ruminants. *Proceedings of the Nutrition Society of Australia*, 10, 55-62.

Williams, Y.J. (2003) Ingestive processes and digestion of highly digestible pastures by strip-grazing dairy cows. PhD thesis, The University of Melbourne, Australia.

Williams, Y.J., G.P. Walker, P.T. Doyle, C.R. Stockdale & A.R. Egan (2005). Pasture type and allowance affects intake and rumen fermentation in grazing dairy cows. *Australian Journal of Experimental Agriculture*, 45 (in press).

Williams, Y.J., G.P. Walker, W.J. Wales & P.T. Doyle (2000). The grazing behaviour of cows grazing persian clover or perennial ryegrass pastures in spring. *Asian-Australasian Journal of Animal Sciences*,13 [Supplement Vol. A], 509-512.

New insights into the nutritional value of grass

D.H. Rearte
Instituto Nacional de Tecnología Agropecuaria. EEA Balcarce. 7620 Balcarce. Argentina, Email: drearte@balcarce.inta.gov.ar

Key points

1. The rumen environment in cattle grazing high quality forage is different to that reported for cattle fed indoors with diets based on processed feedstuffs.
2. Temperate pasture is an excellent source of nutrients for ruminants but a high energy:protein imbalance can occur when it is offered at the stage of optimal digestion.
3. Beef and milk produced on grass in temperate regions have a composition with nutritional advantages over beef or milk produced in indoor systems based on concentrate.
4. Increasing water soluble carbohydrate content of grasses would diminish environmental contamination by reducing the excretion of urea through the urine.
5. Increasing the digestibility of the forage would reduce methane production per unit of animal product.

Keywords: grazing, temperate pasture, pasture quality, environment

Introduction

Animal productivity on grazing is not always satisfactory and individual performance expressed as body weight gain or milk yield is generally lower than that obtained on intensive systems based on conserved forage and concentrate feeding. Pasture availability and forage quality are the main constraints affecting animal productivity. The nutritional limitation occurring on grazing depends on the activity involved in the ruminant production process. In the cow-calf systems where the main objective is to produce calves efficiently, forage availability from pasture and natural grassland are the main limitation. Nutritional limitations are generally a consequence of a poor forage budgeting program, drought, or a limited amount of conserved forage as hay or silage, more than to forage quality.The situation is totally different in cattle rearing and fattening systems and for grazing dairy cows where nutrient requirements are generally high and cannot always be satisfied by the nutrient content of available forage.

In temperate pasture, grazing management is an important tool to maintain pasture in a vegetative stage for as long as possible. As soon as pasture becomes mature, cell wall content and ligninfication increase affecting forage digestibility and consequently forage intake. Even when high quality temperate pastures are an optimum and cheap source of nutrients for ruminants they not always supply the correct amount of nutrients that high performance cattle require.

Dry matter and organic matter (OM) digestibility, cell wall content, fiber degradability and concentration and degradability of protein are the main parameters used to define the nutrient value of temperate pasture, but these do not always explain the production response obtained. For a better understanding of cattle performance on temperate pasture new insights into the nutritional value of fresh forage will be discussed.

Nutritional value of grasses

Pasture quality has always been considered a limiting factor for animal production in grazing systems. Nutritive value of pasture depends mainly on its chemical composition but other factors may also alter its quality, affecting animal productivity. Quality of pasture in temperate regions is influenced by management, environment and by species and cultivars (Sheaffer *et al.,* 1998). Stage of growth at grazing is very significant due to the negative relationship between grass maturity and forage quality (Cherney *et al.,* 1993). Grazing management becomes an important tool to maintain forage quality, however, factors such as sward height, fertilization program and mixture composition and stability are also important in determining forage quality.

Grass and legume species and cultivars differ in chemical composition therefore, their contribution to the pasture mixture will affect forage quality. This generally occurs because the quality of legumes is superior to grasses, which is attributable to the lower cell wall and higher CP concentrations of legumes compared with cool-season grasses (Buxton, 1996). Nutrient content of forage from temperate pasture is generally higher when grazed than when offered to the animals in conserved form as hay or silage (Glenn, 1994). Ruminal digestibility of dry matter and ruminal degradability of protein are also normally high in grazed forage relative to non-pastured forages.

For years nutritive value of pasture was defined as the product of the dry matter intake, digestibility, and efficiency of utilization of end products of rumen digestion. Even when that definition is maintained new parameters on nutrient composition should be considered. Due to the proved negative relationship between fiber content of the forage and dry matter intake and digestibility, concentration of fiber expressed as NDF or ADF has been the main parameter to define temperate pasture quality. Protein is also an important nutrient provided by pasture to grazing ruminants but generally only subtropical pastures are considered to be limiting in this nutrient. Temperate pasture grazed at an immature stage is generally high in protein, and its degradability in the rumen may be more important than its total content as a measure of nutrient availability. Even when fiber digestibility is still the main factor limiting animal productivity on most grazing systems throughout the world, other nutritional limitations may appear when feeding cattle with higher requirements.

Rumen digestion of temperate pasture

Forage digestibility depends not only on its stage of maturity, its mean fiber and lignin contents, but also on the digestion process that occurs in the rumen. Fiber digestion in the rumen will depend on the digestion rate, which will be affected by bacterial activity, and the retention time in that compartment. Because of that, the rumen environment generated by a specific pasture is fundamental to the study of the nutritional value of that forage.

Crawford *et al.,* (1983), working with conserved forage, reported an optimum rumen fluid pH of 6.6-6.8 for maximum fiber digestion and bacterial yield in the rumen. When rumen pH decreased from the optimum, fiber digestion could be affected. Hoover *et al.,* (1984) pointed out that cellulolytic activity was reduced at pH values lower than 6.0 and fiber digestion ceased at pH values less than 5.0. Efficiency of microbial protein synthesis (MPS) would also be reduced at low pH values (Strobel & Russell, 1986). Studies carried out at INTA Balcarce, in Argentina proved that the rumen environment in cattle grazing high quality forage is different to that reported for cattle fed indoors with diets based on processed feedstuffs like hay, silage and concentrate (Rearte & Santini, 1993). Similar results were obtained by other authors working also with cattle grazing high quality pasture (van Vuuren *et al.*, 1986; Carruthers *et al.*, 1996). Only on pasture of lower quality like wheatgrass (*Agropirum elongatum*), tall fescue (*Festuca arundinacea*), or ryegrass (*Lolium perenne*) at the mature stage, rumen pH was 6.3-6.4. On forages of higher quality like oats or perennial ryegrass in the early vegetative stage, rumen pH was 5.9-6.0. Legumes like alfalfa, even when they are supposed to have a higher buffering capacity, have also caused low rumen pH when grazed at the vegetative stage. In addition to a low rumen pH, a high concentration of volatile fatty acids (90-120 mmol/l) with a low acetate:propionate ratio were measured in rumen fluid of cows grazing high quality pasture. Carruthers *et al.* (1996) and other studies reported a lower pH in animals grazing high quality pasture than the optimum mentioned earlier even though there was no evidence that a depression in fiber digestion or a lower microbial efficiency resulted.

De Veth & Kolver (2001) studying the effect of rumen pH on fiber digestion in temperate pasture observed that digestion and microbial protein synthesis were largely insensitive to pH across a broad range (5.8 to 6.6), and a large reduction in both occurred only when pH was 5.4 or lower. These authors concluded that short periods (4h) of suboptimal pH (5.4) reduced the digestibility of DM, OM, and NDF by approximately 4 percentage units. Longer periods (>8h) of suboptimal pH were required before microbial protein synthesis was compromised. These results suggest that the period of time that pH is below optimal pH may be more critical for digestion than the relationship between mean daily pH and optimal pH. The low ruminal pH in grazing cattle is not consistent with the fiber content of pasture but it could be associated with the high ruminal concentration of VFA or the high buffering capacity of fresh pasture (Erdman, 1988). It is also possible that with grazing, salivation rate is lower than expected due to the low content of physically effective fiber in high quality pasture (Allen, 1995).

Related to dietary N metabolism occurring in grazing cattle, ammonia (NH_3) concentration in rumen fluid is high, well above the minimum required for an optimum microbial synthesis owing to the high protein content and degradability of fresh forage. When NH_3 production rate in the rumen is higher than the rate at which it is utilized by the microbial population, its concentration at ruminal level increases, a high proportion is absorbed through the rumen wall, part of it is recycled to the rumen via saliva, and the remainder is converted into urea in the liver and finally excreted in the urine (Siddons *et al.*, 1985). This would be expected to reduce the efficiency of utilization of the dietary N. Studies with growing cattle consuming fresh forage carried out by Beever *et al.* (1986), showed that seasonal changes in N content can lead to different efficiencies for microbial utilization of ruminally degraded protein affecting the non-ammonia N (NAN) flows to the duodenum. Much of the N consumed by cows grazing high quality temperate pasture never reaches the duodenum because of high losses of degraded N due to NH_3 absorption from the rumen, conversion to urea in liver and excretion through urine (Beever, 1993). In a trial carried out at INTA Balcarce (Elizalde *et al.*, 1994; Elizalde *et al.*, 1996), the efficiency of microbial protein synthesis in cows grazing winter oats (WO) (*Avena sativa*) at five different maturity stages, autumn, early winter, winter, spring and late spring, was studied. The effects of date of harvest on N metabolism showed that microbial N production was 24.6 and 32.6 g/kg organic mater digested in the rumen and N loss was 44 and 7% of total N intake for autumn and spring pasture, respectively. These differences were associated with differences in the total protein and soluble carbohydrate contents of the forage at different times of the year. While in autumn the CP content of grass is very high, the amount of soluble carbohydrate is low compared to the concentration that temperate grasses have in spring. It is clear that an imbalance of energy:protein occurs in the rumen of cattle grazing WO in autumn. This imbalance is reflected in a higher NH_3 concentration in the rumen, well above the minimum required for optimum bacterial activity.

Sugar and protein imbalance

The water soluble carbohydrate (WSC) content of temperate grasses is variable but normally too low to balance the high content of highly degradable protein. Consequently, the sugar:protein ratio has become very important in defining the nutritional value of fresh forage in temperate pasture. Supplementation with starchy concentrate could be a way to increase the energy content of the diet but it is not the ideal energy source because its energy, in the form of adenosine triphosphate (ATP), is not liberated in the rumen at the same time that maximum rate of protein degradation occurs. The optimal energy source to balance pasture protein would be WSC contained within the forage.

Conventional plant breeding techniques and gene manipulation have been applied lately to improve the energy: protein ratio of temperate grasses (Miller *et al.*, 2001; Moorby *et al.*, 2001). Promising results obtained through grazing management and grass production techniques could also be alternatives to improve the nutritive value of grasses. Nitrogen fertilization, regrowth age, and time of the day at grazing, were observed to affect the chemical composition of grasses especially those components related to its energy: protein ratio. Several trials have proved that lowering N fertilization level decreased the protein content of pasture, increased its WSC content and improved dietary N utilization by animals (Peyraud *et al.*, 1997; Delaby *et al.*, 1996; van Vuuren *et al.*, 1992). The increase in WSC content by lowering N fertilization can be attributed to a lower use of WSC for protein synthesis and plant growth, which would be influenced in turn by the levels of N available to the plant. Other studies have shown that nutritive value of grass could also be affected by age of regrowth. Delagarde *et al.* (2000) observed that biomass and DM content were increased by increasing regrowth age from 21 to 28 and 35 days, while CP content fell sharply and WSC increased strongly in the same period. The decrease in CP content with ageing may be linked directly to the increase in total WSC content mentioned above.

The chemical constituents of grass also vary at different times of the day due mainly to changes in the photosynthesis and gas exchange rates with the atmosphere. Delagarde *et al.* (1997) and Orr *et al.* (2001) observed that WSC content increased from morning to evening due to the accumulation of photosynthesized simple sugars (especially saccharose), and then fell during the night due to respiration, protein synthesis and export towards storage organs. However, even when the level of N fertilization, days of regrowth at grazing or time of the day at which pasture was offered to the animals have been managed to increase the sugar content of the grass and improve efficiency of N utilization by the animals, responses in animal productivity were not consistent.

Rearte *et al.* (2003) obtained high sugar grass by combining low N fertilization (40kg vs 80kg N), a long regrowth period (28 vs 21 days) and evening instead of morning grazing. Crude protein degradability was reduced on the high sugar grass in line with values reported by Delaby (2000), Peyraud *et al.* (1997) and Delaby *et al.* (1996) for grasses receiving different levels of N fertilization. Dry matter intake was not affected by the sugar:protein ratio of the grass but cows on high sugar grass consumed more soluble carbohydrate and less protein than cows on control grass. However milk yield was not improved by feeding high sugar grass and averaged 22.3 kg/day. Digestion data may explain the lack of a production response. Rumen fluid pH was not affected by the different diets but fluctuation of pH throughout the day was higher in cows fed high sugar grass. Ammonia concentration was far lower in that diet with a relatively constant concentration throughout the day at between 1 and 3 mmol/l. There was no significant effect of treatment on OM digestibility but total tract digestibility and ruminal digestibility of NDF and ADF were significantly lower on cows offered the high sugar grass. Cellulolytic activity was numerically five percentage units lower in cows on high sugar grass. A low NH_3 concentration in rumen fluid would have decreased the capacity for fiber digestion in those cows. Similar results were reported by Delagarde *et al.* (1997) when feeding cows with low N fertilized grass.

Fiber degradation measured in-situ in dry cows fed a standard hay:concentrate diet was similar in both grasses; therefore, the rumen environment would appear to have been responsible for the observed NDF digestibility depression with the high sugar grass. The NH_3 concentration measured in the rumen with high sugar grass was 2.06 mmol/l. This value would not be at the limit suggested by several authors for depressing the digestion rate of forage cell-wall (Orskov, 1992; Satter & Slyter, 1974). However, there is still some ambiguity about the minimum NH_3 concentration required and data on fiber digestibility in dairy cows consuming highly fermentable feeds and with low ruminal N are scarce. Most of the information on reduced fiber digestion caused by ruminal N deficiency was obtained with low quality forage lacking available ruminally degradable carbohydrates (Orskov, 1992). In the early in-vitro trial of Satter and Slyter (1974), the optimal NH_3 concentration to support maximal microbial growth was estimated to be 50 mg/l (2.94 mmol/l) and the addition of non protein N (NPN) to ruminant rations would be justified only if ruminal NH_3 was lower than that concentration. However, Ruiz *et al.* (2002) reported an improvement in ruminal NDF digestibility by increasing ruminal NH_3 concentration from 45 mg/l to 100 mg/l (2.64 to 5.88 mmol/l) by the addition of urea to the ration of dairy cows containing highly degradable carbohydrates. Delagarde *et al.* (1997) also reported a low NH_3 concentration in the rumen (2.7 mmol/l) and a depression in fiber digestion when feeding cows grasses containing 106 g/kg of CP. Similar results were obtained Elizalde *et al.* (1994) evaluating autumn versus spring grazed reygrass. In their trial NDF digestion in cows offered spring grass containing 117 g/kg DM of CP and 205 g/kg DM of WSC, was 10 percentage units lower than that measured on autumn grazing cows consuming pasture containing 230 g/kg DM of CP and 38 g/kg DM of WSC. Ruminal NH_3 concentrations were 3 and 19 mmol/l in cows grazing spring and autumn grass, respectively. It seems that the NH_3 concentration for optimal fermentation and bacterial yield depends on the substrates and the minimum required to maximise digestion is a function of the fermentability of the diet. High quality forage with high digestible fiber would support a greater growth of fiber-digesting bacteria that require NH_3 resulting in the minimum required for the optimal fermentation being greater. In Rearte (2003) the higher ingestion of WSC with the high sugar grass, would compensate for the decrease in ruminal fiber digestion thus maintaining the total quantity of OM that was fermented.

The lack of response to increasing the WSC content of grasses in the efficiency of microbial protein synthesis is in contrast to what numerous authors suggest with respect to supplying extra readily available energy to the rumen microbial population (Nocek & Russell, 1988; Rooke *et al.*, 1987). Stern *et al.* (1978), in contrast to our finding, reported that readily fermentable carbohydrates are more effective than other energy sources in increasing microbial growth at similar VFA production and digestibility for a range of diets. It is important to note that in all the reported studies (Elizalde *et al.*, 1996; Beever *et al.*, 1978;

Peyraud *et al*., 1997; Lee *et al*., 2002) grasses with high WSC had a CP content no higher than 110 g/kg DM and the NH_3 concentration in the rumen was always less than 3 mmol/l. Even when Clark *et al.* (1992) suggested that a mean value of 1.4 mmol/l of NH_3 was the minimum required for maximizing microbial protein synthesis in dairy cows, this value could vary with the fermentable energy of the diet. If a deficiency of N from the diet or recycling occurs, which could be the case in our and the other studies, carbohydrate fermentation and microbial growth will become uncoupled leading to a futile cycle of bacterial energy metabolism and a consequent reduction in the efficiency of microbial protein synthesis.

Effect of pasture composition on product quality

As well as production efficiency, product composition and quality may also be affected by the nutrient composition of pasture offered to grazing ruminants. An excess of highly degradable protein in the diet could affect milk processing quality due to the decrease in the casein fraction and the increase in urea and NPN in milk. Rearte & Santini (1993) reported a lower secretion of urea-N in the milk of cows consuming high sugar grass than would be explained by a lower intake of highly degradable protein. Also the nutritional value of milk and beef is a very important aspect to consider because consumers are becoming more concerned about healthy diet and there is a consensus against the consumption of animal fats. Ruminant products have become unpopular foods due to their high levels of saturated fat and cholesterol, which has been seen by most medical opinion as increasing the risk of the development of certain coronary heart diseases.

But not all beef and milk have the same fat content and composition. Milk from cows fed high sugar grass of longer regrowth had a lower fat content than cows on control grass (Rearte, 2003). This lower milk fat agrees with the findings of Gonda *et al.* (1992), Bauchart *et al.* (1984) and Dewhurst *et al.* (2001) who reported a decrease in milk fat content and a greater degree of fatty acid saturation with longer regrowth periods. Not only is the fat content of milk affected by pasture composition but also its fatty acids (FA) composition is affected. Milk fat of cows fed high sugar grass contained a higher proportion of short chain FA and a lower proportion of long chain FA than milk fat from control cows. It can be seen that the milk fat reduction occurring in cows fed high sugar grass is due to a decrease in the quantity of long chain FA absorbed by the mammary gland and secreted in milk which could not be compensated for by the increase in the quantity of short chain FA and palmitic acid synthesized within the mammary gland.

High quality temperate pasture at the vegetative stage is a rich source of long chain unsaturated FA (up to 30 g/kg of FA on a dry matter basis), of which proportionally about 0.9 are unsaturated C18 acids (Murphy *et al*., 1995). As the growth stage of grasses advance the lipid content decreases and FA become more saturated (Bauchart *et al.*, 1984). Dewhurst *et al.* (2001) reported a decrease of proportionally 0.17, 0.25, 0.34 and 0.45 in the concentration of C18:0, C18:1, C18:2 and C18:3, respectively as regrowth length of perennial rye grass increased from 20 to 38 days. The amount of long chain unsaturated FA consumed was therefore, much higher for cows fed normal control grass harvested at an earlier vegetative stage than for high sugar grass. Long chain unsaturated FA could affect fatty acid synthesis, by their inhibitory effect on Acetyl-CoA carboxylase, a key enzyme involved in the metabolic pathway for de-novo synthesis of short chain FA from acetate and 3-hydroxybutyrate in the mammary gland (Barber *et al.*, 1997). Thus, a higher contribution of long chain unsaturated FA from normal grass compared to high sugar grass, would increase their concentration in milk fat but simultaneously would inhibit de novo synthesis decreasing the percentage of short chain (< C14) and C16:0 FA in milk fat.

Among the unsatured FA in milk the concentration of conjugated linoleic acid, (cis-9 trans-11 CLA, referred to as CLA subsequently) has become an important matter in the last few years due to its proved nutraceutical properties, including anticarcinogenic activity (Ip *et al.*, 1991) and some inhibition of the development of atherosclerosis in animals (Lee *et al.*, 1994). It is well known that milk produced in temperate regions by grazing cows is higher in CLA content compared to that produced on intensive systems based on conserved forage and concentrate (Jahreis *et al.*, 1997; Kelly *et al.*, 1998, Dhiman *et al.*, 1999; Stene *et al.*, 2002). In Rearte (2003) cows fed the high sugar grass produced milk with a lower content of CLA than cows on the normal grass diet with values that were similar to those reported by Chilliard *et al.* (2000), Dhiman *et al.* (1999), and Kelly *et al.* (1998), for cows consuming fresh young grasses as the main component of the diet. The concentration of trans-11 C18:1 fatty acid (vaccenic acid) in milk was higher in cows fed the normal grass. Conjugated linoleic acid is an intermediate product in the biohydrogenation of linoleic acid, after lipolysis of dietary lipid has occurred. Initial isomerisation is followed by the saturation of the cis-9 double bond resulting in the production of vaccenic acid, the major trans isomer of ruminant tissues. Vaccenic acid is important not only because of its high concentration in ruminant tissues but because of its positive correlation with CLA concentration in milk. Conjugated linoleic acid is not an intermediate in linolenic acid biohydrogenation pathways but vaccenic acid is one of the final products. Corl *et al.* (2000) reported that approximately 0.75 of CLA in milk comes from endogenous synthesis in the mammary gland via the desaturation of vaccenic acid by the action of hepatic microssomal Δ^9-desaturase (Griinari *et al.*, 1997). Considering the increase in lipid intake with normal grass and assuming that approximately 0.5 consists of FA a much higher CLA concentration in cows fed that grass compared to high sugar grass could be expected. Here again a compensatory effect of both diets on CLA content in milk could be occurring. A higher Δ^9-desaturase activity in mammary gland of cows fed high sugar grass could be compensating for the favorable effect of the higher supply of CLA precursor with the normal grass. A trend for a lower trans-11 C18:1/CLA ratio observed in milk produced with high sugar grass (2.08 vs 2.34) would support this hypothesis. Nutritional and hormonal control of Δ^9-desaturase activity, involving insulin as the main positive agent, was reported by Ntambi (1995); therefore, it is possible that the higher WSC intake of cows fed high sugar grass increased circulating insulin levels, which would explain the observed increase in Δ^9-desaturase activity. Although intake of unsaturated FA and Δ^9-desaturase activity in mammary gland seem to be the main variables determining the CLA content of milk, production of CLA and CLA precursors in the rumen is also important. The rate of CLA and vaccenic acid production depends on the rumen environment and microbial fermentation, which could be affected by changes in the dietary carbohydrate source. It was suggested that the high content of sugar and soluble fiber found in young immature pasture may create a rumen environment favorable for CLA production or reduced bacteria utilization compared to more mature grasses or conserved forage (Kelly *et al.*, 1998).

Effect of pasture composition on environment pollution

For years plant breeders tried to improve pasture quality aiming only to improve the production response. The objective was to maximize milk yield, daily weight gain and feed conversion efficiency without taking the environment into account. Actually, production systems on grazing still aim to satisfy the nutrient requirements of high producing animals but now also try to minimize negative environmental impact. Temperate pasture is a good source of nutrients for ruminants but its high content of highly degradable protein and the excessive use of N fertiliser with intensive grazing have made N excretion higher than desired. The high NH_3 concentration in the rumen as a result of the high degradability of pasture protein not only affects the efficiency of N utilization and animal performance, but also contributes to environmental pollution due to N excreted in the urine. Reducing degradation of dietary protein in temperate pasture has become a new goal for plant breeders and advances in this are expected to be achieved in the near future. In the meanwhile the use of varieties with high sugar:protein ratios achieved by plant breeding and gene manipulation or using appropriate grazing management and grass production techniques may contribute to the reduction of N excretion to the environment.

Studying N balance in cows fed grasses of different sugar:protein ratios, Rearte *et al.* (2003) observed that faecal N excretion and N secretion in milk were similar with both grasses but urine N output was significantly lower in cows on high sugar grass. Of the total N excreted in urine, urea N was the main N component affected by diet, being significantly higher in cows fed normal grass of low sugar content. Considering the proportion of digestible N that is not excreted in urine as a measure of biological value (BV) of ingested N, high sugar grass had a N component of higher BV than control normal grass. Expressed as a proportion of N intake, cows on high sugar grass had a higher excretion of N through faeces and milk and a lower excretion through the urine than cows fed normal grass. In that study N in milk of cows on high sugar grass was proportionally 0.52 higher and N in urine was proportionally 0.42 less than cows on control grass. The gross efficiency of use of dietary N for milk production by animals offered the high sugar grass (proportionally 0.35 of dietary N excreted in milk) and the control grass (proportionally 0.23 of dietary N excreted in milk) were similar to values quoted by Miller *et al.* (2001) and Peyraud (1997). The lower excretion of urea-N in urine with high sugar grass may reflect a difference in net N balance across the rumen together with protein catabolism. Protein catabolism was apparently hardly modified in this study but the difference in ruminal N balance between the two diets was in line with the difference in urea-N output which was actually reduced by 131 g/day with the high sugar grass compared to normal control grass.

Beside N contamination of the environment, methane (CH_4) production is another ruminant byproduct of concern in relation to the environmental impact of different production systems. Methane and CO_2 are the main gases contributing to the warming of the atmosphere and ruminants are identified as one of the main sources of methane gas. This gas is a byproduct of the enteric fermentation occurring in the rumen and its production is highly correlated with diet digestibility. Increasing the sugar content of grass improves its digestibility and therefore the production of CH_4 per production unit (kg of beef or l of milk) is reduced. Rearte (2003) estimated that production of CH_4 could be reduced from 644 to 379 g /kg of beef or 24 to 15 g /l of milk by improving diet digestibility by 8 percentage units.

Conclusion

Improvement of the nutritive value of temperate pasture continues to be a priority in grazing systems, but new quality parameters should be incorporated in the aims and objectives of plant breeders and pasture management specialists. Dry matter content, fiber digestibility, metabolizable protein and mineral content, among others, are still important parameters to consider but new analysis should be incorporated for a better understanding of the process that explains animal performance. Temperate pasture is an excellent source of nutrients for ruminants but a high energy:protein imbalance can occur when it is offered at the stage of optimal digestion. Increasing WSC content by breeding or grazing management could potentially not only improve animal performance but could also diminish environmental contamination by reducing the concentration of NH_3 in the rumen and consequently the excretion of urea through the urine. Increasing the digestibility of the forage would also reduce CH_4 production per unit of animal product.

Even when animal productivity, expressed as weight gain or milk yield, is still the main objective of using high quality pasture, ruminant product composition has to be considered in order to satisfy the present market demands. Beef and milk produced on grass in temperate regions have a composition with nutritional advantages over beef or milk produced in indoor systems based on concentrate. Beef and milk produced on temperate pasture have a lower content of saturated FA reducing the incidence and risk of arterial coronary diseases. It has also been proven that beef and milk from pasture are the richest natural dietary sources of CLA, which has been shown to have anti-cancer properties. Milk and beef quality and nutrient composition with potential health benefits will be the main characteristics demanded by the consumer in the near future. Grass breeding and grazing management should be applied in a way that will maintain or even improve the benefits and the properties of the product obtained from cattle grazing temperate pastures.

References

Allen, M.S. (1995). Relationship between ruminally fermented carbohydrate and the requirement for physically effective NDF. *Journal of Dairy Science,* 78 (Suppl.1), 265.

Barber, M.C., R.A. Clegg; M.T. Travers & R.G. Vernon (1997). Lipid metabolism in the lactating mammary gland. *Biochimica Biophysica Acta*, 1347, 101-126.

Bauchart, D., R. Verite & B. Remond (1984). Long-chain fatty acid digestion inlactating cows fed fresh grass from spring to autumn. *Canadian Journal of Animal Science*, 64 (Suppl. 1) 330-331.

Beever, D.E., R.A. Terry; S.B. Cammell & A.S. Wallace (1978). The digestion of spring and autumn harvested perennial ryegrass in sheep. *Journal of Agricultural Science*, Cambridge, 90, 463-470.

Beever, D.E., Dhanoa, M.S., Losada, H.R., Evans, R.T., Cammell S.B. & J. France (1986). The effect of forage species and stage of harvest on the processes of digestion occurring in the rumen of cattle. *British Journal of Nutrition*, 56, 439-454.

Beever, D.E. (1993). Ruminant animal production from forages. Present position and future opportunities. *Proceedings of the XVII International Grassland Congress*. Palmerston North, New Zealand, 535-542.

Buxton, D.R. (1996). Quality related characteristics of forages as influenced by plant environment and agronomic factors. *Animal Feed Science and Technology*, 59, 37-49.

Carruthers, V.R., P.G. Neil, & D.E. Dalley (1996). Microbial protein synthesis and milk production in cows offered pasture diets differing in non-structural carbohydrate content. *Proceedings New Zealand Society of Animal Production*, 56, 255-259.

Cherney, D.J.R., J.H. Cherney & R.F. Lucey. (1993). In vitro digestion kinetics and quality of perennial grasses as influenced by forage maturity. *Journal of Dairy Science*, 76, 790-797.

Chilliard, Y., A. Ferlay, R.M. Mansbridge & M. Doreau. (2000). Ruminant milk fat plasticity: nutritional control of saturated, polyunsaturated, trans and conjugated fatty acids. *Annales de Zootechnie*, 49, 181-205.

Clark, J.H., T.H. Klusmeyer & M.R. Comeron. (1992). Microbial protein synthesis and flows of nitrogen fractions to the duodenum of dairy cows. *Journal of Dairy Science*, 75, 2304-2323.

Corl, B.A., L.H. Bau,gard, D.E. Bauman & J.M. Griinari (2000). Role of Δ9-desaturase in the synthesis of the anticarcinogenic isomer of conjugated linoleic acid and other milk fatty acids. In: Proceedings of the Cornell Nutrition Conference for Feed Manufacturers, Cornell University, Ithaca, New York, 203-212.

Crawford, R.J., B.J. Shriver, G.A. Varga & W.H. Hoover (1983). Buffer requirements for maintenance of pH during fermentation of individual feeds in continuous cultures. *Journal of Dairy Science*, 66, 1881-1890.

Delaby, L. (2000). Effet de la fertilisation minérale azotée des prairies sur la valeur alimentaire de l'herbe et les performances des vaches laitières au pâturage. *Fourrages*, 164, 421-436.

Delaby, L. J.L. Peyraud, R.Vérité & B.Marquis (1996). Effect of protein content in the concentrate and level of nitrogen fertilization on the performance of dairy cows in pasture. *Annales de Zootechnie* 45, 327-341.

Delagarde, R., J.L. Peyraud & L. Delaby (1997) The effect of nitrogen fertilization level and protein supplementation on herbage intake, feeding behaviour and digestion in grazing dairy cows. *Animal Feed Science and Technology*, 66, 165-180.

Delagarde, R., J.L. Peyraud, L. Delaby & P. Faverdin. (2000). Vertical distribution of biomass, chemical composition and pepsin-cellulase digestibility in a perennial ryegrass sward:interaction with month of year, regrowth age and time of day. *Animal Feed Science and Technology*, 84, 49-68.

de Veth M.J. & E.S. Kolver. (2001). Digestion of ryegrass pasture in response to change in pH in continuous culture. *Journal of Dairy Science*, 84, 1449-1457.

Dewhurst, R.J., N.D. Scollan, S.J. Youell, J.K. S. Tweed & M.O. Humphreys (2001). Influence of species, cutting date and cutting interval on the fatty acid composition of grasses. *Grass and Forage Science*, 56, 68-74.

Dhiman, T.R., G.R. Anand, L.D. Satter & M.W. Pariza (1999). Conjugated linoleic acid content of milk from cows fed different diets. *Journal of Dairy Science*, 82, 2146-2156.

Elizalde, J.C., F.J. Santini & A.M. Pasinato (1994). The effect of stage of harvest on the processes of digestion in cattle fed winter oats indoors. I. Digestion of organic matter, neutral detergent fiber and water-soluble carbohydrates. *Animal Feed Science and Technology*, 47, 201-211.

Elizalde, J.C., F.J. Santini & A.M. Pasinato (1996). The effect of stage of harvest on the processes of digestion in cattle fed winter oats indoors. 2. Nitrogen digestion and microbial protein synthesis. *Animal Feed Science and Technology*, 63, 245-255.

Erdman, R.A. (1988). Dietary buffering requirements of the lactating dairy cows: A review. *Journal of Dairy Research*, 71, 3246-3266.

Glenn, B.P. (1994). Grasses and legumes for growth and lactation. In: Proceedings Cornell Nutrition Conference for Feed Manufacturers, Cornell University, Ithaca, New York, 1-18

Gonda, H.L., D.H. Rearte, P.T. García, J.J. Santini & M. Maritano (1992). Efecto del contenido de lípidos de la pastura sobre lacomposición de la gasa de la leche. *Revista Argentina de Producción Animal*, 12, 235-251.

Griinari, J.M., P.Y. Chouinard & D.E. Bauman (1997). Trans fatty acid hypothesis of milk fat depression revised. In: Proceedings of the Cornell Nutrition Conference for Feed Manufacturers, Cornell University, Ithaca, New York, 208-216.

Hoover, W.H., C.R. Kincaid, G.A. Varga, W.V. Thayne, & L.L. Junkins, Jr. (1984). Effects of solids and liquid flows on fermentation in continuous cultures. IV. pH and dilution rate. *Journal of Animal Science* 58, 692–699.

Ip, C., S.F. Chin, J.A. Scimeca & M.W. Pariza. (1991). Mammary cancer prevention by conjugated linoleic acid. *Cancer Research*, 51, 6118-3124.

Jahreis, G., J. Fritsche & H. Steinhart (1997). Conjugated linoleic acid in milk fat: high variation depending on production system. *Nutrition Research*, 17, 1479-1484.

Kelly, M.L., E.S. Kolver, D.E. Bauman, M.E. Van Amburgh & L.D. Muller (1998). Effect of intake of pasture on concentration of conjugated linoleic acid in milk of lactating cows. *Journal of Dairy Science*, 81, 1630-1636.

Lee, K.N., D. Kritchevsky & M.W. Pariza (1994). Conjugated linoleic acid and atherosclerosis in rabbits. *Atherosclerosis*, 108, 19-25.

Lee, M.R.F., L.J.Harris, J.M.Moorby, M.O.Humphreys, M.K.Theodorou, J.C.MacRae & N.D.Scollan (2002). Rumen metabolism and nitrogen flow to the small intestine in steers offered Lolium perenne containing different levels of water-soluble carbohyydrate. *Animal Science*, 74, 587-596.

Miller, L.A., J.M. Moorby, D.R. Davis, M.O. Humphreys, N.D. Scollan, J.C. MacRae & M.K. Theodorou (2001). Increased concentration of water-soluble carbohydrate in perennial ryegrass (Lolium perenne L.): milk production from late-lactation dairy cows. *Grass and Forage Science,* 56, 383-394.

Moorby, J.M., L.A. Miller, R.T.Evans, N.D. Scollan, M.K. Theodorou & J.C. MacRae (2001). Milk production and N partitioning in early lactation dairy cows offered perennial ryegrass containing a high concentration of water soluble carbohydrates. *Proceedings of the British Society of Animal Science*, 6.

Murphy, J.J., J.F. Connolly & G.P. McNeill (1995). Effects on cow performance and milk fat composition of feeding full fat soyabeans and rapeseeds to dairy cows at pasture. *Livestock Production Science*, 44, 13-25.

Nocek, J.E. & J.B. Russell (1988). Protein and energy as an integrated system. Relationship of ruminal protein and carbohydrate availability to microbial synthesis and milk production. *Journal of Dairy Science* 71, 2070-2107.

Ntambi, J.M. (1995). The regulation of stearoyl-CoA desaturase (SCD). *Proceedings of Lipid Research*, 34, 139-150.

Orskov, E.R. (1992). Host animal protein requirement and protein utilization. In: Protein Nutrition in Ruminants, 2nd Edition, Harcourt Brace Jovanovich, San Diego, CA, 85-136

Orr, R.J., S.M. Rutter, P.D. Penning & A.J. Rook (2001). Matching grass supply to grazing patterns for dairy cows. *Grass and Forage Science*, 56, 352-361.

Peyraud, J.L., L. Astigarraga & P. Faverdin (1997). Digestion of fresh perennial ryegrass fertilized at two levels of nitrogen by lactating dairy cows. *Animal Feed Science and Technology*, 64, 155-171.
Rearte, D.H., & F.J. Santini 1993. Rumen digestion of temperate pasture: Effects on milk yield and composition. *Proceedings XVII International Grassland Congress New Zealand*, 562-563.
Rearte, D.H., J. L. Peyraud, J.L. & C. Poncet (2003). Increasing the water soluble carbohydrate / protein ratio of temperate pasture affects the ruminal digestion of energy and protein in dairy cows. *Proceedings VI International Symposium on the Nutrition of Herbivores*, Merida, Mexico, Tropical and Subtropical Agroecosystems, 3, 251-254
Rearte, D.H. (2003). Sustentabilidad de los sistemas ganaderos de carne y leche en el marco del actual proceso de transformación de las empresas. Memorias Congreso AACREA-Centro, Córdoba, 24-32.
Rooke, J.A., N.H. Lee & D.G. Armstrong (1987). The effects of intraruminal infusions of urea, casein, glucose syrup, and a mixture of casein and glucose syrup on nitrogen digestion in the rumen of cattle receiving grass-silage diets. *British Journal of Nutrition*, 57, 89-94.
Ruiz, R., L.O. Tedeschi, J.C. Marini, D.G. Fox, A.N. Pell, G. Jarvis & J.B. Russell (2002). The effect of a ruminal nitrogen (N) deficiency in dairy cows: Evaluation of the Cornell net carbohydrate and protein system ruminal N deficiency adjustment. *Journal of Dairy Science*, 85, 2986-2999.
Satter, L.D. & L.L. Slyter (1974). Effect of ammonia concentration on rumen microbial protein production in vitro. *British Journal of Nutrition*, 32, 199-208.
Sheaffer, C.C., P. Seguin & G.J. Cuomo (1998). Sward characteristics and management effects on cool-season grass forage quality. In: Cherney, J.H. and D.J.R.Cherney (eds.) Grass For Dairy Cattle, Cab International, 73-99.
Siddons, E.C., J.V. Nolan, D.E. Beever & J.C.Mac Rae (1985). Nitrogen digestion and metabolism in sheep consuming diets containing contrasting forms and levels of N. *British Journal of Nutrition*, 54, 175-187.
Stene, O., E. Thuen, A. Haug & P. Lindstad. (2002). The effects of grazing versus indoor feeding on cow milk fatty acid composition. In: J.L. Durand, J.C. Emile, C. Huyghe and G. Lemaire (eds.) Multi-Function Grasslands. Quality Forage, Animal Products and Landscapes. 600-601.
Stern, M.D., H. Hoover, C.J.Sniffen, B.A. Crooker & P.H. Knowlton (1978). Effect of non structural carbohydrate, urea and soluble protein levels on microbial protein synthesis in continuous culture of rumen contents. *Journal of Animal Science*, 47, 944-956.
Strobel, H.J. & J.B. Russell (1986). Effect of pH and energy spilling on bacterial protein synthesis by carbohydrate limited cultures of mixed rumen bacteria. *Journal of Dairy Science*, 69, 2941-2947.
van Vuuren, A.M., C.J. Van der Koelen & J. Vroons de Bruin (1986). Influence of level and composition supplements on rumen fermentation patterns of grazing cows. *Netherlands Journal of Agricultural Research*, 34, 457-467.
van Vuuren, A.M., F. Krol-Kramer, R.A. van Der Lee & H. Corbijn (1992). Protein digestion and intestinal amino acids in dairy cows fed fresh Lolium perenne with different nitrogen contents. *Journal of Dairy Science*, 75, 2215-2225.

Genetic characteristics required in dairy and beef cattle for temperate grazing systems

F. Buckley[1], C. Holmes[2] and M.G. Keane[3]
[1]Dairy Production Research Centre, Teagasc, Moorepark, Fermoy, Co. Cork, Ireland, Email: fbuckley@moorepark.teagasc.ie
[2]IVABS, Massey University, Palmerston North, New Zealand
[3]Beef Production Research Centre, Teagasc, Grange, Dunsany, Co. Meath, Ireland

Key points

1. Only about 10% of the world's milk is produced from grazing systems. Consequently the majority of dairy cattle have not been selected under grazing, nor on seasonal systems. This is not true for beef cattle, for which the majority, especially the dams, are managed under seasonal grazing systems.
2. In grazing systems daily feed intake is limited to lower levels than are achievable on concentrate plus conserved forage rations. Consequently, cows most suited to grazing environments are likely to have a lower genetic potential for milk production than cows selected in high concentrate systems, to minimise their relative energy deficit.
3. The traits required under grazing will include those for other systems; high yields of milk with high milk solids, efficient converters of feed to product, functionality, good fertility, health and longevity. Successful grazing systems require dairy cows that are adapted to achieving large intakes of forage relative to their potential milk yields, and therefore able to meet production potential exclusively from forage. Grazing cattle must also be able to walk long distances, and in seasonal systems, must be able to conceive and calve once every year. The ability to be productive when milked once daily may also be desirable in low cost grazing systems in the future.
4. Intensive selection for milk production within the Holstein-Friesian breed on high concentrate diets has generally resulted in a genotype that is not well suited to grazing (high forage) systems, in which these cows exist in permanent energy deficit. This unsuitability is particularly true for seasonal systems, for which good fertility is an essential trait.
5. There is now strong evidence for the existence of interactions between genotype of dairy cattle and feeding system, where the genetics and the systems differ widely. Therefore mutual compatibility between the cow and the system must be optimised for production and profit.
6. The New Zealand Friesian and New Zealand Jersey, and crosses between them, or with other dairy breeds, including the North American Holstein-Friesian, have been shown to be well suited to grazing systems. Increasing evidence suggests that genetics from some Scandinavian breeding programs, e.g. the Norwegian Red, may also be suited to grazing systems, where good fertility is essential.
7. Developments in international sire evaluation (adaptations of multiple across country evaluation (MACE)) that enable differences in management systems to be taken into account will provide different breeding values for different conditions. Until then, sires should be proven in the same general management conditions in which the daughters are to be managed.
8. There is little evidence of important G×E interactions in beef cattle, for growth rate, food intake or carcass traits.
9. But beef cows from large, late maturing breeds are relatively more restricted by inadequate nutrition than smaller early maturing types. As a result both their fertility and milk production can be impaired leading to a lower weaning percentage and lighter weaning weight.

10. Interactions between the effects of genotype and nutrition on carcass composition can occur where the level of energy intake is above the muscle deposition capacity of some breeds.

Keywords: grazing, cattle, dairy, beef, genetics, G×E

Introduction

Until recently, in the world of dairy cattle breeding, the term "high genetic merit" was synonymous with high milk production potential. Now it is acknowledged that the complete index for high genetic merit should reflect as many characteristics as are required to reflect total economic profitability. In particular, due to the decline in reproductive efficiency within the Holstein, many countries have diversified their breeding goals to include measures of survivability or functionality (Philipsson *et al.,*, 1994; Visscher *et al.,*, 1994; Veerkamp *et al.,*, 2002). However, economic and physical conditions, and production systems differ widely between countries, so that the optimum complete selection index must also differ widely between countries. World-wide, cattle are farmed under a wide range of environments and management systems. Even within temperate conditions, these can range from grazing on lush temperate pastures or on low quality range-lands, to totally non-grazing or confinement systems, fed on concentrates and conserved roughages. Only about 10% of the world's milk comes from grazing systems (World Animal Review 1995), consequently the majority of dairy cattle have not been selected under grazing. Cattle on grazing systems must be able to graze effectively and to walk long distances, abilities that are not required in confinement systems. In seasonal grazing systems cattle must also conceive and calve at the right time every year. Cattle on non-grazing systems on the other hand can achieve higher total daily feed intakes and consequently higher growth rates or milk yields than those on grazing systems. There is now strong evidence to show that the cattle that are genetically best suited to non-grazing systems are not best suited to grazing systems, an interaction between genotype and feeding system.

In contrast to dairy cattle, the majority of beef cows are managed under seasonal grazing systems, even if their progeny are grown under more intensive systems in some cases. Therefore it is less likely that interactions between genotypes and feeding systems will be found in beef cattle. Pasture finished cattle have a tendency towards yellow fat (Priolo *et al.,* 2001). Some consumers find this undesirable and require that cattle have a period of feedlot finishing to whiten the fat. Other consumers regard fat yellowness as an indicator of more extensive production systems with better animal welfare and more naturally produced beef. There are clear regional differences between consumers on this matter. Pastured finished cattle have lower levels of carcass and intramuscular fat, higher mono and polyunsaturated fatty acids and higher omega 3 and conjugated linoleic acids (Keane & Allen, 1998; French *et al.*, 2000).

New Zealand and Australia have for many years competed profitably at low world market prices, exporting the majority of dairy produce and meat with no subsidies or incentives. Within Europe, the continued reform of the Common Agricultural Policy (CAP) in order to be more market-focused and future WTO agriculture negotiations, suggests a more unstable and unpredictable time ahead. The potential for dairy farmers to secure higher prices for their output to compensate for their increasing costs and downward pressure in product prices as a result of policies at EU level is very limited. Compliance with Directives on the environment and food safety (Nitrogen Vulnerable Zones, Water Framework and Strategic Environmental Assessment) and with International Agreements (Kyoto Protocol and Gothenburg Protocol) will be required. Worldwide, farming systems will have to focus on costs of production, and be sustainable in terms of the environment, people and animal welfare.

Features of intensive grazing systems

Dairy production

Milk production in countries such as New Zealand and Ireland is based on the efficient conversion of high quality grazed grass to milk (Penno, 2000; Dillon *et al.*, 1995). Pasture based systems are capable of low cost milk production with high milk output per hectare. Hence, countries like New Zealand and Ireland have developed widely used production systems in which the herd's pattern of feed demand and milk supply is harmonized to the seasonal production of grazed grass. The principal characteristic of the system is that the entire herd is calved (including the heifers which calve at 2 years) over a short period of time, usually 10 to 14 weeks, at the beginning of the grass-growing season, so that the increasing feed demand of lactation coincides with the increasing pasture growth of spring. In Ireland, but less commonly in New Zealand, supplementation, in the form of conserved forage/hay and/or concentrates/by-products, is offered until grass supply meets cow demand early in the grazing season. Supplementation may also be offered in late lactation to maintain yield and extend lactation. Less supplementary feed is generally offered in New Zealand, and cows tend to be dried off in late summer/autumn before the slower pasture growth of winter, in order to prevent a feed (grass supply) deficit and excessive loss of body condition. Consequently lactation lengths are relatively shorter in New Zealand (230 to 260 days), than in Ireland (260 to 300 days). In total, grazed grass/conserved forage will usually account for at least 90% of the diet. The differences in management practices between New Zealand and Ireland reflect differences in grass growth patterns between the two countries. Grazing systems generally carry lower direct costs than most other more intensive systems (International Farm Comparison Network Report, 2004).

Beef production

In regions where there is grass growth in winter but drought in summer (e.g. Australia) calving is typically in autumn, whereas, in regions where grass growth ceases in winter but continues throughout the summer (e.g. Ireland) calving is typically in spring. In both situations the objective is the same, namely to have the calves weaned and have cows dry with low feed requirements when grass availability is least. Calf growth rate is closely related to cow milk production. This is a function of grass supply and quality. When these are adequate calf growth rate exceeds 1 kg/day form birth to weaning resulting in weanling weights in excess of 300 kg (McGee *et al.*, 1995). As with dairy cows, the objective with beef cows is also compact calving close to the start of the grazing season. At pasture, cows regain body condition quickly and conceive readily when the breeding season commences, thus maintaining a 365-day calving interval. Cows continue to increase in body condition throughout the grazing season resulting in a large reserve at the start of the following winter. This is then depleted over the winter (or drought period) thus minimising expensive feed inputs (Drennan, 1993). In range conditions weaned calves are generally sold for finishing elsewhere either on pasture or in feedlots and males may be castrated or left entire. In the more productive grassland areas the males are generally castrated and both males and females are either finished entirely on pasture or remain at pasture until 3-5 months before slaughter and are then feedlot finished. Subsequently, they exhibit compensatory growth at pasture and generally two thirds or more of the short-fall in weight gain is compensated for by the end of the grazing season (Keane, 2002). Where light grass-finished carcasses are required animals are slaughtered at the end of this grazing season with none or a low level of concentrate supplementation over the final 2-3 months. Where heavier carcasses and/or higher levels of fatness are required animals receive about a 5-month finishing period. Alternatively, they may be fed for moderate gains again in their second winter and finished off pasture the following grazing season. Where animals are slaughtered at 19-21 months of age, towards the end of their second grazing season, grazed grass provides 75-80% of their lifetime feed dry matter intake and half of the remainder comes from grass silage conserved as a necessary component of grassland management. Where animals are finished indoors in their second winter, grazed grass comprises about 50% of their total lifetime dry matter intake with a further 30% coming from grass silage. Animals retained for slaughter until their third season consume about 70% of their lifetime dry matter intake as grazed grass and 25% as grass silage (Keane, 1996).

Traits required by grazing dairy cattle

The need to farm profitably and be sustainable in the future will continue to be a main aim of all future grass-based farming systems. To this end the type of cattle farmed are required to be compatible with the system. In broad terms the definition of a "high merit animal" will continue to be the animal that can produce the largest quantities of high-value milk solids and/or meat most efficiently and sustainably, from the smallest amounts of physical and financial inputs (including feed, and the various costs of labour, health and reproduction). In Ireland, and probably more generally, maximum profitability in grazing systems is achieved by minimising costs and increasing the proportion of grazed grass in the diet of the lactating dairy cow (Shalloo *et al.*, 2004). The traits required under grazing will include those for other systems; high yields of milk with high contents of fat and protein, functional (easily milked) udders, efficient conversion of feed to product, good health, easy care and long, productive lives. Systems based on grazed grass can limit daily feed intake, for example by 20% in Holstein-Friesians (Kolver & Muller, 1998; Kolver *et al.*, 2002), which may be due to a combination of slower rates of intake when grazing, and slower rates of digestion on roughage diets. Increasing the energy concentration of the diet through concentrate supplementation can reduce these physical limitations. Successful grazing systems require dairy cows that are capable of achieving large intakes of forage relative to their genetic potential for milk production so that they are able to meet the requirements almost entirely from grazing. This should also increase the likelihood of survival in the seasonal grazing systems, for which the maintenance of a 365-day calving interval, and good fertility are essential to optimal financial performance (Lopez-Villalobos *et al.*, 2000). This limit to intake when grazing also suggests that cows most suited to grazing environments are likely to have lower genetic potentials for milk production and live weight, than cows best suited to more intensive diets. A requirement to walk long distances is another basic requirement for cows in grazing systems. Some aspects of the cows' legs and feet, which are associated with lameness and the "walkability trait", are under genetic control (Boelling & Pollott, 1998; Goddard & Wiggans, 1999). Therefore, selection under grazing may have lead to an increased ability to walk long distances. The ability to be productive when milked only once daily is likely to be an important characteristic in cows managed in large, seasonal grazing systems, focussed on low-cost profitable farming (Dalley & Bateup, 2004).

The proportion of North American Holstein-Friesian (NAHF) genetics has increased dramatically in Ireland from 9% in 1990 to 63% in 2001 (Evans *et al.*, 2004), and in New Zealand too, from 2% in 1978 to 50% (estimated) in 2002 (Harris & Winkelman, 2000). The sires used have generally been selected for high milk production in a predominantly confined environment (Rauw *et al.,* 1998). There is increasing evidence that intensive selection for milk production within the NAHF breed has generally resulted in a genotype that is less suited to low input (high forage) and grazing systems. This inappropriateness is compounded when seasonality or a requirement for good reproductive efficiency is demanded. The dramatic decline in reproductive performance that has occurred within the NAHF has been well documented (Hoekstra *et al.*, 1994; Pryce & Veerkamp, 2001; Lucy, 2001). In all production systems, but particularly in a seasonal pasture based system, characteristics other than milk production such as reproductive performance and animal health are very important (Veerkamp *et al.*, 2002).

Results from an on-going collaborative study involving Dexcel, Livestock Improvement and Massey University (in New Zealand), and Teagasc, Moorepark (in Ireland), comparing New Zealand-Friesians (NZHF) and the NAHF under grazing but with different systems of supplementation (in New Zealand: Macdonald *et al.,* 2005; Kolver *et al.,* 2004; in Ireland: Horan *et al.*, 2004; Horan *et al.,* 2005a; Horan *et al.*, 2005b; Horan *et al.*, 2005c; Linnane *et al.*, 2004), clearly demonstrate that cows selected under intensive seasonal grazing, NZHF, are more adapted to that environment. Daily milk yield (and total lactation milk yield under Irish conditions, and when given large amounts of supplementation in New Zealand) is higher for the NAHF highly selected for milk production. However, reproductive efficiency and survival is substantially higher for the NZHF cows (Table 1).

Table 1 Effect of strain of Holstein-Friesian on reproductive performance (Horan *et al.*, 2004; Horan *et al.*, 2005a) (and raw means from the New Zealand study in brackets)

	NAHF	NZHF
Milk yield (kg)	6958 (5130)	6141 (4970)
Fat (kg)	279 (216)	275 (236)
Protein (kg)	241 (175)	224 (179)
Pregnancy rate to 1st service (%)	47 (39)	60 (46)
6 week in calf rate (%)	59 (54)	73 (69)
Empty rate after 14 weeks (%)	21 (13)	9 (7)

The final results in New Zealand have not been analysed yet, but some average data for 2003/04, with 2, 3 and 4 year old cows, are presented in Figure 1. The NZ70 and NZ90 (NZ90 identical to the NZHF in Ireland) strains represent high New Zealand Breeding Worth (BW) cows of the 1970s and 1990s, with at least proportionately 0.9 NZ genetics. The OS90 cows represent high BW cows with at least proportionately 0.9 overseas genetics of the 1990s (identical to the NAHF in Ireland).

As expected, the NZ90 cows achieved higher yields per cow and per hectare than the NZ70 cows at all levels of feed allowance. The NZ90 cows produced much more milk solids than the OS90 cows, when both were fed at the two moderate feed allowances (5.5 and 6.0 t DM/cow), mainly because the OS90 were dried off earlier after shorter lactations at these allowances, because of their thinner body condition. However, at their two higher feed allowances, the OS90 cows produced similar yields to the NZ90 cows. Nevertheless, the NZ90 were more profitable than the OS90 cows at all allowances, because of the high cost of the supplements relative to the price for milk (Kolver *et al.,* 2004).

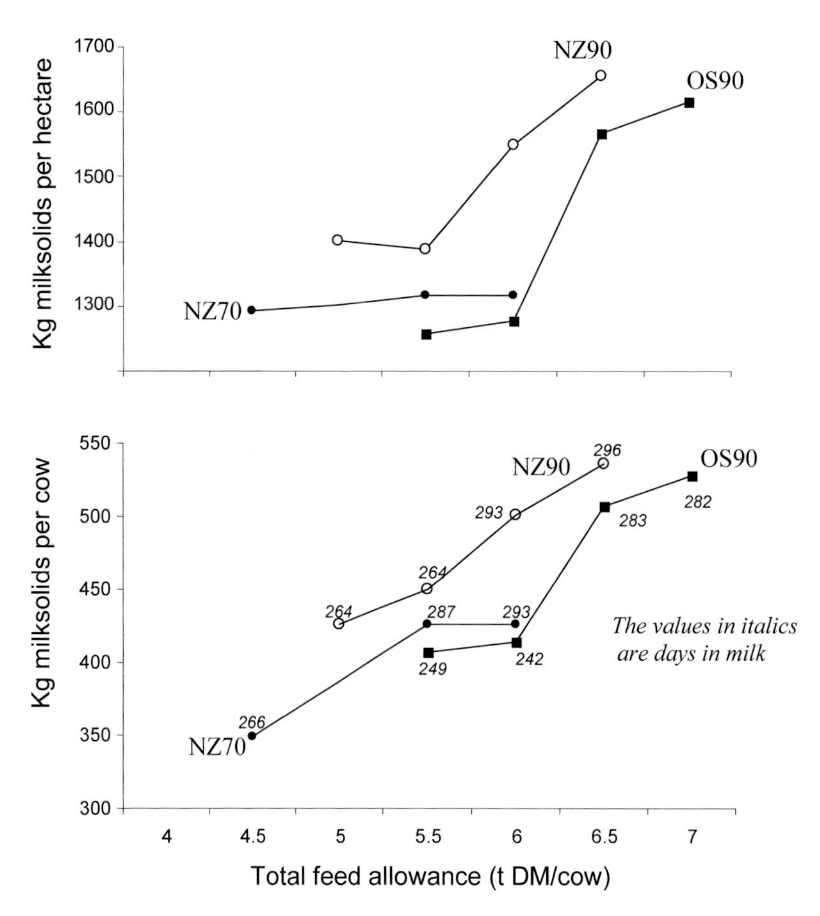

Figure 1 Mean values for yields of milk-solids per cow and per hectare in 2003/04, by three strains of Holstein Friesian cows, on grazing with different levels of supplementation in Hamilton, NZ (Macdonald, 2004).

Although not fully analysed yet, these data from New Zealand and Ireland, do indicate the presence of an interaction between genotype×feeding level or system, which would be of great practical importance. Similar evidence was provided by Fulkerson (2000) for two strains of HF cows grazed on subtropical pastures, with three levels of concentrate feeding.

Body condition and live weight

The importance of body condition score (BCS) in achieving good reproductive performance has been highlighted by many studies (Villa-Godoy *et al.*, 1988; Butler & Smith, 1989; Nebel & McGilliard, 1993; Senatore *et al.*, 1996; Domecq *et al.*, 1997; Buckley *et al.,* 2003; Berry *et al.,* 2003). Other studies suggest that high genetic merit (for yield) cows may have higher energy requirements for maintenance when expressed per $kg^{0.75}$, because one kg of their live weight contains higher proportions of metabolically active tissues (eg:- lean; digestive tract) (Agnew & Yan, 2000; Ferrell & Jenkins, 1985). Historic breeding schemes of the NAHF almost worldwide have focussed solely on increased milk production, which has increased the gap between energy intake and output in early lactation, a gap that is exacerbated on pasture. The correlated response in feed intake in early lactation, due to selection for milk yield, can cover only 40-48% of the extra requirement (Van Arendonk *et al.,* 1991), resulting in a greater mobilization of body reserves (body condition). Experiments comparing cows of high and moderate genetic merit for milk production, associate high genetic merit with more severe negative energy balance (NEB) in early lactation (Holmes, 1988; Gordon, 1996; Veerkamp & Emmans 1995; Oldenbroek, 1984; Buckley *et al.*, 2000, Kennedy *et al.*, 2003). Holmes (1988) also showed that at any given level of BCS at parturition, higher merit cows (for fat yield), lost more body condition (or gained less) post partum, as recently suggested by Dechow *et al.,* (2002). These problems caused by selection for yield, "angularity" (or sharpness, and thinness) and large size, were reviewed by Hansen (2000).

In the New Zealand study outlined above (Kolver *et al.,* 2004; Macdonald *et al.*, 2005) all cows calved in condition scores above 5 (scale 1-10), but by mid to late lactation the OS90 cows were thinnest (about 3.5), with the NZ90 cows intermediate (about 4.0) and the NZ70 cows fattest (about 4.5). The live weights before calving were about 520, 550 and 590 kg for the NZ70, NZ90 and OS90, respectively.

Just as in the New Zealand study, Horan *et al.,* (2005a), have shown that in Ireland the NZHF cows tended to be lighter than the NAHF cows, but tended to have a higher BCS at all times. The NZ strain had BCS values of 3.37 (Scale 1-5) immediately post partum, 2.84 at nadir and 3.13 at drying-off. The BCS of the high production NAHF strain tended to be lower at 3.17, 2.45 and 2.68, respectively, while that of the more moderately selected 'high durability' NAHF strain was intermediate at 3.24, 2.65 and 2.93, for the same times, respectively.

A recent analysis of 100,000 first lactation cows from sire proving herds under seasonal grazing in New Zealand from 1987 to 1999 (Harris & Winkelman 2000) measured the main differences between the NZHF and OSHF two-year olds. Those with a high proportion of OS genetics: were heavier (+43 kg), produced the same fat yield but more protein (8 kg) and more milk (390 litres); but were less likely to conceive to AI (47% vs. 61%), in a 5 to 7 week mating period; and had lower survival rates from first to second lactation (78% vs. 89%), and especially from first to fifth lactation (33% vs. 60%).

The lower body weight (BW) and higher BCS of the NZ strain is likely to be a consequence of more than 50 years of selection for higher yields of milk fat, and later protein also, in cows managed under grazing conditions, with the associated restrictions on maximum daily feed intake and milk yield. Maintenance of these characteristics will probably be helped by the recently included negative weighting on milk volume and BW within the New Zealand Breeding Worth Index (Harris *et al.*, 1996). However, the superior BCS of the NZHF is also likely due to historic culling of relatively infertile cows that failed to meet the targets for seasonal calving, in New Zealand herds, and the moderate to strong positive genetic association that exists between BCS and reproductive efficiency (Pryce *et al.,* 2001; Berry *et al.,* 2003).

Meeting energy demands from grass

Kennedy *et al.*, (2003) showed that with a high proportion of grass in the diet, NAHF cows highly selected for milk production were not capable of eating much more than NAHF cows of a lower production potential. On high concentrate diets however, high producing NAHF cows achieved higher DM intakes. Similar conclusions can be drawn from the results of Kolver *et al.,* (2002), for NZHF and NAHF on either pasture or TMR. Hence the difference in milk yield between genotypes which differ in genetic potential for milk yield are smaller under grazing because intake is limited by the constraining factors discussed by Kolver & Muller (1998).

Dry matter intake, estimated using the n-alkane technique as modified by Dillon & Stakelum (1989), differed by only 0.4kgDM/day between the high production NAHF and the NZHF strains (17.9 v. 17.5) on grass only (Horan *et al.*, 2005c). A greater differential in total DM intake (1.9kg) was observed with cows offered a daily allowance of 3.7kg concentrate DM (20.8 v. 18.9 kgDM/day) while grazing. This is supported by the findings in New Zealand (Kolver *et al.,* 2004). The NZ70 is able to maximise yield per cow on grazing only, with no supplementary feed; the NZ90 is superior to the NZ70 with or without supplementary feed, but shows increases in yields with supplementary feed; whereas the OS90 is no better than the NZ70 without supplement but with generous supplementation it is almost equal to the NZ90, in milk yields but not in profit. This is in agreement with the results of Kolver *et al.,* (2002), who reported values for daily DMI of 16.6 and 20.4 for grazed pasture and 17.3 and 24.0 on TMR, for New Zealand HF cows or Overseas HF cows respectively. For both strains, intakes were lower on pasture than on TMR, but on TMR the OS cows showed a much bigger increase in intake (3.6) than the New Zealand cows (0.7). This feature of a high grass-concentrate substitution rate (resulting in a low response to concentrate supplementation), coupled with a higher BCS in the NZHF cows suggests that these animals achieve a greater proportion of their potential milk production on grass alone than do the high production potential NAHF cows. Linnane *et al.* (2004) concluded that the grazing appetite of the NZHF is compromised by the provision of supplementary food. Energy balance calculations also revealed that the energy balance of the NZHF was more positive than that of the NAHF in early lactation. When adjusted for BCS, Horan *et al.,* (2005c) found that the lighter NZHF had the highest grass DM intake per kg live-weight, in agreement with the results of Caicedo-Caldas *et al.,* (2001) for a lighter and heavier strains of NAHF cows, and of Grainger & Goddard (2004) who compared Jerseys and NZHF. Faverdin *et al.,* (1991) showed that substitution rate is lower with high yielding cows when energy requirements are not being met. In such situations, concentrate supplementation only slightly reduces herbage intake and appreciably improves animal performance. The high producing NAHF cows put the extra energy from the concentrate-based diet into milk rather than reducing their energy deficit, and showed lower substitution rates than the NZHF cows (Horan *et al.,* 2005c). The decreased energy balance during early lactation maybe primarily the result of genetically controlled energy partitioning rather than the result of feed intake not keeping up with yield, and that the physiological processes may be similar to those normally associated with under nutrition (Veerkamp *et al.,* 2003). It appears that the NZ strain can achieve a greater proportion of it's energy requirement from a grass-only diet, and increased intake of concentrates reduces the intake of grass; on the other hand, increased concentrate supplementation resulted in higher energy intakes with the high producing NAHF and appear to be necessary to achieve it's genetic production potential. The key common factor is the size of the "relative energy deficit", between the cow's potential yield and energy demand, and its actual energy intake (Penno *et al.,* 2001).

Convincing evidence of a G×E interaction was also provided by Kolver *et al.,* (2002), who compared widely different feeding systems and two distinctly different genetic strains of Holstein-Friesian cows. For yields of milk and milk solids, the New Zealand strain was slightly superior on pasture, but the overseas strain was much superior on TMR. This was due, at least in part, to the inability of the OS strain to eat enough pasture to meet the needs of its heavier live weight and extra lactose synthesis (in higher milk volumes), which resulted in thinner body condition. In terms of overall merit, the OS strain would have been a complete failure on pasture because of its high proportion of non-pregnant cows (63%). However, current studies are showing that the OS (NAHF) strain may be suitable for extended lactations in non-seasonal systems with generous pasture and concentrates, where they can maintain high yields for longer periods of time (over 500 days up to the end of 2004; Eric Kolver, unpublished data).

Lactation profiles

Horan *et al.,* (2005b) showed that the high producing NAHF production profile is characterised by a steeper incline from calving to peak and a greater decline from peak to the end of lactation (i.e. lower persistency). The persistency of the NZ strain was superior at low concentrate levels. However, the difference in persistency between the strains was reduced by adding supplement to the diet. Kolver *et al.,* (2000) showed that on a total mixed ration diet there was no difference in the persistency between NAHF and NZHF animals, but Kolver *et al.,* (2002) reported that the genotype×feeding system interactions tended to increase as lactation progressed. This is likely to have implications for survival on seasonal pasture-based systems, because very high milk yield at the beginning of lactation (steep lactation curve) followed by reduced persistency exposes the cow to greater physiological stress, often increasing the frequency of reproductive disorders or metabolic diseases (Madsen, 1975).

Other breeds and cross breeding

To date the use of non-additive genetic variance has been uncommon in dairy cattle breeding compared with other domestic species. This is most likely because past experience showed that in general crosses did not exceed the best parental breed for milk volume (Willham, 1985), namely the Holstein-Friesian, and because of inadequate attention to performance components other than milk volume that influence the life-cycle efficiency. However, due to genetic improvement for milk production in other dairy breeds and a decline in fertility/survival within the Holstein-Friesian, crossbreeding may now be more attractive.

The use of alternative breeds or crossbreeding to counter the decline in reproductive efficiency and health within NAHF populations is now being considered as a solution by some farmers in many countries. Lopez-Villalobos (1998) concluded that breeds other than NAHF, or crossbreds involving NAHF might be superior in providing a higher net farm income. The use of crossbreeding to counter the decreases in reproductive abilities, health and survival, and inbreeding, was discussed by McAllister (2002) and VanRaden & Sanders (2003). Lopez-Villalobos *et al.,* (2000) showed a dual effect of increased survival on profitability through lower replacement rates and through higher milk yields with higher proportions of mature animals in a simulation study of seasonal calving herds in New Zealand.

The lack of an across-breed evaluation procedure is an obstacle to the efficiency of this strategy in many countries, paradoxically due to a lack of data on which an evaluation system could be based. However, in New Zealand, with 25% of Jersey×Friesian cows, an across-breed evaluation has existed for the last 15 years, and within-breed selection (mainly Jersey and Friesians) has been carried out successfully for over 50 years. A recent study by Harris *et al.,* (1999) showed that crossbreeding brought considerable merits in terms of fertility and survival. Heterosis values of 5-6% for production traits and up to 18% for reproduction and health traits were observed. Twenty percent more crossbred cows survived to 5th lactation than did Holstein-Friesians, with the largest effect of heterosis in the NAHF×Jersey cross, and smallest in the NAHF×NZ HF cross. The Jersey breed, with its small size, represents the extreme of the dairy type, but its small size limits its value for beef although it enables it to eat more per kg live weight than other breeds (Grainger & Goddard, 2004). Under grazing in New Zealand, Jerseys produced less milk and solids per cow than Holstein Friesians but more per ha when both breeds were stocked at the same live weight/ha, or for maximum milk solids/ha (Ahlborn & Bryant, 1992). Slightly higher maximum net incomes were predicted for the Jerseys, at stocking rates of 3.7 Jerseys and 3.0 Holstein Friesians per ha. The Jersey has a slightly shorter period from calving to first mating but a slightly lower conception rate, and slightly lower survival to 5th lactation than the NZHF (Harris & Winkelman, 2000). Jerseys had significantly higher yields of milk solids on once-a-day milking in grazing systems, than NZHF cows (Dalley & Bateup, 2004; R. Gale unpublished report 2005).

Holstein Friesians and Jerseys were managed either on confined feeding systems or on grazed pasture plus haylage and concentrates in the USA, and calved in spring and autumn (White *et al.,* 2002; Washburn *et al.*, 2002). The authors concluded that "maintaining seasonal reproduction appears more feasible with Jerseys than with Holsteins, regardless of production system".

Scandinavian countries, particularly Norway and Sweden have included health and fertility traits, in addition to yield, in their total merit index for the past 30 years. As a result, fertility performance has not decreased in these breeds over time (Lindhe & Philipsson, 2001). These cattle (Norwegian Red (NRF)) are now the subject of considerable interest in Ireland (Buckley *et al.,* 2004; Ferris *et al.,* 2004) as potentially suitable for cross breeding with the NAHF. Results presented by Buckley *et al.,* 2004, highlighted the benefits of the Norwegian selection program, attributes that are similar to those observed with the NZHF (Table 2).

Table 2 Effect of breed of dairy cow on reproductive performance (Buckley *et al.,* 2004)

	NAHF	NRF
Pregnancy rate to 1st service (%)	42	59
6 week in calf rate (%)	56	75
Empty rate after 14 weeks (%)	19	10

Dillon *et al.,* (2003) reported higher BCS and superior fertility/survival with the French dual-purpose Montbeliarde and Normande breeds compared to NAHF when evaluated under pasture-based systems of milk production in Ireland. Breeding schemes for dual-purpose dairy breeds in France are designed to improve dairy attributes and functional traits in addition to beefing qualities, including growth and conformation. Heifers on seasonal grass based systems must typically calve at 2 years of age, and must therefore reach puberty by 12 to 14 months of age. However, in France the age and live weight at which puberty occurs is 10 months of age and 260 kg, and 16 months of age and 340 kg, for the Holstein-Friesian and Montbeliarde, respectively (Coulon *et al.,* 1997). No data are available for puberty in crosses between these breeds. In New Zealand, Holstein-Friesians with higher proportions of North American genetics generally reached puberty at older ages and heavier weights than NZHF

(Garcia-Muniz *et al.,* 1997; McNaughton *et al.,* 2002). These differences may have implications for the suitability of various breeds on seasonal grazing systems.

Cross breeding in beef production

The advantages of heterosis for reproduction and maternal traits in beef cows have been widely demonstrated and there are further advantages in progeny performance from using a sire from a third breed. Ideally therefore, beef cows should be crossbred and should be mated to a bull from a third breed (Cundiff *et al.,* 1992). This happens in controlled, well managed, large herds where it is practical to allocate a proportion of the herd and use separate sires or artificial insemination specifically for the production of replacements. In practice, many beef cow herds are small and have a requirement for only one bull. In such circumstances a choice has to be made between use of a bull to produce replacements or to produce slaughter animals. The latter option is usually preferred with the intention of sourcing replacements from outside the herd. In the past, such replacements were available as beef crosses from the dairy herd (i.e. Hereford or Limousin cross Friesians). These were very suitable as suckler cows with hybrid vigour, good fertility and good milk production (Drennan, 1993). When crossed with a bull of late maturing breed type the resultant progeny had high overall performance.

Genetic evaluations across countries and systems

Currently Interbull routinely provides across country evaluations for production and some linear traits. This 'globalization' of dairy cattle breeding allows the provision of proofs from sires all around the world to be compared on a country's own base through the Multiple Trait Across Country Evaluation procedure (MACE) (Schaeffer, 1994). To date Interbull does not routinely apply MACE procedures (or Multiple-Trait Herd Cluster techniques) to fertility information. However initial studies (van der Linde & de Jong, 2003) have reported low (–0.05) genetic correlations between longevity across some countries; the average genetic correlation between countries for direct longevity was 0.60. Nevertheless, such low correlations are partially attributable to the diversity in definition of longevity related traits between the different countries. Similarly, the European countries that use survival analysis to measure longevity exhibited genetic correlations between 0.56 and 0.88 (Van der Linde & de Jong, 2003). Despite this van der Linde & de Jong (2003) concluded that MACE for longevity traits is feasible. Recent studies (Weigel & Rekaya, 2000; Zwald *et al.*, 2003) have been published which employ cluster analysis techniques or associated procedures to group herds of similar characteristics together thereby facilitating a borderless genetic evaluation of dairy cattle. Such techniques, if adopted, could increase genetic progress through improved accuracy of genetic evaluations for each management category.

Traits required by grazing beef cattle

Many of the comments on dairy systems apply equally to beef systems. On temperate grassland, beef production, like milk production, is based on the efficient conversion of grass (grazed and conserved) to meat. Biological efficiency is optimised when beef cows calve in late winter/early spring and increasing herd feed demand at least to mid season is matched by increasing grass growth and herbage supply. In winter when cows are dry and for a short period after calving they mobilise body reserves to meet a portion of their nutrient requirements, and then replenish these reserves during the grazing season when herbage supply is abundant. Growing cattle can be "stored" in winter and subsequently exhibit compensatory growth at pasture.

Unlike, Holstein-Friesian dairy cows, different strains of individual beef breeds have, either not evolved from selection in different production environments, or they have not spread outside of their own environment or geographical area. Thus, experiments like those described earlier where various Holstein-Friesian strains were compared in different production environments do not exist for beef breeds. In beef production, genotype×environment interactions usually refer to different breed types compared in different production environments. Geay & Robelin (1979) recommended that comparisons of different genotypes of cattle should include various feeding levels, as variability in performance is enhanced when animals are fed a high energy diet *ad libitum* which allows them express their growth potential, particularly for muscle. Since then, many studies have investigated genotype×environment (including nutrition) effects. Most of these studies have compared different breeds, with very few comparing cattle from one breed with different genetic merits, and few have included a total grazing system.

Reproductive and maternal traits

One report showed small but significant interactions between breed and grazing environment in New Zealand, for weight of calf weaned per cow mated (Morris *et al.,* 1993). In general the large European breeds grew faster to heavier mature weights, but reached puberty at older ages and had lower reproductive efficiency, especially in less favourable conditions. Clearly, if cow breeds differing in mature size (and hence in maintenance requirements) are compared in an environment where nutrition is limiting, those with the highest requirement will be most adversely affected. This may result in impaired fertility and consequent knock-on effects on calf production. Similarly, if cow breeds differing in milk production potential are compared, inadequate nutrition will cause a greater reduction in milk yield in those of higher potential with consequent effects on calf performance. When nutrition is not limiting all breeds can perform to their genetic potential for reproduction and milk production.

Weight gains

There is general agreement that large late maturing continental breeds (e.g. Charolais, Belgian Blue) have higher live weight gains than smaller late maturing breeds (e.g. Limousin, Piedmontese), early maturing breeds (e.g. Hereford, Angus) and dairy breeds (e.g. Friesian, Normand). However, the extent to which differences between breed types depend on plane of nutrition is unclear. There were no significant genotype by level of nutrition interactions in the studies of Lanholz (1977), Ferrell *et al.,* (1978), Ferrell & Jenkins (1998) or Steen and Kilpatrick (1995), which included a range of non-grazing diets and breeds. Similar conclusions can be drawn from studies which did include pasture in the system (Baker, 1977, Liboriussen *et al.,* 1977), and from a study of progeny from bulls of 11 breeds, grazing at three locations in New Zealand, which differed in their ability to support high levels of animal production (Baker *et al.,* 1990). Liboriussen *et al.,* (1977) studied the progeny from four sires, which were different in genetic merit for growth rate, and reared on four planes of nutrition. There was no significant interaction for live weight gain but there was for carcass gain. The ranking of the sires did not differ significantly for the different feeding levels but the superiority of one sire decreased with a decrease in feeding level.

Feed intake

In most of the studies where different breed types were subjected to different levels of nutrition, the feeding levels were controlled, so no interactions between genotype and nutritional level for feed intake were possible. Furthermore breed type and weight were often confounded. It is generally accepted that Holstein-Friesians, other dairy breeds and perhaps also dual purpose breeds have a higher intake capacity than beef breeds, but differences between beef breeds are small and there was no evidence of interactions between genotype and feeding level (Geay & Robelin, 1979; Steen, 1995; Keane *et al.*, 1989; Keane, 1994).

Carcass composition

An interaction between genotype and nutrition for carcass composition was described by Geay & Robelin (1979). A reduction of 17% in ME intake of early maturing Salers bulls between 9 and 15 months of age, reduced the proportion of fat in the carcass without a significant effect on rate of body weight gain. A similar reduction in the intake of Charolais reduced body weight gain but had no effect on body composition. The authors suggested that the effects of a reduction in energy intake depends on the protein deposition capacity of the animal relative to its energy intake. When energy intake is greater than protein deposition capacity, restriction reduces lipid deposition only and thus changes composition. Conversely, when energy intake just matches protein deposition capacity, restriction reduces both lipid and protein deposition with little change in composition. The rate of live weight gain in Friesian bulls declined linearly after about 300kg live weight and protein deposition followed the same pattern. In contrast, the rates of gain in live weight and protein in Charolais bulls increased up to about 480kg live weight before then declining. At the same feed intake, Friesians have a higher maintenance requirement than Charolais so the latter have more energy for growth which they use to deposit more protein and associated water and less lipid (Geay & Robelin, 1979). The intake capacity of Charolais matches their protein deposition capacity but Friesians, with their higher intake capacity have energy in excess of their protein deposition capacity, which is deposited as lipid, (Geay & Robelin, 1979). Growth of muscle was affected by level of feeding to a greater extent in Angus than in Friesians (Fortin *et al.*, 1981), and the interaction was significant, but a similar study by Steen & Kilpatrick (1995) showed no significant interaction. A lower level of feeding increased the proportions of muscle and bone, but decreased the proportion of fat in Angus and Holstein cattle (Fortin *et al.*, 1981), and in Friesian, Limousin×Friesian and Belgian Blue×Friesian cattle (Steen & Kilpatrick, 1995). In the former study, the interaction was significant because the lower feeding caused a decrease in the weight of muscle in the Angus but not in the Holstein. The interaction was not significant in the latter study, or in two other similar studies (Ferrell *et al.*, 1978; Ferrell & Jenkins, 1998). There were no interactions between the effects of breed and location for 13 months weight, slaughter weight or slaughter traits of the progeny (up to 13 months), of 11 sire breeds, including 6 European beef breeds, plus Angus, Hereford, Friesian and Jersey, and grazed at three different sites, ranging from favourable to moderately hard grazing conditions in NZ (Baker *et al.*, 1990).

Conclusions and implications

Genetic selection for yields of milk and solids has generally had similar effects regardless of production system; higher yields, thinner cows, and a decline in longevity. However, with intensive grazing systems where seasonal calving is required, good reproductive efficiency/survival is essential, and potential daily intake is lower than for cattle on concentrated rations. Because only about 10% of the world's milk production is from grazing systems the majority of dairy cattle have not been selected under grazing. Until recently most experimental results have indicated little or no important breed/strain×feeding system interactions in temperate dairying systems (Holmes, 1995). However, there is increasing evidence to suggest that the highly selected (for milk production) NAHF is unable to express its full genetic potential for milk production in a grass-based environment. Cows selected under grazing conditions do appear to exhibit characteristics that make them suited to production from a grass only diet; a lower genetic potential for milk yield, an ability to consume high intake of herbage relative to their potential energy demand, an ability to maintain body condition (energy balance), a minimal requirement for concentrate supplementation, and an ability to reproduce and survive within the constraints of the seasonal system. Crossbreeding provides a simple method to increase the health and efficiency of strains such as the NAHF through the introduction of favourable genes governing some of these characteristics, and through heterosis.

In future, improvement programmes should use a selection index that combines all the economically important traits appropriately for the local conditions and systems. Genetic proofs should be based on the performance of daughters managed under these local conditions, although this may become less necessary as further and more detailed genetic information about individuals becomes available in a range of systems. In the future, however, the cows must be compatible with the system used, and prediction of the phenotypic performance of dairy cattle must be based on knowledge of the cow's genotype as well as the environment in which they are managed. Polarisation of dairy cattle breeding is therefore likely in the coming years as the selection criteria chosen are refined to best reflect the profitability of various different systems of milk production.

In marked contrast, almost all beef cows and most growing cattle are managed under grazing systems. Therefore there has been much less tendency to select cattle for a particular system of feeding or management and selection has probably not resulted in such extremely high genetic potential for energy required per day, as in dairy cows. The relatively limited amount of research has shown few important interactions between genotype×feeding systems.

Where beef cow breeds of different mature size are managed in a sub-optimal nutritional environment, the larger breeds with higher requirements are more adversely affected by inadequate nutrition with consequent effects on calf performance. Aside from fertility and milk production, there is little evidence of genotype×environment in interactions for growth rate, weight for age, slaughter weight or the common slaughter traits. The ranking of widely different breeds for these traits is consistent across a wide range of production environments. There is also little evidence for a genotype×environment interaction for intake, but the issue has not been studied in detail.

Genotype x environment interactions for carcass composition depend on the animals genetic potential capacity for muscle deposition relative to its dietary intake of energy. When energy intake is higher than the capacity to deposit protein, surplus energy is deposited as fat; this will occur more readily, and at lower intakes in animals with lower protein deposition capacities, due to genetics and/or to age and growth phase.

References

Agnew, R.E. & T. Yan (2000). Impact of recent research on energy feeding systems for dairy cattle. *Livestock Production Science*, 66, 197-215.

Ahlborn, G. & A.M. Bryant (1992). Production, economic performance and optimum stocking rates of Holstein-Friesian and Jersey Cows. *Proceedings of the New Zealand Society of Animal Production*, 52, 7-10.
Baker, H.K. (1977). Notes on genotype - nutrition interactions. In "Crossbreeding Experiments and Strategy of Beef Utilisation to Increase Beef Production". Published by Commission of the European Communities, 414-426.
Baker, R.L., A.M. Carter, C.A. Morris & D.L. Johnson (1990). Evaluation of 11 cattle breeds for crossbred beef production: performance of progeny up to 13 months of age. *Animal Production,* 50, 63-77
Boelling, D. & G.E. Pollott (1998). Locomotion, lameness, hoof and leg traits in cattle. 2. Genetic relationships and breeding values. *Livestock Production Science,* 54, 203-213
Berry, D.P., F. Buckley, P. Dillon, R.D. Evans, M. Rath & R.F. Veerkamp (2003b). Genetic relationships among body condition score, body weight, milk yield, and fertility in dairy cows. *Journal of Dairy Science,* 86, 2193-2204.
Buckley, F., K. O' Sullivan, J.F. Mee, R.D. Evans & P. Dillon (2003). Relationships among milk yield, body condition, cow weight, and reproduction in spring-calved Holstein-Friesians. *Journal of Dairy Science,* 86, 2308-2319.
Buckley, F., P. Dillon, M. Rath & R.F. Veerkamp (2000). The relationship between genetic merit for yield and live weight, condition score, and energy balance of spring calving Holstein Friesian dairy cows on grass based systems of milk production. *Journal of Dairy Science,* 83, 1878-1886.
Buckley, F., J.F. Mee, N. Byrne, M. Herlihy & P. Dillon. (2004). A comparison of reproductive efficiency in four breeds of dairy cow and two crossbreeds under seasonal grass-based production systems in Ireland. *Journal of Dairy Science,* 87, Supplement 1, Abstract 526.
Butler, W.R. & R.D. Smith (1989). Interrelationships between energy balance and postpartum reproductive function in dairy cattle. *Journal of Dairy Science,* 72, 767-783.
Caicedo-Caldas, A., V. Lemus-Ramirez, C.W. Holmes & N. Lopez-Villalobos (2001). Feed intake capacity in Holstein-Friesian cows which differed genetically for body weight. *Proceedings of the New Zealand Society of Animal Production,* 61, 207-209.
Coulon, J.B., A. Hauway, B. Martin & J.F. Chamba (1997). Pratiques d'elevage, production laitiere et caracteristiques des fromages dans les Alpes du Nord. *INRA Productions Animales,* 10, 195-205.
Cundiff, L.V., R. Nunez-Dominguez, G.E. Dickerson, K.E. Gregory & R.M. Koch (1992). Heterosis for lifetime production in Hereford, Angus, Shorthorn and crossbred cows. *Journal of Animal Science,* 70, 2397-2410.
Dalley, D. & N. Bateup (2004). Once-a-day-milking; is it opportunity knocking. Proceedings of Dairy 3. 1-6; Massey University and Dexcel Ltd.
Dechow, C.D., G.W. Rogers & J.S. Clay (2002). Heritability and correlations among body condition score loss, body condition score, production and reproductive performance. *Journal of Dairy Science,* 85, 3062-3070.
Dillon, P., S. Snijders, F. Buckley, B. Harris, P. O'Connor & J.F. Mee (2003). A comparison of different dairy cow breeds on a seasonal grass-based system of milk production: 2. Reproduction and survival. *Livestock Production Science,* 83, 35-42.
Dillon, P., S. Crosse, G. Stakelum & F. Flynn (1995). The effect of calving date and stocking rate on the performance of spring-calving dairy cows. *Grass and Forage Science,* 50, 286-299.
Dillon, P. & G. Stakelum (1989). Herbage and dosed alkanes as a grass measurement technique for dairy cows. *Irish Journal of Agricultural Research,* 28 , 104 (Abstract).
Domecq, J.J., A.L. Skidmore, J.W. Lloyd & J.B. Kaneene (1997). Relationship between body condition scores and conception at first artificial insemination in a large dairy herd of high yielding Holstein cows. *Journal of Dairy Science,* 80, 113-120.
Drennan, M.J. (1993). Planned Suckler Beef Production. Beef Series No 4, Published by Teagasc, ISBN 0948321709, 38-47.
Evans, R., P. Dillon, F. Buckley, M. W allace, V. Ducrocq & D.J. Garrick (2004). Trends in milk production, fertility and survival of Irish dairy cows as a result of the introgression of Holstein-Friesian genes. *Animal Science,* Submitted November 2004.
Faverdin, P., J.P. Dulphy, J.B. Coulon, R. Verite, J.P. Garel, J. Rouel & B. Marquis (1991). Substitution of roughage by concentrates for dairy cows. *Livestock Production Science,* 27, 137-156.
Ferrell, C.L. & T.C. Jenkins (1985). Cow type and the nutritional environment. *Journal of Animal Science,* 61, 725 – 741.
Ferrell, C.L. & T.G. Jenkins (1998). Body composition and energy utilisation by steers of diverse genotypes fed a high concentrate diet during the finishing period. 1. Angus, Belgian Blue, Hereford and Piedmontese sires. *Journal of Animal Science,* 76, 637-646.
Ferrell, C.L., R.H. Kohlmeier, J.D. Crouse & H. Glimp (1978). Influence of dietary energy, protein and biological type of steer upon rate of gain and carcass characteristics. *Journal of Animal Science,* 46, 225-270.

Ferris, C.P., D.C. Patterson & J.A. McKeague. (2004) A comparison of the first lactation performance of Holstein-Friesian and Norwegain dairy cows on Northern Ireland dairy farms. *Proceedings of the British Society of Animal Science*, Winter Meeting, York. Paper No.47.

Fortin, A., J.T. Reid, A.M. Maiga, D.W. Sim & G.H. Wellington (1981). Effect of energy intake level and influence of breed and sex on the physical composition of the carcass of cattle. *Journal of Animal Science* 51:331-339.

French, P., C. Stanton, F. Lawless, E. O'Riordan, F.J. Monahan, P.J. Caffrey & A.P. Moloney (2000). Fatty acid composition including conjugated linoleic acid of intramuscular fat from steers offered grazed grass, grass silage or concentrate based diets. *Journal of Animal Science,* 78, 2849-2855.

Fulkerson, W.J. (2000). The productivity of Friesian cows; effects of genetic merit and level of concentrate feeding. Final Report DAN 082; Wollongbar Agricultural Institute, New South Wales Agriculture, xxpp.

Garcia-Muniz, J.G., C.W. Holmes, D.J. Garrick & B. Wickham (1997). Growth and onset of puberty in two genetically different lines of Holstein Friesian heifers selected for either heavy or light liveweight. *Proceedings of the New Zealand Society of Animal Production,* 57, 46-47.

Geay, Y. & J. Robelin (1979). Variation of meat production capacity in cattle due to genotype and level of feeding: genotype - nutrition interaction. *Livestock Production Science,* 6, 263-276.

Goddard, M.E. & G.R. Wiggans (1999). Chapter 18. In: R.Fries & A. Ruviusky (eds.) The Genetics of Cattle, C.A.B.I. Publishing.

Gordon, I. (1996). Controlled reproduction in cattle and buffaloes. Controlled Reproduction in Farm Animals, Volume 1, CAB International, UK.

Grainger, C. & M.E. Goddard (2004). A review of the effects of dairy breed on feed conversion efficiency. *Proceedings of the Australian Society of Animal Production,* 25,

Hansen, L.B. (2000). Consequences of selection for milk yield from a geneticist's viewpoint. *Journal of Dairy Science,* 83, 1145-1150.

Harris, B. L. & A. M. Winkelman (2000). Influence of North American Holstein genetics on dairy cattle performance in New Zealand. *Proceedings of the Australian Large Herds Conference,* 122-136.

Harris, B.L., C.W. Holmes, A.M. Winkelman, Z.Z. Xu (1999). Comparisons between fertility and survival of strains of Holstein-Friesian cows, Jersey cows and their crosses in New Zealand. Fertility in the High-Producing Dairy Cow. British Society of Animal Science Occasional Publication No. 26 Volume 2, 491-493.

Harris, B.L., J.M. Clark & R.G. Jackson (1996) Across breed evaluation of dairy cattle. *Proceedings of the New Zealand Society of Animal Production* Vol. 56, 12-15.

Holmes, C.W. (1988). Genetic merit and efficiency of milk production by the dairy cow. In: P.C. Garnsworthy (ed.) Nutrition and lactation in the dairy cow. Published by Butterworths, UK., xxpp.

Holmes, C. W. (1995). Genoype x environment interactions in dairy cattle: A New Zealand perspective. In: T.L.J. Lawrence, F.J. Gordon and A. Carson (eds.) Breeding and Feeding the High Genetic Merit Dairy Cow. Occasional Publication No 19, British Society of Animal Science, 51-58.

Hoekstra J, A.W. Van der Lugt, J.H.J. Van der Werf, W. Ouweltjes (1994). Genetic and phenotypic parameters for milk production and fertility traits in upgraded dairy cattle. *Livestock Production Science,* 40, 225-232.

Horan, B., J. F. Mee, M. Rath, P. O'Connor & P. Dillon (2004). The effect of strain of Holstein-Friesian cow and feed system on reproductive performance in seasonal-calving milk production systems. *Animal Science,* 79, 453 – 468.

Horan, B., P. Dillon, P. Faverdin, L. Delaby, F. Buckley &M. Rath (2005a). Strain of Holstein-Friesian by pasture-based feed system interaction for milk production; bodyweight and body condition score. *Journal of Dairy Science,* (in press).

Horan, B., P. Dillon, D.P. Berry, P. O'Connor & M. Rath (2005b). The Effect of Strain of Holstein-Friesian, Feeding System and Parity on Lactation Curves Characteristics of Spring Calving Dairy Cows. *Livestock Production Science,* (in press).

Horan, B., P. Faverdin, L. Delaby, F. Buckley, M. Rath & P. Dillon (2005c). The effect of Strain of Holstein-Friesian Dairy Cow on Grass Intake and Milk Production in Various Pasture-based Systems. *Animal Science,* (submitted).

International Farm Comparison Network (2004). In: T. Hemme (ed.) Dairy Report, www.ifcndairy.org

Keane, M.G. (1994). Productivity and carcass composition of Friesian, Meuse-Rhine-Issel (MRI) x Friesian and Belgian Blue x Friesian steers. *Animal Production,* 59, 197-208.

Keane, M.G., G.J. More O'Ferrell & J. Connolly (1989). Growth and carcass composition of Friesian, Limousin x Friesian and Blonde d'Aquitaine x Friesian steers. *Animal Production,* 48, 353-365.

Keane, M.G. (1996). Beef Systems in Ireland - Role of Forage. In "Grass and Forage for Cattle of High Genetic Merit". *Proceedings British Grassland Society Winter Meeting,* 6 pages.

Keane, M.G. & P. Allen (1998). Effects of production system intensity on performance, carcass composition and meat quality of beef cattle. *Livestock Production Science,* 56, 203-214.

Keane, M.G. (2002). Response in weanlings to supplementary concentrates and subsequent performance. Occasional Series No 4. Grange Research Centre, Published by Teagasc, ISBN 12 184170 3117, 36pp.

Kennedy, J., P. Dillon, K. O' Sullivan, F. Buckley & M. Rath (2003). The effect of genetic merit for milk production and concentrate feeding level on the reproductive performance of Holstein-Friesian cows in a grass-based system. *Animal Science,* 76, 297-308.

Kolver, E.S. & L.D. Muller (1998). Performance and nutrient intake of high producing Holstein cows consuming pasture or a total mixed ration. *Journal of Dairy Science,* 81, 1403-1411.

Kolver,E.S., Napper,A.R., Copeman,P.J.A., and Muller,L.D. 2000. A
Comparison of New Zealand and Overseas Holstien Friesian heifers. *Proceedings of the New Zealand Society of Animal Production,* 60, 265-269.

Kolver, E.S., J.R. Roche, M.J. de Veth, P.L. Thorne & A.R. Napper (2002). Total mixed rations versus pasture diets: Evidence for a genotype x diet interaction in dairy cow performance. *Proceedings of the New Zealand Society of Animal Production,* 62, 246-251.

Kolver. E., B. Thorrold, K. Macdonald, C. Glassey, & J.R. Roche (2004). Black and White answers on the modern dairy cows. South Island Dairy Event, Lincoln University and Dexcel, 168-185.

Langholz, H.J. (1977). Interaction between growth potential and feeding level in beef production. In "Crossbreeding Experiments and Strategy of Beef Utilisation to Increase Beef Production. Published by Commission of the European Communities, 445-463.

Liboriussen, T., A. Sorensen & B. Andersen, (1977). Genotype x Environment interactions in beef production. In: "Crossbreeding Experiments and Strategy of Beef Utilisation to Increase Beef Production. Published by Commission of the European Communities", 427-444.

Lindhe, B. & J. Philipsson (2001). The Scandinavian experience of including reproductive traits in breeding programmes. Fertility in the High-Producing Dairy Cow. British Society of Animal Science Occasional Publication No. 26, 251-261.

Linnane, M., B. Horan, J. Connolly, P O'Connor, F. Buckley & P. Dillon (2004). The effect of strain of Holstein-Friesian w and feed system on reproductive performance in seasonal-calving milk production systems. *Animal Science,* 78, 169 – 178.

Lopez-Villalobos, N., D.J. Garrick, H.T. Blair & C.W. Holmes (2000). Possible effects of twenty five years of selection and crossbreeding on the genetic merit and productivity of New Zealand dairy cattle. *Journal of Dairy Science,* 83, 154-163

Lopez Villalobos, N., D.J. Garrick, C.W. Holmes (2002). Genetic opportunities to improve milk value in New Zealand. *Proceedings of the New Zealand Society of Animal Production,* 62, 90-94.

Lopez Villalobos, N. (1998). Effects of crossbreeding and selection on the productivity and profitability of the New Zealand dairy industry. PhD Thesis, Massey University, Palmerston North, New Zealand.

Lucy, M.C. (2001). Reproductive loss in high-producing dairy cattle: where will it end? *Journal of Dairy Science,* 84, 1277-1293.

McAllister, A.J. (2002). Is crossbreeding the answer to questions of dairy breed utilization? *Journal of Dairy Science,* 85, 2352-2357.

Macdonald, K. (2004). Unpublished report on the Holstein Friesian Strain Trial; Dexcel Ltd, Hamilton.

Macdonald, K., B. Thorrold, C. Glassey, J. Lancaster, G. A. Verkerk, J.E. Pryce & C. W. Holmes (2005). Farm performance of Holstein-Friesian cows of three genetic strains on grazed pasture. International Grassland Conference, Satellite Workshop at Cork; July 2005, (in press)

Madsen, O. (1975). A comparison of some suggested measures of persistency of milk yield in dairy cows. *Animal Production*, 20, 191-197.

McGee, M., M.J. Drennan & P.J. Caffrey (1995). Suckler cow milk production and calf performance. In: P. O' Kiely, J.F.Collins & T. Storey (eds.), *Proceedings Agricultural Research Forum,* Published by Teagasc, Dublin, 87-88,

McNaughton, J. R., S. R. Morgan, P. Gore, G. A. Verkerk, C. W. Holmes & T. J. Parkinson (2002). Monitoring onset of puberty in three genetic strains of Holstein Friesian dairy cattle. *Proceedings of the New Zealand Society of Animal Production,* 62, 30-33

Morris, C.A., R.C. Baker, S.M. Hickey, D.L. Johnson, N.G. Cullery & J.A. Wilson (1993) Evidence of a genotype by environment interaction for reproductive and maternal traits. *Animal Production,* 56, 69-83

Nebel, R.L., M.L. Mc Gilliard (1993). Interactions of high milk yield and reproductive performance in dairy cows. *Journal of Dairy Science*, 76, 3257-3268.

Oldenbroek, J. K. (1984). A comparison of Holstein Friesian, Dutch Friesians and Dutch Red and Whites. Production characteristics. *Livestock Production Science*, 11, 69-81.

Penno, J. W., K. A. Macdeonald & C. W. Holmes (2001). Towards a predictive model of supplementary feed response from grazing dairy cows. *Proceedings of the New Zealand Society of Animal Production,* 61, 229-233.

Penno, J.W. (2000). Profitable milk production from pasture. *Irish Grassland and Animal Production Association Journal,* 34, 78-86.

Philipsson, J., G. Banos & T. Arnason (1994). Present and future uses of selection index methodology in dairy cattle. *Journal of Dairy Science,* 77, 3252-3261.
Priolo, A., D. Micol & J. Agabriel (2001). Effects of grass feeding systems on ruminant meat colour and flavour. A review. *Animal Research,* 50, 185-200.
Pryce, J.E., R.F. Veerkamp & G. Simm (1998). Expected correlated responses in health and fertility traits to selection on production in dairy cattle. Proceedings of the 6th World Congress on Genetics applied to Livestock Production, 383-386.
Pryce, J.E. & R.F. Veerkamp (2001). The incorporation of fertility indices in genetic improvement programmes. Fertility in the High-Producing Dairy Cow. British Society of Animal Science Occasional Publication No. 26, 237-249.
Pryce, J.E., M.P. Coffey & G. Simm (2001). The relationship between body condition score and reproductive performance. *Journal of Dairy Science,* 84, 1508-1515.
Rauw, W.M., E. Kanis, E.N. Noordhuizen-Stassen & F.J. Grommers (1998). Undesirable side effects of selection for high production efficiency in farm animals: a review. *Livestock Production Science,* 56, 15-33.
Senatore, E.M., W.R. Butler & P.A. Oltenacu (1996). Relationships between energy balance and post-partum ovarian activity and fertility in first lactation dairy cows. *Animal Science,* 62, 17-23.
Schaeffer, L. R. (1994). Multiple-country comparison of dairy sires. *Journal of Dairy Science,* 77, 2671-2678.
Shalloo, L., P. Dillon, M. Rath & M. Wallace (2004). Description and validation of the Moorepark Dairy Systems Model. *Journal of Dairy Science,* 87, 1945-1958.
Steen, R.W.J. (1995). The effect of plane of nutrition and slaughter weight on growth and food efficiency in bulls, steers and heifers of three breed crosses. *Livestock Production Science,* 42, 1-11.
Steen, R.W.J. and D.J. Kilpatrick (1995). Effect of plane of nutrition and slaughter weight on the carcass composition of serially slaughtered bulls, steers and heifers of three breed crosses. *Livestock Production Science,* 43, 205-213.
Van Arendonk, J.A.M., G.J. Nieuwhof, H. Vos & S. Korver (1991). Genetic aspects of feed intake and efficiency in lactating dairy heifers. *Livestock Production Science,* 29, 263-275
Van der Linde, C. & G. de Jong (2003). MACE for longevity traits. INTERBULL Bulletin No. 30, 3-9.
VanRaden, P.M. & A.H. Sanders (2003). Economic merit of crossbred and purebred US dairy cattle. *Journal of Dairy Science,* 86, 1036-1044.
Veerkamp. R.F & G.C. Emmans (1995). Sources of genetic variation in energetic efficiency of dairy cows. *Livestock Production Science,* 44, 87-97.
Veerkamp, R.F., P. Dillon, E. Kelly, A.R. Cromie & A.F. Groen, (2002). Dairy cattle breeding objectives combining yield, survival and calving interval for pasture-based systems in Ireland under different milk quota scenarios. *Livestock Production Science,* 76, 137-151.
Veerkamp, R.F. (1998). Selection for economic efficiency of dairy cattle using information on live weight and feed intake: A review. *Journal of Dairy Science,* 81, 1109-1119.
Veerkamp R.F., B. Beerda & T. Van der Lende (2003). Effects of genetic selection for milk yield on energy balance, level of hormones, and metabolites in lactating cattle, and possible links to reduced fertility. *Livestock Production Science,* 83, 257-275.
Villa-Godoy, A., T.L. Hughs, R.S. Emery, L.T.Chapin & R.L. Fogwell (1988). Associations between energy balance and luteal function in lactating dairy cows. *Journal of Dairy Science,* 71, 1063-.
Washburn, S.P., S.L. White, J.T.Green & G.A.Benson, G.A. (2002). Reproduction, mastitis and body condition of seasonally calved Holstein and Jersey cows in confinement or pasture systems. *Journal of Dairy Science,* 85, 105-111
Weigel, K.A. & R. Rekaya (2000). A multiple-trait herd cluster model for international dairy sire evaluation. *Journal of Dairy Science,* 83, 815-821.
White, S.L., G.A. Benson, G.A., S.P. Washburn, & J.T. Green (2003). Reproduction, mastitis and body condition of seasonally calved Holstein and Jersey cows. *Journal of Dairy Science,* 85, 95-104
Willham, R.L. & Pollak, E. (1985). Theory of Heterosis. *Journal of Dairy Science*, 68, 2411-2417
World Animal Review (1995). A classification of livestock production systems. 84/85, (3-4), 83-93
Zwald, N.R., K.A. Weigel, W.F. Filkse & R. Rekaya (2003). Application of a multiple-trait herd cluster model for genetic *evaluation* of dairy sires from seventeen countries. *Journal of Dairy Science,* 86, 376-382.

Grass growth modelling: to increase understanding and aid decision making on-farm

P.D. Barrett[1,3] and A.S. Laidlaw[2]
[1]*Agricultural Research Institute of Northern Ireland, Hillsborough, Lisburn, BT26 6DR, UK*
Email: p.barrett@vionville.datanet.co.uk
[2]*Department of Agriculture and Rural Development for Northern Ireland, Plant Testing Station, Crossnacreevy, Belfast, BT6 9SH, UK*
[3]*Current address; TIMAC ALBATROS Ltd., Arena House, Sandyford, Dublin 18*

Key points

1. Crop and grass growth models have been developed over the last 50 years, or so, but general appreciation of their benefits and potential has been recognised only relatively recently. The most popular application of grass growth models has traditionally been for knowledge understanding.
2. There is growing awareness of the potential of models in decision support systems (DSS) applications to aid pasture management and grassland budgeting on dairy farms.
3. Although some models have been developed for DSS, their widespread uptake in industry has been slow; challenges still exist which need to be addressed in order to improve their precision and user-friendliness.

Keywords: simulation, grassland budgeting, forecasting, decision support

Introduction

A mathematical model may be defined as a concise mechanism for providing a numerical description of a process or an object (Sheehy & Johnson, 1988). In simple terms, a model is a representation of a real life system, although often it is an artificial and highly simplified representation. Modelling has led to the accelerated knowledge of many commonly accepted agricultural processes and systems and provides an irreplaceable mechanism for predicting and preparing for future scenarios and offering decision support, even at a global level. Models traditionally have been used to advance understanding of the system while more recently they have been adopted into on-farm DSS including those for grazing management.

Grass growth is highly variable, even under standard management conditions. For example, in Northern Ireland in variety trials managed to a strict and consistent protocol at one site, mean growth rate over 15 years between simulated grazing in early-mid April and late April-early May was 60 kg DM/ha/day but the range was from 21 to 100 kg DM/ha/day (T. G. Gilliland, personal communication). If this level of variation can be expected under 'standard' management conditions, clearly variation of sward growth for a given time of year on-farm is likely to be even more pronounced as sward age, grazing and fertiliser management vary. This variability in sward growth rate is one of the factors which results in poor or variable utilisation of herbage produced on-farm, as farmers are unable to manage grazing with precision. For example, in a survey of utilisation by grazing of herbage grown on five dairy farms in southwest England, efficiency of utilisation varied from 51 to 83%. By increasing predictability of grass growth and animal requirement, feed budgets can be drawn up with confidence. Taking this a stage further, decision support systems can be designed, based on growth models, the interaction between the herbage produced and the animals' intake, to be a grazing management aid. An important component of such a system is the herbage growth model. While its output will be influenced by the presence of the grazing animal (briefly documented later), a major objective in building DSS for grazing management is to produce a reliable model of grass growth based on relatively simple environmental variables so that it can be used on-farm.

This paper will highlight the types of models used for the modelling of grassland production and their various applications and purposes. A number of existing grass models will be highlighted, as will the benefits of using models and their associated limitations.

Classification of models

In practice, grass growth models help to rationalise the complex interacting effects of weather and soil components on grass growth. Although they describe a complex system, models vary in their complexity; the more complex the model, the greater is the requirement for inputs. Empirical models are, in general, the simplest, based on experimental data and often consisting of simple regression relationships between input and observable output variables. However, empirical models do not lend themselves to describe complex systems, such as the physiology of grass production, due to their many contributing and interacting components. Equally, empirical models can rarely offer any insight into the biological mechanism at work within the system. Their simplicity, however, ensures that they have been used extensively to estimate crop yields (Jame & Cutforth, 1996) including grass production. For example, Corrall (1988) ambitiously collected grass growth and weather data from a series of trials around 30 locations across Europe. Others include a simple model to aid grassland management in England (Rook *et al.*, 2001), while two developed in Ireland (Brereton *et al.*, 1996; Han *et al.*, 2003a) provide relatively good estimates of herbage production and are based on meteorological parameters.

However, empirical models are specific to the circumstances under which they were developed. A model developed in the UK could not readily be applied in, for example, Australia, North America, Sweden or Spain, where prevailing climatic conditions, and probably soil type and local topography, are very different. The level of restriction is in practise much greater than this and micro-elements rather than macro-elements are generally responsible for precluding the widescale uptake of any empirical model. For example the model of Han *et al.* (2003a) was produced in Ireland based on a permanent grassland pasture and therefore would only be applicable to a permanent pasture sward. Likewise, empirical models such as linear regression equations describing the relationship between light interception and crop growth are often reasonable under average weather conditions but fail to deliver accurate outputs in adverse conditions when good predictions are particularly required (Monteith, 1981).

Mechanistic models, as alternatives to empirical models, are process-based as they describe the underlying processes described by theory and knowledge of the main biological principles. Mechanistic models are generally transparent (Thornley, 1998) and facilitate the scrutiny of the underlying processes of the system. Mechanistic models, while more complex, may contain empirical elements or sub-models. Such models tend to be more readily adopted in grassland modelling because of their superior robustness, accuracy and flexibility of application. For grassland systems, the main models are mechanistic and include the Hurley Pasture Model (Thornley, 1998), the Australian GrassGro model (Moore *et al.*, 1997) and the EU GrazeGro model (Barrett *et al.*, 2005), itself based on the mainly mechanistic Wageningen LINTUL model (Spitters & Schapendonk, 1990). Most of the models discussed further in this paper are mechanistic.

Table 1 A selection of recent models developed for grass growth prediction

Reference	Purpose/objective	Type	Application
Barrett *et al.*, 2005	Grassland production and quality	Dynamic/mechanistic	European GrazeMore DSS and Northern Irish GrassCheck grass prediction service
Topp and Doyle, 2004	Policy	Dynamic/mechanistic	Forecasts of yield productivity in different agro-climatic zones in N Europe
Han *et al.*, 2003	Grassland production and quality	Linear empirical	Silage production in Irish permanent pasture pasture swards
Rook *et al.*, 2001	Grassland production	Empirical	Improved grassland management for grazing in England
Woodward, 2001	Grassland production and quality	Dynamic/mechanistic	Improved grassland management in New Zealand
Brereton *et al.*, 2001	Grassland production	Empirical	Improved grassland management under Irish grazing conditions
Rodruegez *et al.*, 1999	Climate change	Dynamic/mechanistic	Assessing elevated atmospheric CO_2 concentration
Thornley, 1998*	Biological understanding	Dynamic/mechanistic	Improved understanding
Schapendonk *et al.*, 1998	Grassland production	Dynamic/mechanistic	Land use evaluation, crop yield forecasting and climate change impact
Moore *et al.*, 1997	Grassland production	Dynamic/mechanistic	Grass growth for GrazePlan decision support system in Australia
Mohtar *et al.*, 1996	Grassland production	Dynamic/mechanistic	GRASIM decision support system in USA
Topp and Doyle, 1996	Effects of climate change	Dynamic/mechanistic	Grassland (perennial ryegrass and white clover) production in Scotland
Gustavsson *et al.*, 1995	Grassland production	Dynamic/mechanistic	Improve management of timothy silage swards in Sweden
Parsons *et al.*, 1988	Biological understanding	Dynamic/mechanistic	Understanding dynamics of grazed grass/clover swards
Corrall, 1988	Grassland production	Empirical	Pan-european grass production for growth prediction of irrigated and non-irrigated perennial ryegrass and timothy swards.

*The Hurley Pasture Model has been developed over many years and encompasses many publications and sub-models and routines (overviewed in Thornley, 1998).

Application of models

The application of grass production models is vast. During their early development, they were often used for knowledge synthesis to gain greater scientific insight (de Wit, 1965). More recently, however, emphasis has shifted to more practical and operational use. Some recently published models and their application are presented in Table 1.

Models are useful for research and development, education and training. They can be more useful than experimentation when experimentation is not feasible or when hypotheses need to be framed before an experiment is designed. Models can also be used to transcend time and to speculate over future or historic events. Grass growth models have been used to predict the impact of climate change on grassland productivity and other vegetation (Holden & Brereton, 2002). Rodreguez *et al.* (1999) used an adaptation of LINGRA (Schapendonk *et al.*, 1998) to investigate the impact of climate change and elevated levels of atmospheric CO_2 on grass growth, while Topp & Doyle (1996) developed a model to predict the specific impact of global warming on milk and forage production in Scotland. Models are also used for land and vegetation-zonation programmes. Topp & Doyle (2004) have adapted their earlier model to compare productivity and profitability of a number of grasses and legumes for silage production in different agro-climatic zones across northern Europe, while LINGRA grass growth model has been used in the Crop Growth Monitoring Scheme (Bouman *et al.,* 1996) and also used by the Joint Research Centre of the EU for crop yield forecasting in the EU (Vossen & Rijks, 1995). Potential impact of restrictions imposed by the EU on nitrogen application have also been investigated (Topp & Doyle, 2004). In recent years, grass growth models have been used most frequently as predictive tools to aid decision making by reducing uncertainties about grass production systems.

Model comparison

Due to the wide range of models available, testing may be necessary within the range of circumstances and conditions for which the models will be used. Brereton and O'Riordan (2001) and Barrett *et al.* (2004) have tested the suitability of models, varying in their degree of complexity, for prediction of seasonal grass growth curves, based mainly on meteorological inputs. The models tested in both studies were adaptations of those of Brereton *et al.* (1996), Schapendonk *et al.* (1998) and Johnson and Thornley (1985). As Brereton and O'Riordan (2001) were mainly interested in a model to predict grass growth under Irish conditions, they tested output against data for 5 years from a research centre in a grass growing region in Ireland. Barrett *et al.* (2004) were primarily interested in finding a model which would predict grass growth under the range of conditions encountered in western Europe and tested the models against data from two centres (south east England and Northern Ireland) covering a total of 28 centre seasons. In the comparison of Brereton and O'Riordan (2001), while LINGRA produced the closest fit to actual data, the least mechanistic of the three i.e. Brereton *et al.* (1996) could be easily parameterised for the specific site and so predict growth adequately. The comparison of Barrett *et al.* (2004) supports the suitability of the model of Brereton *et al.* (1996) for specific sites, in this instance the site in SE England (Figure 1, Table 2) but taking the two sites together, LINGRA, modified to take account of the reproductive phase, provides the best prediction of growth. Indeed, on the basis of this latter comparison, the modified LINGRA was used as a basis for development of a grass growth model for the EU Grazemore DSS.

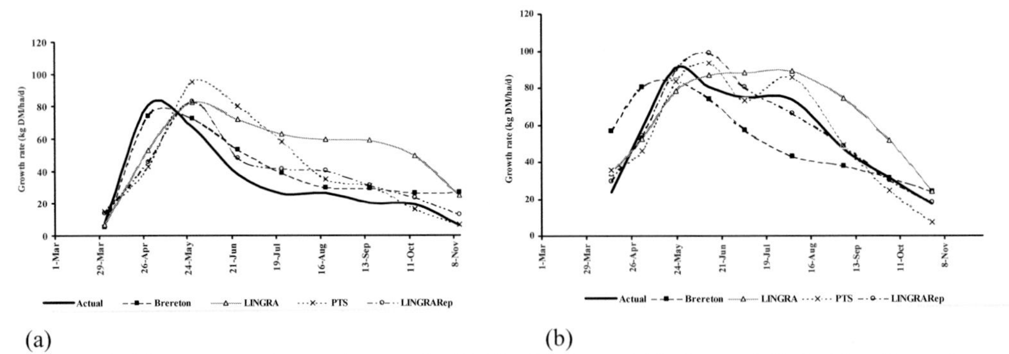

(a) (b)

Figure 1 Mean seasonal growth rates (kg DM/ha/day) at (a) Hurley and (b) Belfast (legends shown on graph)

Table 2 Comparison of THE precision (kg DM/ha/day) of models tested (actual growth was 34.5 and 56.7 for Hurley and Belfast, respectively; MPE is mean prediction error)

Model	Hurley			Belfast		
	Bias	R^2	MPE	Bias	R^2	MPE
Brereton *et al.* (2001)	+5.5	0.67	0.55	-1.1	0.20	0.55
LINGRA	+19.4	0.27	1.00	9.3	0.54	0.42
Johnson & Thornley (1985)	+11.5	0.34	0.79	0.5	0.62	0.33
LINGRA (reproductive)	+5.7	0.48	0.70	1.3	0.65	0.32

These comparisons raise the issue of complexity and suitability. The most complex (those based on Johnson and Thornley, 1985) are not necessarily the most suitable for predicting daily grass growth. They may also have a requirement for inputs, which are not generally available. On the other hand, the least mechanistic was less robust when tested under contrasting conditions. So parameterising it for one set of circumstances could weaken its ability to predict satisfactorily for other conditions.

Model construction

Models applied to grassland can be categorised in more than one way as they have been built mainly to satisfy regional requirements for prediction of herbage growth and quality. For example, while models developed in western Europe and New Zealand tend to be primarily built for perennial ryegrass with or without white clover e.g. Thornley (1998), Topp & Doyle (1996), Wu & McGechan (1999), Barrett *et al.* (2005), Brereton *et al.* (1996), Parsons *et al.* (1988), Groot & Lantinga (2004) and Woodward (2001), those developed in Northern Europe have focussed on timothy and meadow fescue (Gustavsson *et al.*, 1995) and for the drier south of Europe, tall fescue and cocksfoot (Duru & Ducrocq, 2002). The multispecies swards containing a high proportion of annual species in southeastern Australia have influenced the form of herbage production models in that region e.g. the pasture growth submodel in the decision support system GRAZPLAN (Moore *et al.*, 1997). However, the development of a model in a specific area or region need not confine its use to that geographical zone. Variants of the Hurley Pasture Model or its simpler precursors are used in many temperate grassland regions throughout the world far removed from southeast England!

In recent times modellers seem more willing to re-develop and adapt existing models for their own requirements than develop entirely new models, serving to improve the development of tried and tested models. For example, Topp & Doyle (1996) based their model on Johnson & Thornley (1985) but introduced some adaptations to meet their specific requirements, particularly in relation to elevated carbon dioxide levels, and with some other modifications related to nitrogen uptake and moisture stress. In turn, Wu & McGechan (1998) adapted the Topp and Doyle (1996) model, including introducing the Swedish SOILN model. The chain has been continued with further development of these models (McGechan & Topp, 2004). There are other examples such as modification of the original LINTUL model (Spitters & Schapendonk, 1990) to LINGRA (Schapendonk *et al.*, 1998) for perennial ryegrass swards, modification at other centres to meet specific requirements (e.g. adapted for Timothy, Höglind *et al.*, 2001), to study climatic change (Rodruegez *et al.*, 1999) and to be more generally applicable to grazing by introducing further functions (Barrett *et al.*, 2005).

On-farm requirements of grass growth models

The main objective of many grass growth models is to improve grassland management at farm-level by predicting grass production, ranging from origins in the Southern Hemisphere (Moore *et al.*, 1997; Woodward, 2001) to Northern Europe (Höglind *et al.*, 2001; Gustavsson *et al.*, 1995). Some primarily predict grass production and, in some instances, quality for grazing (Barrett *et al.*, 2005) and for silage production (Gustavsson *et al.*, 1995: Han *et al.*, 2003b; Groot & Lantinga, 2004) while others predict both (e.g. Moore *et al.*, 1997).

Synchronising herbage supply with herbage demand is the fundamental objective of grassland management for dairy farmers operating pasture-based systems. Grassland budgeting is simple but as it must precede production of herbage, its accuracy is severely limited by the uncertainty of future herbage supply. Therefore, grass growth prediction models clearly have a role to play in the management of pasture and paddock planning on dairy farms by limiting the uncertainty of grass supply figured into the calculations.

Traditionally, farmers can estimate paddock and farm grass covers by measuring sward height with a sward stick (Barthram, 1986) or rising plate meter (Mitchell, 1982) or making a visual estimate by 'eye-balling' herbage mass (Stockdale, 1984). Other more sophisticated methods have been developed such as the pasture probe capacitance meter (Vickery *et al.*, 1980). Otherwise, destructive methods such as grass clipping and weighing from a known area can be employed (Frame, 1993). All these methods however, are time consuming and also have a relatively low level of precision (O'Donovan *et al.*, 2002; Frame, 1993). O'Donovan *et al.* (2002) concluded from a comparison of methods that the most accurate estimation of herbage mass was by visual assessment. However, this assumed the operator was well experienced and had sufficient opportunity to 'calibrate' their eye against real data from cut herbage, which is difficult on-farm. The rising plate meter is the most popular method of herbage mass measurement on-farm. In a recent study, the rising plate was found to require considerable re-calibration throughout the season (Barrett & Dale, 2005) to the point that it would make it impractical on a farm.

Models, however, can be used to replace such laborious methods, while also having the advantage of being predictive. They can be used for both strategic and tactical planning in advance of growth and can be used to run endless examples of 'what if' scenarios, quantifying the outcome when inputs are varied. There is evidence that grassland farmers are favourably disposed to the use of models in management. In a preliminary survey of 80 farmers throughout Europe to gauge their attitude to a grassland management decision support system (Mayne *et al.*, 2004) they cited prediction of grass growth (and herbage intake) as a priority requirement from a DSS but considered prediction of grass quality to be unnecessary for grazed swards

Linking into a DSS

The construction of models in their basic forms often would not be appropriate for use in DSS without the addition of a user-friendly interface. The interface must facilitate the input of required parameters, allow necessary degree of manipulation of data and must provide output information, all in a user-friendly and intuitive way. For complete system programs, grass growth models must be linked to herbage intake models to facilitate both the production and utilization elements of the system. This represents a major challenge but two examples are the GrassGro model incorporated into the GRAZPLAN DSS and the GrazeGro model used in the

GrazeMore DSS. GRAZPLAN is a commercially available program developed in Australia but has also been used to good effect in Canada (Cohen *et al.*, 2003). Secondly the GrazeGro model has been interfaced in the GrazeMore DSS together with the INRA produced Grazeln model (Delegarde *et al.*, 2004).

A simple example of farmer decision support

The GrazeGro model is currently being used in Northern Ireland in a farmer-funded programme, GrassCheck, that monitors and reports grass growth and quality on a weekly basis to industry via the farming press (Barrett & Laidlaw, 2005). It is a prime example of the direct on-farm benefits that can be gained from the application of a grass growth model. The previous 3-week's growth is monitored every week on plots across the Province and growth for the next two weeks is predicted using the GrazeGro model. Farmers can, therefore, make management decisions with relatively high confidence based on current and future estimates of grass production. Removing uncertainties from the management system is of great importance to the dairy farmer and this is a pioneering instance where the output from a modern model has been made directly available to the public for application. This project was supported by farmer demand (and farmer funding) and farmers have reported the usefulness of monitored growth rates, but in particular, the prediction of growth.

Figure 2 shows actual and predicted growth rates for 2003 and 2004. GrazeGro predicted growth very closely, indicated by high goodness of fits, with R^2 = 0.85 and R^2 = 0.89, for 2003 and 2004, respectively. This considerably exceeds the value suggested by Woodward (2001) (i.e. R^2 = 0.50 or above) to be sufficient for models for on-farm growth prediction. However during local validation, considerable differences in measured growth rates were found between sites that were managed identically and were situated close to each other. Regressions of weekly estimates of growth of each of the three sites at ARINI on each other were poor considering the expected similarities in growth rates due to similar managements. For Site 1 vs. Site 2, R^2 = 0.73; Site 1 vs. Site 3, R^2 = 0.57 and for Site 2 vs. Site 3, R^2 = 0.89. All macro difference are accounted for, as all sites experienced the same fertiliser and cutting regime and were within 0.5 km of one another, therefore climatic conditions were similar, as was topography and soil classification. However, there were sufficient additional, genotypic, environmental or historical management differences to cause relatively large differences in growth rates.

This reinforces the difficulty in providing grass growth prediction as a service for a wide range of pastures. In this case, the system was simplified, as growth was determined from small plots that were cut on the same day after 21 days regrowth. In reality, on the farm, circumstances are more complex and additional factors influence the sward. In paddocks there are problems of poaching damage (important in Ireland), rejected herbage, different soil types and differences in pH and nutritional status, even within paddocks, as demonstrated by modern soil sampling techniques using GIS and GPS technology (Bailey *et al.*, 2000; Jordan *et al.*, 2003). A model capable of operating under such circumstances is optimistic but must be a realisable target. However, as always, a balance has to be struck between inputs, which are appropriate, readily available to the farmer and easy to upload against output, which is feasible, if not strictly accurate, and in which the farmer has confidence. However, simplicity of the software presenting the model must be realistic. Arnold & de Wit (1976) recognised that intuitive user-friendliness is essential.

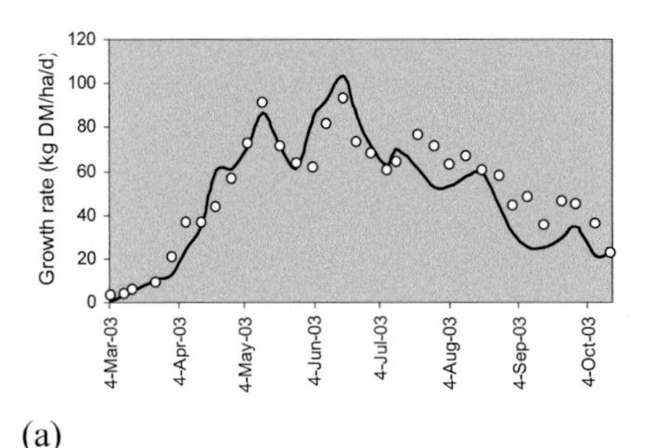

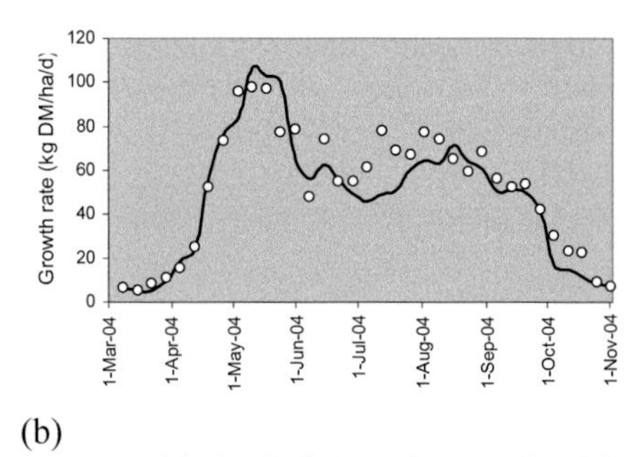

(a) (b)

Figure 2 Grass growth rate measured from plots on a weekly basis in Northern Ireland (○) and the predicted growth rate as determined from the GrazeGro model (——) in (a) 2003 and (b) 2004

Problems with grass growth decision support models

While the use of DSS is gathering in popularity, they are still underutilised as they are perceived to be more appropriate for researchers or other specialists than general on-farm users. Also, as grass growth models depend on weather measurements their predictive power is limited by reliability of weather forecasts. Some efforts can be made to circumvent this by using average weather conditions to estimate average growing conditions. Models are often based on data from plot or even glasshouse experiments and applied to fields which may have many unknown or unquantifiable variables and limiting factors. Addressing this creates difficulties for both the modeller and the user, as input parameters must be increased to account for as many of these factors as possible. The many input parameters must be balanced against the difficulty and, often, expense of input collection. Parsons *et al.* (2001) have clearly outlined the problems with increased complexity required when scaling theoretical models up to field level with the associated increase in spatial variability and heterogeneity, usually associated with grazing. Also, while defoliation of grazed swards has been well quantified and incorporated into sub-models of herbage production, detailed modeling of the other components (excretal return and treading) have been less well quantified. Reference has already been made to the modeling of the consequence of deposition of excreta on N transformation in the soil (McGechan & Topp, 2004) for dairy cows. In GRAZPLAN, primarily developed for sheep grazing, treading is taken into account in determining the rate of loss of standing dead material into the litter pool, through stocking rate (Moore *et al.*, 1997). Many other functions may be required to make a grass growth model fully applicable to the grazing environment

Conclusion

The various problems with grass growth models, and particularly their decision support application, hinder their progress and widespread uptake. The perfect decision support model must be complex and all-encompassing yet robust and simple to use. All models are simplifications but for developers the decision needs to be made as to how simple the model can be while still being appropriate for the intended application. For decision support easy input with at least moderate accuracy is required. Whilst major gaps still exist in some of the knowledge for grass growth modelling, particularly in the area of plant quality, problems with a deterministic model capable of representing the complexities of the typical on-farm conditions still require much greater development and remain a major challenge for scientists, agronomists and modellers. Equally, promotion and education about the benefits of grass models and DSS should be a key priority to help promote their uptake and drive further progress in their development and accuracy.

References

Arnold G.W. & C.T.de Wit (eds.) (1976). Critical evaluation of systems analysis in ecosystems research and management. Simulation Monographs, PUDOC, Wageningen, The Netherlands, 108pp.

Bailey, J.S., A. Higgins & C. Jordan (2000). Empirical models for predicting the dry matter yield of grass silage swards using plant tissue analyses. *Precision Agriculture* 2, 131-145.

Barrett, P.D. & A.S. Laidlaw (2005). GrassCheck: Monitoring and predicting grass production in Northern Ireland. XX International Grassland Congress, Cork Satellite meeting.

Barrett, P.D. & A.J. Dale (2005). Assessment of rising plate meter calibration on Northern Ireland dairy farms. *Irish Grassland and Animal Production Association* (submitted).

Barrett, P.D., A.S. Laidlaw & C.S. Mayne (2004). An evaluation of selected perennial ryegrass growth models for development and integration into a pasture management decision support system. *Journal of Agricultural Science,* 142, 327-334.

Barrett, P.D., A.S. Laidlaw & C.S. Mayne (2005). GrazeGro: A European herbage growth model to predict pasture production in perennial ryegrass swards for decision support. *European Journal of Agronomy* (in press).

Barthram, G.T. (1986). Experimental techniques: the HFRO sward stick. *Biennial Report Hill Farming Research Organisation, 1984-85*, 29-30.

Bouman, B.A.M., H. van Keulen, H.H. van Laar & R. Rabbinge (1996). The 'school of de Wit' growth simulation models: a pedigree and historical overview. *Agricultural Systems,* 52, 171-198.

Brereton, A.J., S.A. Danielov & D. Scott (1996). Agrometeorology of Grass and Grasslands for Middle Latitudes. Technical Note No. 197. World Meteorological Organisation. Geneva, 36pp.

Brereton, A.J. & E.G. O'Riordan (2001) A comparison of grass growth models. In: N.M. Holden (ed.) Agrometeorological Modelling, Agmet, Dublin, 136-154.

Cohen, R.D.H., J.P. Stevens, A.D. Moore & J.R. Donnelly (2003). Validating and using the GrassGro decision support tool for a mixed grass/alfalfa pasture in western Canada. *Canadian Journal of Animal Science,* 83, 171-182.

Corrall, A.J. (1988). Prediction of production from grassland. Information Bulletin of the FAO European Cooperative Network on Pastures and Field Crops. Herba, 1, Agronomy Institute of Florence, Italy, 25-28.

Delagarde, R., P. Faverdin, C. Barratte, M. Bailhache & J.L. Peyraud, (2004). The herbage intake model for grazing dairy cows in the EU Grazemore project. In: Proceedings of the 20th General Meeting of the European Grassland Federation, Luzern, 650-652.

de Wit, C.T. (1965). Photosynthesis of leaf canopies. Agricultural Research Report 663, PUDOC, Wageningen, The Netherlands, 57pp.

Duru, M. & H. Ducrocq (2002). A model of lamina digestibility of orchardgrass as influenced by nitrogen and defoliation. *Crop Science*, 42, 214-223.

Frame, J. (1993). Herbage Mass. In: Davies, A., Baker, R.D., Grant, S.A., Laidlaw, A.S. (Eds.), The Sward Measurement Handbook (Second Edition), 39-68.

Groot, J.C.J. & E.A. Lantinga (2004). An object-oriented model of the morphological development and digestibility of perennial ryegrass. *Ecological Modelling* 177, 297-312.

Gustavsson, A.-M., J.F. Angus & B.W.R. Torssell (1995). An integrated model for growth and nutritive value of Timothy. *Agricultural Systems*, 47, 73-92.

Han, D., D.W. Sun & P.O. O'Kiely (2003a). Linear models for the dry matter yield of the primary growth of a permanent grassland pasture. *Irish Journal of Agricultural and Food Research*, 42, 17-38.

Han, D., D.W. Sun & P.O. O'Kiely (2003b). Linear models for the in vitro dry matter digestibility of the primary growth of a permanent grassland pasture. *Irish Journal of Agricultural and Food Research,* 42, 39-54.

Höglind, M., A.H.C.M. Schapendonk & M. Van Oijen (2001). Timothy growth in Scandanavia: combining quantitative information and simulation modelling. *New Phytologist,* 151, 355-367.

Holden, N.H. & A.J. Brereton (2002). An assessment if the potential impact of climate change on grass yield in Ireland over the next 100 years. *Irish Journal of Agricultural and Food Research*, 41, 213-226.

Jame, Y.M. & H.M. Cutforth (1996). Crop growth models for decision support systems. *Canadian Journal of Plant Science*, 76, 9-19.

Johnson, I.A. & J.H.M. Thornley (1985). Dynamic model of the response of the vegetative grass crop to light, temperature and nitrogen. *Plant Cell and Environment,* 8, 485-499.

Jordan, C., Z. Shi, J.S. Bailey & A.J. Higgins (2003). Sampling strategies for mapping 'within-field' variability in the dry matter yield and mineral nutrient status of forage grass crops in cool temperate climes. *Precision Agriculture*, 4, 69-86.

Mayne C.S., A.J. Rook, J.L. Peyraud, J. Cone, K. Martinsson & A. Gonzalez (2004). Improving sustainability of milk production systems in Europe through increasing reliance on grazed pasture. Proceedings of the 20th General Meeting of the European Grassland Federation, Luzern, 584-586.

McGechan, M. B. & C.F.E. Topp, (2004). Modelling environmental impacts of deposition of excreted nitrogen by grazing dairy cows. *Agriculture Ecosystems & Environment*, 103, 149-164.

Mitchell, P. (1982). Value of a rising plate meter for estimating herbage mass of grazed perennial rye-grass-white clover dairy pastures. *Proceedings of the New Zealand Grassland Association*, 49, 117-122.

Mohtar, R.H., D.R. Buckmaster & S.L. Fales (1996). A grazing simulation model: GRASIM I: Model development. *Transactions ASAE*. (in review).

Monteith, J.L. (1981). Epilogue: themes and variations. *Plant and Soil*, 58, 305-309.

Moore, A.D., J.R. Donnelly & M. Freer (1997). GRAZPLAN: Decision support systems for Australian grazing enterprises. III. Pasture growth and soil moisture submodels, and the GrassGro DSS. *Agricultural Systems*, 55, 535-582.

O'Donovan, M., P. Dillon, M. Rath & G. Stakelum (2002). A comparison of four methods of herbage mass estimation. *Irish Journal of Agricultural and Food Research*, 41, 17-27.

Parsons, A.J., I.R. Johnson & A. Harvey (1988). Use of a model to optimise the interaction between the frequency and severity of intermittent defoliation and to provide a fundamental comparison of the continuous and intermittent defoliation of grass. *Grass and Forage Science*, 43, 49-59.

Parsons, A.J., S. Schwinning & P. Carrere (2001). Plant growth functions and possible spatial and temporal scaling errors in models of herbivory. *Grass and Forage Science*, 56, 21-34.

Rodruegez, D., M. Van Oijen & A.H.M.C. Schapendonk (1999). LINGRA-CC: a sink-source model to simulate the impact of climate change and management on grassland productivity. *New Phytologist*, 144, 359-368.

Rook, A.J., T.M. Martyn & R.O. Clements (2001). Forecasting future herbage growth from current yields and meteorological data. In: Rook A.J. & P.D. Penning (eds) Grazing management: the principles and practice of grazing, for profit and environmental gain, within temperate grassland systems. British Grassland Society Occasional Symposium No. 34. British Grassland Society, Harrogate, 57-58.

Schapendonk, A.H.M.C., W. Stol, D.W.J. Van Kraalingen & B.A.M. Bouman (1998). LINGRA, a source/sink model to simulate grassland productivity in Europe. *European Journal of Agronomy*, 9, 87-100.

Sheehy, J.E. & I.R. Johnson (1988). Physiological models of grass growth. In: Jones, M.B. & A. Lazenby (eds.), The Grass Crop: The Physiological Basis for Production. Chapman and Hall, London, 243-275.

Sheehy, J.E., F. Gastal, P.L. Mitchell, J.L. Durand, G. Lemaire & F.I. Woodward (1996). A nitrogen-led model of grass growth. *Annals of Botany*, 77, 165-177.

Spitters, C.J.T. & A.H.M.C. Schapendonk (1990). Evaluation of breeding strategies for drought tolerance in potato by means of crop growth simulation. *Plant and Soil*, 123, 193-203.

Stockdale, C.R. (1984). Evaluation of techniques for estimating the yield of irrigated pastures intensively grazed by dairy cows. (1) Visual assessment. *Australian Journal of Experimental Agriculture*, 24, 300-304.

Sylvester-Bradley, R. (1991). Modelling and mechanisms for the development of agriculture. The Art and Craft of Modelling in Applied Biology, *Aspects of Applied Biology*, 26, 55-68.

Thornley, J.H.M. & I.R. Johnson (1990). Plant and Crop Modelling. Oxford University Press, Oxford.

Thornley, J.H.M. (1998). Grassland Dynamics: an Ecosystem Simulation Model, CABI Publishing.

Topp, C.F.E. & C.J. Doyle (1996). Simulating the impact of global warming on milk and forage production in Scotland. 1. The effects on dry matter yield of grass and grass-white clover swards. *Agricultural Systems*, 52, 213-242.

Topp, C.F.E. & C.J. Doyle (2004). Modelling the comparative productivity and profitability of grass and legume systems of silage production in northern Europe. *Grass and Forage Science*, 59, 274-292.

Vickery, P.J., I.L. Bennett & G.R. Nicol (1980). An improved electonic capacitance meter for estimating herbage mass. *Grass and Forage Science*, 35, 247-252.

Vossen, P. & D. Rijks (1995). Early Crop Yield Assessment of EU Countries: The System Implemented by the Joint Research Centre. Publication EUR 16318, of the Office for Official Publications of the EU, Luxembourg, pp.180.

Woodward, S.J.R. (2001). Validating a model that predicts daily growth and feed quality of New Zealand dairy pastures. *Environment International*, 27, 133-137.

Wu, L. & M.B. McGechan (1998). Simulation of biomass, carbon and nitrogen accumulation in grass to link with a soil nitrogen dynamics model. *Grass and Forage Science*, 53, 233-249.

Wu, L. & M.B. McGechan (1999). Simulation of nitrogen uptake, fixation and leaching in a grass/white clover mixture. *Grass and Forage Science*, 54, 30-41.

Modelling of herbage intake and milk production by grazing dairy cows

R. Delagarde[1] and M. O'Donovan[2]
[1]*INRA, UMR Production du Lait, 35590 Saint-Gilles, France*
Email: Remy.Delagarde@rennes.inra.fr
[2]*TEAGASC, Moorepark Research Center, Fermoy, Co. Cork, Ireland*

Key-points

1. Models predicting intake and performance of grazing ruminants from animal, sward, grazing and supplements characteristics are rare, but they are now included in several decision support systems (DSS).
2. An evaluation of the performance and accuracy of published models are rarely undertaken by their authors, but this is proposed in this paper.
3. There is still a need for experimental research but also for the development of generic and dynamic models to predict intake and performance over a wide range of grazing conditions.

Keywords: grazing, model, evaluation, pasture intake, dairy cow

Introduction

In many parts of the world, grazed pasture is the main feed available for extensive and intensive ruminant production systems. A grazing system requires short-term and long-term management decisions for adequate herd feeding and pasture budgeting over the grazing season. However, feeding of grazing ruminants is difficult to manage in practice due to the inability of farmers to accurately estimate nutrient intake from grazed pasture. This can be achieved through an accurate prediction of pasture intake. Prediction of voluntary intake of ruminants fed indoors has been investigated over a long period, and many feeding system models have been developed (Ingvartsen, 1994; Forbes, 1995; Faverdin, 1995). Few models have been adapted to grazing, which take account of pasture availability (Pittroff & Kothmann, 2001). However, recent efforts have been made to develop and include such predictive models in decision support tools or simulators for the feeding of grazing cattle or for grazing management (Freer *et al.*, 1997; Herrero *et al.*, 2000; Delaby *et al.*, 2001b; Cros *et al* ., 2003; Heard *et al.*, 2004; Delagarde *et al.*, 2004).

The purpose of this paper is to describe and list the main methods for predicting daily herbage intake and performance of grazing ruminants, especially dairy cows. The performance of the models under varying grazing and feeding managements are tested by simulations and the precision of the predictions of some models are statistically compared.

How to predict herbage intake at grazing

Factors affecting herbage intake are numerous, and can be divided into five classes: animal, sward, grazing management, supplementation and environmental factors. They were reviewed by Faverdin *et al.* (1995) for winter feeding and by Poppi *et al.* (1987), Dove (1996), McGilloway & Mayne (1996), Cherney & Mertens (1998), Peyraud & González-Rodríguez (2000) and Delagarde *et al.* (2001) for grazing. All these factors regulate DM intake through metabolic, digestive and behavioural limitations. The short-term control of feeding behaviour includes the motivation of the animal to eat and the prehensibility of the feed, determining the rate of nutrient intake. At a larger scale, the medium- or long-term regulation of intake includes the digestive and metabolic adaptations of the animal, as well as the management of body reserves and reproductive function (including gestation and lactation), especially for farm animals (Faverdin *et al.*, 1995).

The general structure and complexity of existing models is be extremely variable according to the time scale of prediction, the biological function considered as determinant for intake prediction (metabolism, digestion, behaviour), the possible links with pre-existing feeding systems and the mathematical approach and databases selected for characterisation of the relationships. For simplicity, models predicting intake will be considered either as empirical or mechanistic.

Empirical models

Empirical models relate intake to several known factors that affect intake, generally by way of multiple regressions of compiled experimental data. Such an approach was developed for a long time for grazing dairy cows, but often considered a limited number of factors. The most complete multiple regressions consider animal characteristics (often milk yield and live weight), sward nutritive value (digestibility), grazing management (herbage allowance, herbage mass or sward height) and supplementary feeds (concentrate level) (Stockdale, 1985; Caird & Holmes, 1986; Peyraud *et al.*, 1996; Stockdale, 2000; Maher *et al.*, 2003; Stakelum & Dillon, 2004). Surprisingly, such empirical models predicting animal performance at grazing are scarce, compared to those predicting intake (Delaby *et al.*, 2001a; Maher *et al.*, 2003). A selection of multiple regressions predicting herbage intake and milk production by grazing dairy cows is presented in Table 1.

The main advantage of these empirical models is that intake and performance can be predicted rapidly from a single equation. However, they are limited by the size and range of the database used, the measured experimental factors, and the factors accounted for in the regression analysis. However, many interactions between factors cannot be predicted and the accuracy of the predictions is limited in extreme situations due to the simple mathematical approach of such models. Linear relationships in biological systems are unreliable across a widely differing range of situations.

Mechanistic models

Mechanistic models predict intake from a series of equations describing the main mechanisms regulating intake. They can be derived from many knowledge sources and sometimes only from theoretical concepts. As the mechanisms regulating intake are numerous and can be investigated at different time scales, the structure of these models is very variable. Some models are based on the short-term defoliation processes and on the dynamic estimates of bite mass, time per bite and grazing time according to the constraints of the sward canopy (Demment & Greenwood, 1988; Woodward, 1997; Smallegange & Brunsting, 2002; Baumont *et al.*, 2004). Other models consider that animals graze successively different pools of homogeneous sward quality, the best quality pools being selected before the worse quality pools during the grazing-down process (Sibbald *et al.*, 1979; Freer *et al.*, 1997). Finally, others consider directly an integrated response of the animals to sward structure or pasture availability on a daily basis (Johnson & Parsons, 1985; Herrero *et al.*, 2000; Delaby *et al.*, 2001b; Cros *et al.*, 2003; Heard *et al.*, 2004; Delagarde *et al.*, 2004). In many cases, intake at grazing is calculated relative to voluntary intake determined often from pre-existing winter-feeding models.

The advantages of a mechanistic approach are numerous. Generic models are potentially adaptable to many animal types, swards and management practices. The mathematical approach can be complex, integrating logarithmic or exponential relationships, with asymptotic limits of the predictions for extreme situations. Variations of intake can also be considered as relative to a non-limiting situation rather than absolute, enabling the simple simulation of interactions. Finally, the choice of algorithms or the succession of equations, with the development of some iterative calculations, leads to the increase of the model robustness and the prediction of many interactions between factors.

However, many so called mechanistic models should be considered to be both empirical and mechanistic. Many equations, parameters or assumptions in mechanistic models are based on expertise or on simple literature surveys and can be considered as empirical. Moreover, a truly mechanistic model should imply that all the relationships included in the model are from cause to effect relationships, which are rarely proven. As an example, milk yield can be seen as the result of nutrient intake (output) but genetic merit potential also influences the motivation to eat and the intake capacity of the dairy cow (input). Eating time is sometimes described as a determinant of herbage intake and is stated as a predictive variable in intake models, but it can also be the result of the balance between the motivation of the animal to eat (intake capacity) and sward state driving intake rate. In that case, eating time is a consequence and not a cause of herbage intake and cannot be used as an input in models predicting intake. Describing how the cow eats and digests will not necessarily help in the prediction of how much pasture will be consumed (Kyriazakis, 2003).

Several models for grazing dairy cows

The main characteristics, inputs and outputs of a selection of five published herbage intake models for grazing dairy cows are described in Table 2. Three of these models are simple considering the required inputs and algorithms (Sepatou, Pâtur'IN, Diet-Check). Two models can be considered more complex, with a higher number of required input variables, but also with greater applicability (GrazFeed, Grazeln). All predict daily herbage intake and are included in decision support systems.

Sepatou is a biophysical dairy farm model developed in France to evaluate rotational grazing management strategies. The animal sub model predicting herbage intake is fully described by Cros *et al.* (2003). It is based on a simplified version of the French Feed Unit system (INRA, 1989), considering the intake capacity of the cows, driven by peak milk yield and stage of lactation, and the ingestibility or fill value of the feeds, as influenced by their digestibility. Fill value of either concentrates or forages offered as supplements are different but fixed. This voluntary intake model is then adapted to grazing, considering only a linear effect of herbage allowance on intake below 20 kg DM/day). Herbage allowance is calculated from herbage mass to ground level minus 0.8 t DM/ha, considering that this amount is not grazeable by cows. The specificity of this model is that the OM digestibility of the herbage selected is calculated for each defoliated stratum assuming a theoretical vertical distribution of herbage OM digestibility. The average OM digestibility is then integrated from the top of the sward to the post-grazing sward height, depending on herbage allowance. However, the average OM digestibility of the herbage selected remains difficult to calculate from the published equations and intake predictions seem insensitive to variations in OM digestibility, which is linked to herbage allowance variations.

Table 1 A selection of multiple regressions predicting herbage intake and milk yield of grazing dairy cows

Reference	Meijs & Hoekstra, 1984	Stockdale, 1985	Caird & Holmes, 1986	Caird & Holmes, 1986	Peyraud *et al.*, 1996	Peyraud *et al.*, 1996	Stockdale, 2000	Maher *et al.*, 2003	O'Donovan *et al.*, unpub	Delaby *et al.*, 2001a	Delaby *et al.*, 2003	Maher *et al.*, 2003	O'Donovan *et al.*, unpub
Grazing system	R	R	R	S	R	R	R	R	R	R	R	R	R
Predicted variable	HOMI	HDMI	TOMI	TOMI	HOMI	HOMI	HDMI	HOMI	HDMI	MY	MY	MY	MY
n	117	223	165	144	95	95	-	192	2839	197	550	192	2839
R^2	0.90	0.79	0.67	0.53	0.72	0.60	0.81	0.78	0.52	0.89	0.91	0.74	0.84
r.s.d	0.78	1.30	1.91	2.52	1.52	1.78	0.26	1.12	2.03	1.67	1.47	1.68	2.71
Intercept	-0.61	-7.817	0.32	8.23	-40.3	-2.5	2.34	3.85	-12.5	-1.9	-4.2	-0.835	-9.1
Animal													
LW		1.10	1.00	0.40	0.95	0.74		0.50	1.19				
FCMpre					0.26	0.26		0.14		0.69	0.63	0.66	
MY			0.18	0.21									
FCMmax									0.23				0.59
SL				0.069					0.035				-0.38
Sward and grazing													
Height basis (cm)	4	0	0	0	0	0	0	4	4	5	5	4	4
OM digestibility (0-1)	6			25				10				24	
A^{-1}					-114								
HM		1.1	-1.0		9.4				2.6				-0.9
HM^2					-0.82				-0.0005				
HA	0.98	0.27	0.54			0.33	0.28	0.23	0.17	0.2	0.68	0.30	0.14
HA^2	-0.0140	-0.0018	-0.0060			-0.0033	-0.0073		-0.0017		-0.0133		-0.0013
HA^{-1}													
RPMH				-0.29			0.42						
C supplementation													
C	1.48		1.64	-0.12					-0.48	1.04	0.95		0.76
C^2									-0.037				
C*HA	-0.039		-0.048										
C*RPMH				0.13									

R: Rotational ; S: Set-stocking ; TOMI or HOMI (HDMI): Total or herbage OM (DM) intake (kg); n number of data; R^2 coefficient of correlation; r.d.s. residual standard deviation; LW live weight (100 kg); FCMpre pre-experimental fat-corrected milk; MY milk yield (actual); FCMmax maximum fat-corrected milk at the peak of lactation; SL stage of lactation (weeks from calving); Height basis: for calculations of HM and HA; A daily offered area (m^2/cow/day); HM herbage mass (t DM/ha); HA herbage allowance (kg DM/cow/day), RMPH rising plate meter height (cm), C concentrate intake (kg DM)

Table 2 Main characteristics, input and output variables for five predictive herbage intake models for grazing dairy cows (• factor taken into account, (•) factor partly taken into account, - factor not taken into account).

Model	GrazFeed	Pâtur'IN	Sepatou	DietCheck	GrazeIn
Country	Australia	France	France	Australia	E.U.
Reference	Freer *et al* . 1997	Delaby *et al* . 2001b	Cros *et al* . 2003	Heard *et al* . 2004	Delagarde *et al.* 2004
Type of animals	all ruminants	dairy cows	dairy cows	dairy cows	dairy cows
Type of swards	many	PRG-WC	PRG-WC[a]	several	many
Grazing system	R+S[b]	R	R	R	R+S
Animal					
Peak milk yield	•	•	•	-	•
Live weight	•	•	-	•	•
Body condition score	•	-	-	-	•
Age	•	-	-	-	•
Days in milk	•	•	•	-	•
Stage of gestation	•	-	-	-	•
Sward					
Species	•	-	-	•	•
OM digestibility (offered)	•	-	•	-	•
OM digestibility (selected)	•	-	•	-	-
Crude protein	•	-	-	-	•
Vertical structure	•	-	(•)	-	-
Morphology	•	-	-	-	-
Grazing					
Herbage allowance	•	•	•	•	•
Herbage mass	•	•	(•)	•	•
Daily access time	-	-	(•)	-	•
Supplementation					
Concentrate amount	•	•	•	•	•
Concentrate nature	•	-	-	-	•
Forage amount	•	•	•	-	•
Forage nature	•	-	-	-	•
Interactions					
Animal × Grazing	•	•	-	-	•
Animal × Supplem	•	•	-	-	•
Grazing × Supplem	•	•	•	•	•
Animal × Grazing × Supplem		•	•	-	- •
Outputs of the model					
Herbage intake	•	•	•	•	•
Milk yield	•	-	-	-	•
Weight gain	•	-	-	-	-

[a] PRG-WC perennial ryegrass-white clover; [b] R: rotational, S: set-stocking; Supplem.: supplementation

Pâtur'IN is a tactical decision support tool developed in France to help grazing management of dairy herds (Delaby *et al.*, 2001b). The intake sub model is based on a simplified version of the French Feed Unit system (INRA, 1989), estimating intake capacity of the herd from the average peak milk yield, stage of lactation and live weight of the cows. The fill value of grazed herbage, concentrates and forages offered as supplements are fixed. The relative herbage intake at grazing is calculated as a proportion of voluntary intake considering grazing conditions, mainly sward depletion. This model is designed, in particular, to estimate the day-to-day variation in herbage intake under rotational grazing with several days residency time in each paddock. Each day within a paddock, the increasing negative effect of sward structure on herbage intake is taken into account from an exponential function based on the ratio between the sward depth still available for grazing expressed as a proportion of the initial sward depth available for grazing. The sward depth available for grazing is calculated as the difference between the pre-grazing sward height and a minimum post-grazing sward height, defined as a proportion of the pre-grazing sward height. Under strip grazing, the sward height depletion effect on intake is calculated hourly with the same exponential function. Moreover, low or high pre-grazing herbage masses negatively affect herbage intake.

Diet-Check is a simple tactical DSS developed in Australia to help dairy farmers to estimate nutrient intake by strip-grazing dairy cows. The herbage intake model is fully described by Heard *et al.* (2004). Daily herbage intake (per 100 kg LW) is firstly calculated for unsupplemented cows from herbage allowance (per 100 kg LW), pasture height and sward species. The effect of herbage allowance to ground level is considered with an exponential function, and pasture height with a positive and linear effect on intake. For cows receiving concentrates, the substitution rate is a linear function of herbage intake (of unsupplemented cows), supplement intake, season and sward species. The relationships were developed from a large Australian database. In the decision support tool, the prediction of herbage intake enables the calculation of energy and protein balance, as well as the marginal milk response to supplements from the herbage intake of unsupplemented cows (given their body condition score and the season).

GrazFeed is a commercially available sofware package providing estimates of animal intake and production at grazing, and is part of the decision support tool GrazPlan developed in Australia. The details of the herbage intake model are fully described by Freer *et al.* (1997). It is designed for any ruminant and sward type. The potential herbage intake is firstly calculated from body size, peak milk production and stage of lactation. The relative herbage intake at grazing is thereafter calculated taking into account the effect of green and dead herbage masses for either strip-grazing or continuous grazing, and the herbage allowance effect under strip grazing. The specificity of this model is that herbage mass is arbitrarily divided into six pools of fixed digestibility. During the grazing process, animals select successively these pools from the highest to the lowest digestibility. Under strip-grazing, these calculations are made five times a day, to account for the high rate of sward depletion. However, the herbage allowance effect on intake is not easily simulated from the published equations. Rotational grazing with several days of residency time in a paddock also seems difficult to simulate. For supplemented animals, the model considers that the amount of the supplement offered will not be automatically consumed. The amount of supplement really consumed is calculated through an estimate of the motivation of the animal to eat the supplement. This motivation is a function of the relative digestibility of the supplement compared to the digestibility of the grazed pasture pool. Finally, the substitution rate between supplements and grazed pasture varies with a number of factors, including the availability of ruminal degradable protein and lactation stage. This model also estimates milk production and live weight change.
GrazeIn is a model for predicting herbage intake and milk yield of grazing dairy cows, developed as part of the European Grazemore decision support tool (Mayne *et al.*, 2004). The animal sub model is briefly described by Delagarde *et al.* (2004) and a full description of the model will be published soon. The model first calculates the voluntary herbage intake according to the principles of the French Fill Unit system (INRA, 1989), including the intake capacity of

the cows and the offered feeds ingestibility. Intake capacity is a function of peak milk yield, live weight, condition score, age, stage of lactation and of gestation. Herbage fill value depends mainly on the main sward species, OM digestibility and crude protein content. Concentrate fill value is a function of substitution rate, depending on energy balance of the cows. In a second step, the model estimates the relative intake at grazing taking into account the effects of herbage allowance and herbage mass under strip- or rotational grazing, sward surface height under set-stocking, and daily access time to pasture whatever the grazing system. Exponential equations were developed from a literature review. The specificity of this model is based on the assumption that herbage mass has no effect on intake when swards are compared at similar herbage allowance above 2 cm. Moreover, iterative calculations enable the estimation of all the interactions between animal, sward, grazing conditions and supplement characteristics. This model is designed for dairy cows only, but is easily adaptable for other ruminants. The model also predicts the herd milk production for each grazed paddock.

Predicting grazing conditions and supplementary feeding effects on intake

The influences of animal characteristics and sward nutritive value on intake are not specific to grazing situations, and they will not be investigated. In order to make possible the comparison between predictions, characteristics of the dairy herd (multiparous cows, peak milk yield of 40 kg/cow per day, DIM of 150 days, LW of 600 kg), sward grazed (pure vegetative perennial ryegrass, herbage mass of 4.2 t DM/ha to ground level, i.e. 2.0 t DM/ha above 4 cm, OM digestibility of 0.80, CP of 180 g/kg DM) and season (spring) were fixed in all simulations. Some simulations were not possible to run from the published description of the models but were obtained from direct use of the DSS in which the models are included (GrazFeed version 4.1.5.) or with the help of the authors (Pâtur'IN).

Herbage allowance

The allowance-intake relationship for rotationally grazing dairy cows has been widely researched with linear, curvilinear and more recently exponential relationships developed. A selection of seven published curvilinear or exponential relationships between herbage allowance to ground level and herbage intake of grazing dairy cows are presented in Figure 1.

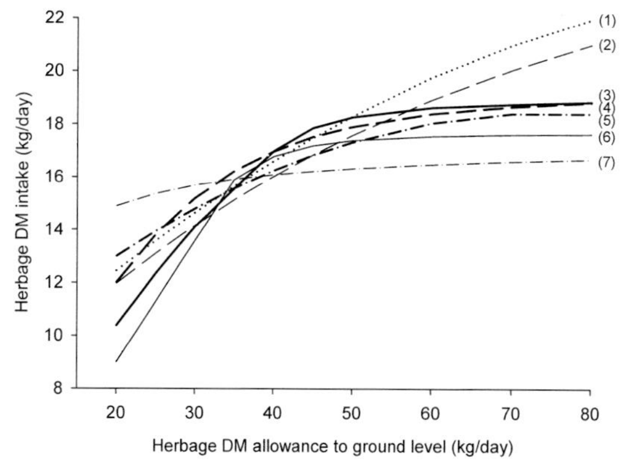

Figure 1 Simulated effect of herbage allowance on herbage intake by grazing dairy cows. The conditions of the simulations are described in the text. Models : (1) Stockdale (2000), (2) Diet-Check (Heard *et al.*, 2004), (3) GrazFeed (Freer *et al* ., 1997), (4) GrazeIn (Delagarde *et al.*, 2004), (5) Stockdale (1985), (6) Pâtur'IN (Delaby *et al.*, 2001b), (7) Peyraud *et al.* (1996)

The average increase in herbage intake is close to 0.20, 0.15 and 0.11 kg DM per kg DM increase in herbage allowance in the ranges 20 to 30, 30 to 40 and 40 to 50 kg DM herbage allowance, respectively. Intake predictions are quite similar between models for medium herbage allowances but predicted intake differences are greatest at low (< 30 kg DM/day) and high (> 50 kg DM/day) herbage allowances. The range in herbage intake predictions between models is close to 6, 1, 4 and 6 kg DM/day for herbage allowances of 20, 40, 60 and 80 kg DM/day, respectively. These discrepancies for extreme herbage allowances may arise from the dataset used to calibrate equations but also from the mathematical approach, which often determines the curvature of the relationship and the robustness of the predictions for atypical situations.

Pre-grazing herbage mass or sward height

In this section, pre-grazing herbage mass and sward height will be considered as similar descriptors of sward state, i.e. a linear relationship between mass and height. In rotational grazing systems, the effect of pre-grazing sward height on herbage intake has not been researched to the same extent as the effect of herbage allowance. Recent reviews or data compilations suggest that intake is less sensitive to sward height than to herbage allowance (Delagarde *et al.*, 2001; Heard *et al.*, 2004). In published multiple regressions, the effect of sward height or herbage mass on intake is linear or quadratic, with a large range in slopes (Table 1). Delagarde *et al.* (2001) highlighted that the slope of the relationship between herbage mass and herbage intake largely depends on the height at which herbage mass and then herbage allowance are considered. In the more complex models, the effect of sward height is often taken into account indirectly (see above models description). The simulated effect of sward height on intake for four recent models is shown in Figure 2. Sward heights of 8, 12, 16, 20 cm are approximately equivalent to 1, 2, 3 and 4 t DM/ha above 4 cm, and 3.0, 4.2, 5.3 and 6.4 t DM/ha to ground level, respectively. Globally, all models predict a positive effect of sward height on intake at similar herbage allowance to ground level, but with very different slopes and interactions with herbage allowance. GrazFeed and GrazeIn predict a curvilinear effect of sward height on intake, with a slightly higher effect at high herbage allowance for GrazFeed and a higher effect at low allowance for GrazeIn. Diet-Check predicts a strong and linear effect of sward height on intake, whatever the herbage allowance. Pâtur'IN predicts an intake reduction for low sward height whatever herbage allowance but also for high sward height at high herbage allowance. These discrepancies of approach and of results between models highlights that the effect of pre-grazing sward height or herbage mass on intake for rotationally grazed dairy cows is not yet clearly established and requires further investigations. Under set-stocking management, the curvilinear relationship between sward surface height and herbage intake is widely known (Penning *et al.*, 1991; Rook *et al.*, 1994) and taken into account in GrazFeed and GrazeIn.

Sward structure

For a given herbage mass, sward structure may be defined by the bulk density (ratio between mass and height), by the vertical distribution of the herbage mass over different strata, and by the morphological composition of the sward, i.e. proportion of leaves, stem, pseudostem and dead material. The influence of these factors on daily herbage intake by grazing ruminants were scarcely studied, and generally are not taken into account in existing models. However, models already taking into account the effects of herbage allowance, herbage mass, sward species, herbage digestibility and CP content probably indirectly account for several aspects of sward structure. GrazFeed is the model integrating most of the sward structure variables. However, simulation with the GrazFeed software shows that herbage intake prediction is not sensitive to sward bulk density *per se* at similar herbage mass, with a decrease of 0.1 to 0.2 kg DM of herbage intake from 250 to 330 kg DM/ha/cm of the above ground bulk density. Herbage intake prediction is more sensitive to dead material proportion in the sward, decreasing by 1.0 kg DM from 10 to 20% of dead material, at similar herbage OM digestibility.

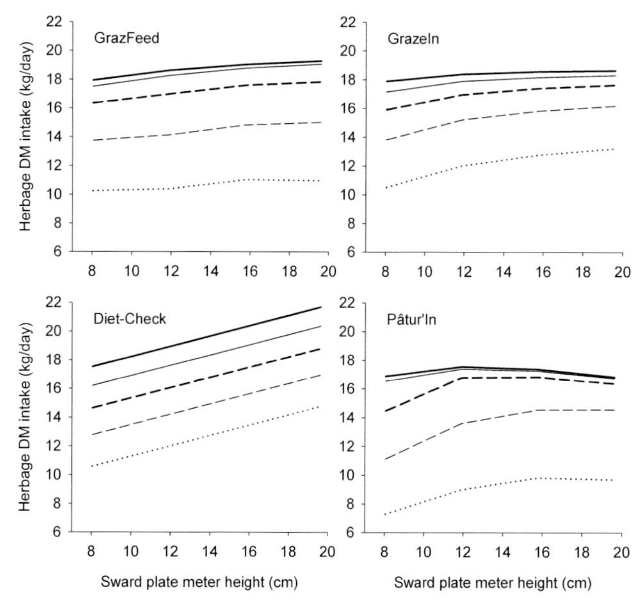

Figure 2 Simulated effect of pre-grazing sward height, measured with a plate meter, on herbage intake by grazing dairy cows according to different herbage allowances measured to ground level. The conditions of the simulations are described in the text. Models : GrazFeed (Freer *et al.*, 1997), GrazeIn (Delagarde *et al.*, 2004), Diet-Check (Heard *et al.*, 2004), Pâtur'IN (Delaby *et al.*, 2001b). Herbage allowance: ····· 20, – – 30, ▬ ▬ 40, —— 50 and ▬▬ 60 kg DM/cow/day.

Daily access time to pasture

In most of the models, herbage intake is predicted only for full daily access to pasture, i.e. approximately 18 to 20 h per day for lactating cows milked twice daily. However, in autumn, winter and early spring, cows have frequently limited access to pasture, for instance between milking times. Buckmaster *et al.* (1997) first tried to take into account the daily access time available for grazing with a simple two linear-phase equation, considering that access time to pasture is not limiting for intake up to 8 h per day (Figure 3). More recently, Delagarde *et al.* (2004) built an exponential relationship between intake and access time from a literature review (Figure 3). The relationship is modulated by the sward height, which determines the potential intake rate by the grazing cows when daily access time is limiting. Compared to the herbage intake of unsupplemented dairy cows with full daily access time to pasture, the herbage intake predicted by GrazeIn is approximately 0.94, 0.90, 0.84 and 0.67 for daily access times of 12, 8, 6 and 4 hours, respectively. The decrease of herbage intake with decreasing access time to pasture is low above 8 hours of daily access because cows generally graze close to 8 hours daily and because they are able to confine their grazing activities during this period of access. Better prediction of herbage intake by ruminants with limited access time to pasture requires extra experimental research, particularly considering the possible interactions with other grazing or supplementary feeding conditions.

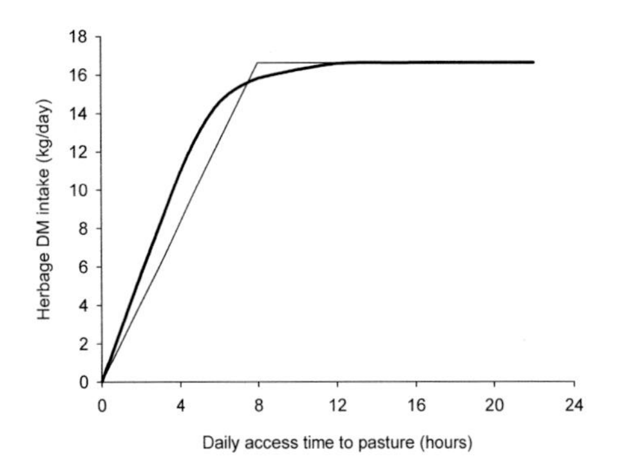

Figure 3 Simulated effect of daily access time to pasture on herbage intake by grazing dairy cows. The conditions of the simulations are described in the text. Models : ▬ GrazeIn (Delagarde *et al.*, 2004), —— Buckmaster *et al.* (1997)

Concentrate supplementation

For dairy cows fed indoors, the marginal substitution rate between roughages and concentrates depends on animal requirements, roughage quality, concentrate quality and finally the energy balance of the animal (Faverdin *et al.*, 1991). At grazing, the same concepts can be applied. Rate of substitution increases with increasing pasture availability, from 0 for high grazing pressure to 0.6-0.8 for low grazing pressure (Stockdale, 2000; Peyraud & Delaby, 2001). The challenge for an accurate prediction of substitution rate at grazing is to account for all of the possible interactions between animal, sward, grazing conditions and supplement characteristics. An empirical approach cannot achieve this. Responses of strip-grazing dairy cows to concentrate intake level in four models are shown in Figure 4. All models show similar trends, and predict increasing substitution rate and decreasing marginal milk response for increasing herbage allowance and concentrate intake level. However, the absolute values of substitution rate and marginal milk response to concentrate are quite different between models. European models predict lower substitution rate and higher milk response than Australian models. This difference can originate partly from different cow production potentials as illustrated in the study of Horan *et al.* (2005). In this study a low milk production response to concentrate and high substitution rate was observed with a New Zealand cow strain, while a Holstein Friesian of high milk production potential exhibited a low substitution rate and a high response to concentrate.

Forage supplementation

As forage (hay, haylage or silage) supplementation at grazing is not as extensively researched as concentrate supplementation, there are no multiple regression equations predicting intake for grazing dairy cows supplemented with forages. Substitution rate between grazed pasture and forage supplements was reviewed by Phillips (1988). All the models presented in Table 2 predict the substitution rate for forage-supplemented dairy cows, with interactions between forage supplements and pasture availability. Substitution rates are higher for forage than for concentrate supplementation due to higher forage fill value.

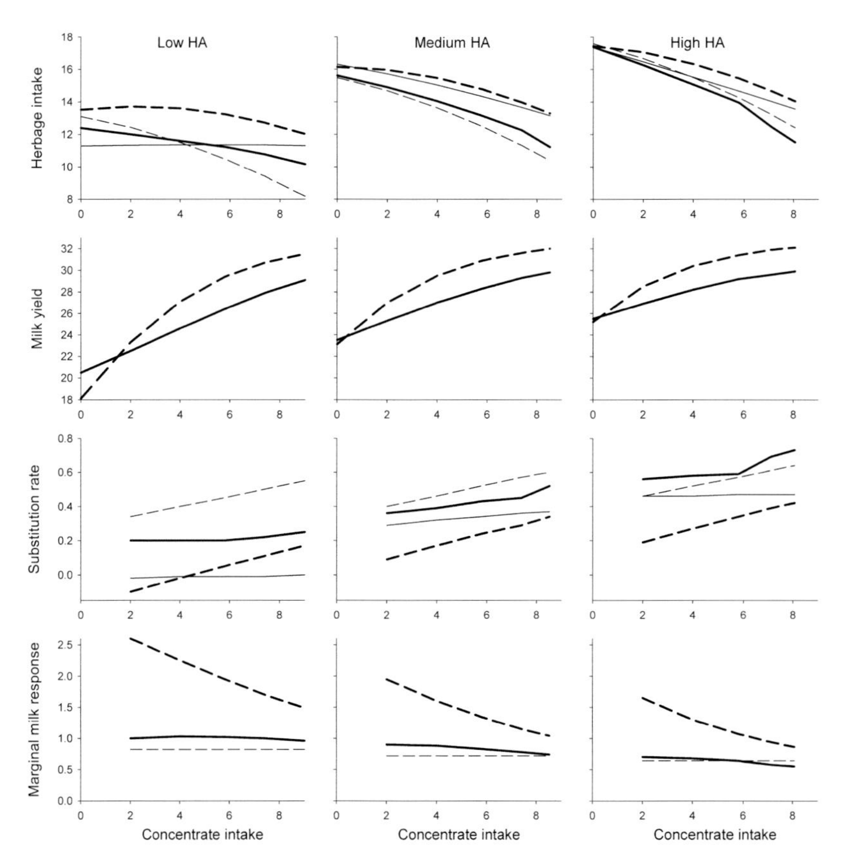

Figure 4 Simulated effect of concentrate supplementation level (kg DM/day) on herbage intake (kg DM/day), milk yield (kg/day), substitution rate (DM basis) and marginal milk response to concentrate (kg milk/kg DM concentrate) by grazing dairy cows. The conditions of the simulations are described in the text. Simulations are done for low, medium and high herbage allowance (HA: 25, 37 and 50 kg DM/day to ground level, respectively). Models: – – Diet-Check (Heard *et al.*, 2004), ▬ ▬ Grazeln (Delagarde *et al.*, 2004); —— Pâtur'IN (Delaby *et al.*, 2001b); ▬▬ GrazFeed (Freer *et al.*, 1997)

Comparison of the accuracy of a number of models

The authors of respective models rarely carry out an evaluation of the accuracy of model predictions. In this section, a global and statistical comparison of the accuracy of some multiple regression equations and models is investigated. Experimental databases coming from TEAGASC (Ireland) and from INRA (France) were used to compare model predictions and actual values of herbage intake by grazing dairy cows. Individual herbage intake was measured from n-alkanes at TEAGASC and from faecal output and herbage digestibility estimates at

INRA. The individual cow characteristics and herbage intake were averaged per grazing herd and per paddock before running the different models. The dataset from TEAGASC includes a total of 20 experiments from 1988 to 2000, representing 190 grazing herds with an average of 15 cows per herd. The dataset from INRA includes 11 experiments from 1988 to 2000, representing 114 grazing herds with an average of 6 cows per herd (Table 3). The accuracy of predictions is investigated through the calculation of the mean prediction error (MPE) and the proportions of the mean square prediction error explained by the mean bias, the line bias and random error (Rook *et al.*, 1990). A relative MPE of 0.15, expressed as a proportion of the mean actual value, means that the model predicts herbage intake with an error of 15%.

Models compared

Five multiple regressions and four herbage intake models taking into account at least some animal characteristics and grazing conditions were selected for this comparison (Table 4). Unfortunately, it was not possible to run a number of models from the published information, for example GrazFeed (Freer *et al.*, 1997). A constant substitution rate of 0.4 was added in the multiple regressions of Stockdale (1985) and Peyraud *et al.* (1996) in order to predict intake for supplemented cows. For the Sepatou model (Cros *et al.*, 2003), the OM digestibility of the herbage offered (above 4 or 5 cm) was used as an input because the estimation of the OM digestibility of the selected herbage cannot be easily estimated. Herbage mass and herbage allowance were measured above 4 cm at TEAGASC and above 5 cm and/or to ground level at INRA. As many regressions and models are calibrated to ground level, herbage mass and thus herbage allowance was calculated to ground level before simulations.

Table 3 Description of the database of 304 experimental herds of grazing dairy cows (190 from TEAGASC and 114 from INRA) used to compare the precision of the prediction of herbage intake between different multiple regressions and models

Variable	Mean (n = 304)	s.d.	Max	Min	TEAGASC (n = 190)	INRA (n = 114)
Actual Intake (kg DM)						
Herbage	15.9	2.09	22.0	9.9	15.6	16.2
Concentrate	0.9	1.23	5.4	0.0	0.9	0.8
Total	16.7	2.28	23.4	10.4	16.6	17.0
Actual milk yield (kg)	21.7	5.51	41.8	8.9	22.4	20.6
Peak milk yield (kg)	33.7	5.73	46.7	21.0	31.1	38.1
Stage of lactation (weeks)	24.2	7.80	39.9	3.8	21.3	29.0
Live weight (kg)	565	39.9	677	487	549	592
Herbage OM digestibility	0.80	0.035	0.87	0.72	0.82	0.78
Herbage CP (g/kg DM)	184	36.9	277	86	200	156
Herbage mass (t DM/ha)						
above 4 cm	2.5	0.85	5.7	0.6	2.3	2.8
to ground level[a]	4.9	1.14	9.1	3.1	4.4	5.8
Area (m^2/cow)	93	38.8	246	21	111	64
Allowance (kg DM/cow)						
above 4 cm	21.3	8.58	61.2	6.6	24.5	15.8
to ground level[a]	43.2	14.91	105.4	16.5	48.6	34.4

[a] measured or estimated

Results

The mean bias between actual and predicted herbage intake ranged from –1.3 to 1.1 kg DM between models considering the 304 experimental herds (Table 4). The overall under- or over-estimation of herbage intake according to the model was generally consistent between the two datasets. The average mean prediction error ranged from 1.44 to 3.82 kg DM between models. The overall precision of the prediction of herbage intake averaged 0.15 and ranged from 0.10 to 0.25 according to the model. This range is similar to that found by Keady *et al.* (2004) evaluating five intake models for dairy cows fed on grass silages (0.10 to 0.20). Rook *et al.* (1990) also reported similar precision for intake models in beef cattle fed on grass silages, ranging from 0.08 to 0.26.

Table 4 Statistical comparison of the accuracy of different multiple regressions and models predicting herbage intake of grazing dairy cows. The TEAGASC, INRA and GLOBAL (TEAGASC+INRA) databases include 190, 114 and 304 experimental herds, respectively

Database	Regression or model	Mean predicted HI	Mean bias (P-A)	R^2	MSPE	Proportion of MSPE			MPE	
						bias %	line %	random %	kg DM	relative
TEAGASC	(1)	16.3	0.6	0.31	3.40	12	2	86	1.84	0.12
	(2)	16.3	0.7	0.01	19.77	2	77	21	4.45	0.29
	(3)	15.3	-0.4	0.39	3.95	4	32	64	1.99	0.13
	(4)	16.9	1.2	0.23	6.76	23	30	47	2.60	0.17
	(5)	15.4	-0.2	0.68	1.38	3	0	97	1.18	0.08
	(6)	17.0	1.4	0.17	5.46	37	0	63	2.34	0.15
	(7)	15.2	-0.5	0.30	3.13	7	0	93	1.77	0.12
	(8)	16.2	0.5	0.23	5.23	5	33	62	2.29	0.15
	(9)	15.8	0.2	0.48	2.21	1	1	98	1.49	0.10
INRA	(1)	16.5	0.3	0.22	4.58	2	24	74	2.14	0.14
	(2)	15.7	-0.5	0.17	5.86	4	34	62	2.42	0.15
	(3)	14.9	-1.3	0.25	5.51	31	9	60	2.35	0.15
	(4)	17.2	1.0	0.19	9.27	10	52	38	3.04	0.19
	(5)	16.3	0.1	0.28	3.22	0	3	97	1.79	0.12
	(6)	16.9	0.6	0.14	4.16	9	1	90	2.04	0.13
	(7)	13.6	-2.6	0.32	10.54	65	7	28	3.25	0.20
	(8)	16.3	0.1	0.21	6.95	0	50	50	2.64	0.17
	(9)	15.5	-0.8	0.39	3.38	18	3	79	1.84	0.12
GLOBAL	(1)	16.4	0.5	0.26	3.84	7	10	83	1.96	0.13
	(2)	16.1	0.2	0.03	14.56	0	71	29	3.82	0.25
	(3)	15.1	-0.7	0.32	4.53	12	23	65	2.13	0.14
	(4)	17.0	1.1	0.21	7.70	17	39	44	2.78	0.18
	(5)	15.7	-0.1	0.53	2.07	1	1	98	1.44	0.10
	(6)	17.0	1.1	0.14	4.97	25	0	75	2.23	0.15
	(7)	14.6	-1.3	0.18	5.91	27	13	60	2.43	0.16
	(8)	16.2	0.4	0.22	5.87	2	42	58	2.43	0.16
	(9)	15.7	-0.2	0.41	2.65	1	2	97	1.63	0.11

Multiple regressions : (1) Stockdale, 1985; (2) Caird & Holmes, 1986; (3) Peyraud *et al.*, 1996; (4) Stockdale, 2000; (5) O'Donovan *et al* ., unpublished
Models: (6) Sepatou (Cros *et al.*, 2003); (7) Pâtur'IN (Delaby *et al.*, 2001b); (8) Diet-Check (Heard *et al.*, 2004); (9) Grazeln (Delagarde *et al.*, 2004)

The lowest precision of intake prediction (0.29) was observed with the multiple regression of Caird & Holmes (1986) in the TEAGASC dataset. The highest precision of intake prediction (0.08) was observed with the multiple regression of O'Donovan *et al.* (unpublished, presented in Table 1) in the TEAGASC dataset, but the same data were used to develop the multiple regression (at the cow level) and to test it (at the herd level). However it also predicted herbage intake well in the INRA dataset. The large size of the TEAGASC database probably explains its high level of accuracy (Table 1). Among models, GrazeIn predictions seem the most precise in both datasets (MPE of 10 and 12%), because more factors are taken into account and because more interactions are estimated compared to the other models or regressions. The simple models are less precise (MPE from 12 to 20%) but can be used more easily.

For each model, the study of the correlations between the herbage intake bias (predicted minus actual) and the main input variables showed that the most significant correlations ($R^2 > 0.30$) were found with herbage allowance. The correlation was negative for the model of Caird & Holmes (1986) and positive for the models of Stockdale (2000) and Heard *et al.* (2004). The quadratic intake/allowance relationship of Caird & Holmes (1986) model clearly under-estimated herbage intake for high herbage allowances. The low curvature of the intake/allowance relationship of the models of Stockdale (2000) and Heard *et al.* (2004) possibly over-estimates intake for high herbage allowances (Figure 1). In the global database, considering only grazing herds offered less than 60 kg DM/cow above ground level (n=272), the accuracy of the three above models was significantly increased (MPE of 0.18, 0.16 and 0.14 for Caird & Holmes (1986), Stockdale (2000) and Heard *et al.* (2004), respectively).

Conclusions

This review has shown that models predicting intake and performance of grazing dairy cows are scarce, particularly regarding models with a mechanistic approach. The different ways of considering grazing conditions and pasture availability leads to large variations of intake predictions for atypical situations. However, the selected multiple regressions or models showed precision comparable to winter feeding models when comparing predicted values to a large set of independent actual values (mean prediction error range 10 to 20%). From a practical point of view, future models should have higher applicability, predicting both intake and performance over a large range of feeding and grazing management practices, and preferably from easy-to-obtain input variables. Moreover, much effort should be made to evaluate these models, testing prediction, robustness and accuracy. Today, even the more complex models are static, intake and performance being predicted from the description of the actual conditions. The challenge for future models is also to predict the dynamic pattern of intake during the gestation-lactation cycle, considering the carry-over effects of previous feeding strategies on subsequent intake and performance. Obviously, the development of such models should not be considered without the development of decision support tools enabling the effective use of such models.

Acknowledgements

Dr. Janna Heard is gratefully acknowledged for the clarifications on the Diet-Check program. Many thanks are due to Dr. Luc Delaby for the provision of equations to evaluate the herbage intake model of Pâtur'IN and to Dr. Gearoid Stakelum for the compilation of the TEAGASC experimental database.

References

Baumont, R., D. Cohen-Salmon, S. Prache & D. Sauvant (2004). A mechanistic model of intake and grazing behaviour in sheep integrating sward architecture and animal decisions. *Animal Feed Science and Technology*, 112, 5-28.

Buckmaster, D.R., L.A.Holden, L.D. Muller & R.H. Mohtar (1997). Modeling intake of grazing cows fed complementary feeds. In: Proceedings of the 18th International Grassland Congress, Winnipeg, Canada, 2, 9.

Caird L., W. Holmes (1986). The prediction of voluntary intake of grazing dairy cows. *Journal of Agricultural Science Cambridge*, 107, 43-54.

Cherney, D.J.R. & D.R. Mertens (1998). Modelling grass utilization by dairy cattle. In: J.H. Cherney & D.J.R. Cherney (eds.) Grass for dairy cattle. CAB International, Wallingford, 351-371.

Cros, M.J., M. Duru, F. Garcia & R. Martin-Clouaire (2003). A biophysical dairy farm model to evaluate rotational grazing management strategies. *Agronomie, Paris*, 23, 105-122.

Delaby, L., J.L. Peyraud & R. Delagarde (2001a). Effect of the level of concentrate supplementation, herbage allowance and milk yield at turn-out on the performance of dairy cows in mid lactation at grazing. *Animal Science*, 73, 171-181.

Delaby, L., J.L. Peyraud & P. Faverdin (2001b). Pâtur'IN: le pâturage des vaches laitières assisté par ordinateur [Pâtur'IN: Computer-assisted grazing of dairy cows]. *Fourrages*, 167, 385-398.

Delaby, L., J.L. Peyraud & R. Delagarde, R. (2003). Faut-il complémenter les vaches laitières au pâturage [Is it necessary to supplement dairy cows at grazing]. *INRA Productions Animales*, 16, 183-195.

Delagarde, R., S. Prache, P. D'Hour & M. Petit (2001). Ingestion de l'herbe par les ruminants au pâturage [Herbage intake by grazing ruminants]. *Fourrages*, 166, 189-212.

Delagarde, R., P. Faverdin, C. Baratte & J.L.Peyraud (2004). Prévoir l'ingestion et la production des vaches laitières: GrazeIn, un modèle pour raisonner l'alimentation au pâturage [Predicting herbage intake and milk production of dairy cows: GrazeIn, a model for the feeding management at pasture]. *Rencontres Recherches Ruminants*, 11, 295-298.

Demment, M.W. & G.B. Greenwood (1988). Forage ingestion: effects of sward characteristics and body size. *Journal of Animal Science*, 66, 2380-2392.

Dove, H. (1996). Constraints to the modelling of diet selection and intake in the grazing ruminant. *Australian Journal of Agricultural Research*, 47, 257-275.

Faverdin, P., J.P. Dulphy, J.B. Coulon, R. Vérité, J.P. Garel, J. Rouel & B. Marquis (1991). Substitution of roughage by concentrates for dairy cows. *Livestock Production Science*, 27, 137-156.

Faverdin, P., R. Baumont & K.L. Ingvartsen (1995). Control and prediction of feed intake in ruminants. In: Proceedings of the IVth International Symposium on the Nutrition of Herbivores, Clermont-Ferrand, 95-120.

Forbes, J.M. (1995). Prediction of voluntary intake. In: J.M. Forbes (ed.) Voluntary food intake and diet selection in farm animals. CAB International, Wallingford, 384-415.

Freer, M., A.D. Moore & J.R. Donnelly (1997). GRAZPLAN: Decision support systems for Australian grazing enterprises. II. The animal biology model for feed intake, production and reproduction and the GrazFeed DSS. *Agricultural Systems*, 54, 77-126.

Heard, J.W., D.C. Cohen, P.T. Doyle, W.J. Wales & C.R. Stockdale (2004). Diet-Check - a tactical decision support tool for feeding decisions with grazing dairy cows. *Animal Feed Science and Technology*, 112, 177-194.

Herrero, M., R.H. Fawcett, V. Silveira, J. Busqué, A. Bernués & J.B. Dent (2000). Modelling the growth and utilisation of kikuyu grass (*Pennisetum clandestinum*) under grazing. 1. Model definition and parameterisation. *Agricultural Systems*, 65, 73-97.

Horan, B., P. Faverdin, L. Delaby, F. Buckley, M. Rath & P. Dillon (2005). The effect of strain of Holstein-friesian dairy cows on grass intake and milk production in various pasture based systems. *Journal of Dairy Science,* (in press).

Ingvartsen, K.L. (1994). Models of voluntary food intake. *Livestock Production Science*, 39, 19-38.

INRA (1989). Ruminant Nutrition: Recommended Allowances and Feed Tables. R. Jarrige (ed.). John Libbey, London, 389pp.

Johnson, I.R. & A.J. Parsons (1985). A theoretical analysis of grass growth under grazing. *Journal of Theoretical Biology*, 112, 345-367.

Keady, T.W.J., C.S. Mayne & D.J. Kilpatrick (2004). An evaluation of five models commonly used to predict food intake of lactating dairy cattle. *Livestock Production Science*, 89, 129-128.

Kyriazakis, I. (2003). What are ruminant herbivores trying to achieve through their feeding behaviour and food intake? In: Proceedings of the 6th International Symposium on the Nutrition of Herbivores, Mérida, México, 153-173.

Maher, J., G. Stakelum & M. Rath (2003). Effect of daily herbage allowance on the performance of spring-calving dairy cows. *Irish Journal of Agricultural and Food Research*, 42, 229-241.

Mayne, C.S., A. Rook, J.L. Peyraud, J.W. Cone, K. Martinsson & A. González-Rodríguez (2004). Improving the sustainability of milk production systems in Europe through increasing reliance on grazed pasture. *Grassland Science in Europe*, 9, 584-586.

McGilloway, D.A. & C.S. Mayne (1996). The importance of grass availability for the high genetic merit dairy cow. In: P.C. Garnsworthy, J. Wiseman & W. Haresign (eds.) Recent advances in animal nutrition. University Press, Nottingham, 135-169.

Meijs, J.A.C. & J.A. Hoekstra (1984). Concentrate supplementation of grazing dairy cows. I. Effect of concentrate intake and herbage allowance on herbage intake. *Grass and Forage Science*, 39, 59-66.

Penning, P.D., A.J. Parsons, R.J. Orr & T.T. Treacher (1991). Intake and behaviour responses by sheep to changes in sward characteristics under continuous stocking. *Grass and Forage Science*, 46, 15-28.

Peyraud, J.L. & L. Delaby (2001). Ideal concentrate feeds for grazing dairy cows – Responses to supplementation in interaction with grazing management and grass quality. In: P.C. Garnsworthy & J. Wiseman (eds.) Recent advances in animal nutrition. University Press, Nottingham, 203-220.

Peyraud, J.L. & A. González-Rodríguez (2000). Relations between grass production, supplementation and intake in grazing dairy cows. *Grassland Science in Europe*, 5, 269-282.

Peyraud, J.L., E. A. Comerón, M.H. Wade & G. Lemaire (1996). The effect of daily herbage allowance, herbage mass and animal factors upon herbage intake by grazing dairy cows. *Annales de Zootechnie*, 45, 201-217.

Phillips, C.J.C. (1988). The use of conserved forage as a supplement for grazing dairy cows. *Grass and Forage Science*, 43, 215-230.

Pittroff, W. & M.M. Kothmann (2001). Quantitative prediction of feed intake in ruminants. II. Conceptual and mathematical analysis of models for cattle. *Livestock Production Science*, 71, 151-169.

Poppi, D.P., T.P. Hughes & P.J. L'Huillier (1987). Intake of pasture by grazing ruminants. In: A.M. Nicol (ed.) Feeding livestock on pasture. New Zealand Society of Animal Production, Occasional Publication n°10, 55-63.

Rook, A.J., M.S. Dhanoa & M. Gill (1990). Prediction of the voluntary intake of grass silages by beef cattle. 3. Precision of alternative prediction models. *Animal Production*, 50, 455-466.

Rook, A.J., C.A. Huckle & R.J. Wilkins (1994). The effects of sward height and concentrate supplementation on the performance of spring calving dairy cows grazing perennial ryegrass-white clover swards. *Animal Production*, 58, 167-172.

Sibbald, A.R., T.J. Maxwell & J. Eadie (1979). A conceptual approach to the modelling of herbage intake by hill sheep. *Agricultural Systems*, 4, 119-134.

Smallegange, I.M. & A.M.H. Brunsting (2002). Food supply and demand, a simulation model of the functional response of grazing ruminants. *Ecological Modelling*, 149, 179-192.

Stakelum, G. & P. Dillon (2004). The effect of herbage mass and allowance on herbage intake, diet composition and ingestive behaviour of dairy cows. *Irish Journal of Agricultural and Food Research*, 43, 17-30.

Stockdale, C.R. (1985). Influence of some sward characteristics on the consumption of irrigated pastures grazed by lactating dairy cattle. *Grass and Forage Science*, 40, 31-39.

Stockdale, C.R. (2000). Levels of pasture substitution when concentrates are fed to grazing dairy cows in northern Victoria. *Australian Journal of Experimental Agriculture*, 40, 913-921.

Woodward, S.J.R. (1997). Formulae for predicting animals' daily intake of pasture and grazing time from bite weight and composition. *Livestock Production Science*, 52, 1-10.

Decision support for temperate grasslands: challenges and pitfalls

J.R. Donnelly[1], L. Salmon[1], R.D.H. Cohen[2], ZL. Liu[3] and XP. Xin[4]
[1]*CSIRO Plant Industry, Canberra, 2601, Australia, Email:john.r.donnelly@csiro.au*
[2]*RDC Ranch Consulting, 313-10033 110 St. NW, Edmonton AB. T5K 1J5, Canada*
[3]*University of Inner Mongolia, Hohot 010021, Inner Mongolia, China*
[4]*Institute of Agricultural Resource and Regional Planning, Chinese Academy of Agricultural Sciences, No.12 South Zhongguancun Ave, Beijing 100081, China*

Key points

1. Successful adoption of decision support tools (DS tools) to address grassland management issues requires careful attention in design to ensure ease-of-use, accuracy in prediction and the flexibility to simulate actual practices.
2. DS tools must handle spatial variability and where possible include facilities for automatic sourcing of essential information for initialisation.
3. Advances in the development of DS tools will depend on resolution of scientific issues in grassland biology including investment in dedicated experiments to determine parameter values for model equations.
4. The use of mechanistic models, the integration of remote sensing technology and cooperation between research groups to develop modular simulation frameworks to share models will enhance the value of DS tools in grassland management.

Keywords: decision support tools, grazing systems, models, GrassGro

Introduction

For more than four decades computer models and decision support tools (DS tools) have been advocated for guiding the management of temperate grasslands. The first uses were for research but subsequently this has extended to farm advice, landscape management and rural policy. At best their adoption has been modest, but given their complexity and diversity this is not surprising. It is worth noting that this is similar to the uptake of many other technologies in agriculture. In a review of technology transfer in the wool industry in Australia, Vizard & Edwards (1992) point to several examples where simple but very beneficial technologies have had low adoption. However, there are some notable examples of productivity improvements with increased profits resulting from the use of DS tools, particularly where variable weather makes the outcomes of decisions uncertain (Donnelly *et al.*, 2002).

Grasslands are complex biological systems. Their optimal management requires a comprehensive, systems-level approach that is best addressed using models. Modelling grasslands is indeed the “big science" recognised by Thornley (2001), but some sections of the scientific and farming communities have yet to appreciate the benefits that it can deliver. Well-tested systems models and DS tools are expensive to develop, as they require expertise in several disciplines such as biology, hydrology, mathematics, economics and computing. David et al. (2002) estimate that costs range from US$15-30M per model, but costs must be weighed against the benefits derived from their use. In Australia in 2002, an independent assessment for CSIRO by the Center for International Economics estimated a return of more than AU$70 for each dollar invested in the development of the GrazFeed DS tool (Freer *et al.*, 1997).

Challenges and pitfalls

Computer-based DS tools provide an integrated framework for farm managers to identify opportunities and quantify risks to profitable livestock production from grasslands. They can help farmers integrate livestock management with other farming enterprises and policy makers can use them to evaluate alternative uses for grasslands. Moore (2005) reviewed briefly key features of a number of DS tools that have contributed to the management of grasslands at the main section of this Congress. However, there are some issues discussed below, about tool design and general requirements that must be addressed to ensure widespread use and to capture the full potential of this powerful technology.

Tool design

In practice, most models and DS tools are used for research and then only by the research groups that built them (Hook, 1997). Early ideas that farmers would be the main users of DS tools are now gradually being replaced with recognition that benefits may be delivered best by those with appropriate expertise (McCown, 2002). The more complex dynamic DS tools or simulators designed for evaluation of strategic management options are more appropriate for use by farm consultants or other professionals. Application of DS tools to day-to-day decisions by farmers may be more likely if they are installed on small, handheld devices suitable for paddock use.

DS tools must be able to represent attributes of the system that the user considers important (such as the legume content of pastures), otherwise they will not be adopted. They must be easy to use with a minimum investment of time to solve a problem and they must facilitate access to in-built or web-accessible databases such as soil properties or climate if these data are required. The interface must guide the user in deconstruction of a problem and in identification of the key controlling variables so that analysis is feasible. The DS tool must help interpret the information generated and produce a report. Analysis of more complex problems may require a factorial design. Such facilities need to be in-built and linked to smart ways to extract critical information from output, which can otherwise be unmanageable. Moore (2005) describes the redevelopment of the GrassGro interface to focus on problem solving, so that pre-configured formats for analyses are available and the simulation results linked with customised reporting. The analyses can also be stored for later use. The objective is to make tailor-made advice more effective and affordable.

Estimating coefficients for equations

Precision and accuracy in prediction is important for widespread acceptance of DS tools. A key issue is the generality and reliability of coefficients in the equations within the underlying models. Since the coefficients can vary with environmental and other conditions, estimating their values is one of the most difficult problems in modelling (Ahuja & Ma, 2002). Estimation by experiment is costly and the lack of standard methodology means it is difficult to achieve. Simplification of biophysical processes in models sometimes frustrates experimental measurement; for example, the artificial partitioning of soil organic matter into recalcitrant and labile pools and the uncertainty in the organic matter turn-over times between pools.

Most pasture plant models have a structure based on a water budget, assimilate production and distribution, reproduction and senescence. A major issue, however, is the lack of a standard approach to quantifying the parameters of the equations for plant growth and development. A relatively large number of parameters are required to discriminate the patterns of growth, reproduction and senescence of different plant species (Moore *et al.*, 1997) and obtaining estimates of these parameters is demanding even when data are available; for many species there are no data. In practice, the relative values for these parameters are far more important than their absolute values as their purpose is to discriminate between species.

This pragmatic approach has enabled continued tool development despite the lack of precise data. As the parameter sets describing pasture species are external to the model, few coding changes are generally required. This model design, has enabled GrassGro to be used in more extreme environments than those of temperate Australia. Rapid parameterisation of 19 species common to the Canadian prairie (Cohen *et al.*, 2003) and descriptions of 7 species common to the steppe of Mongolia and Inner Mongolia (*Leymus chinensis, Stipa grandis, S. krylovii, Agropyron michnoii, A. cristatum, Cleistogenes squarrosa* and *Artemisia frigida*) have been possible (Figure 1).

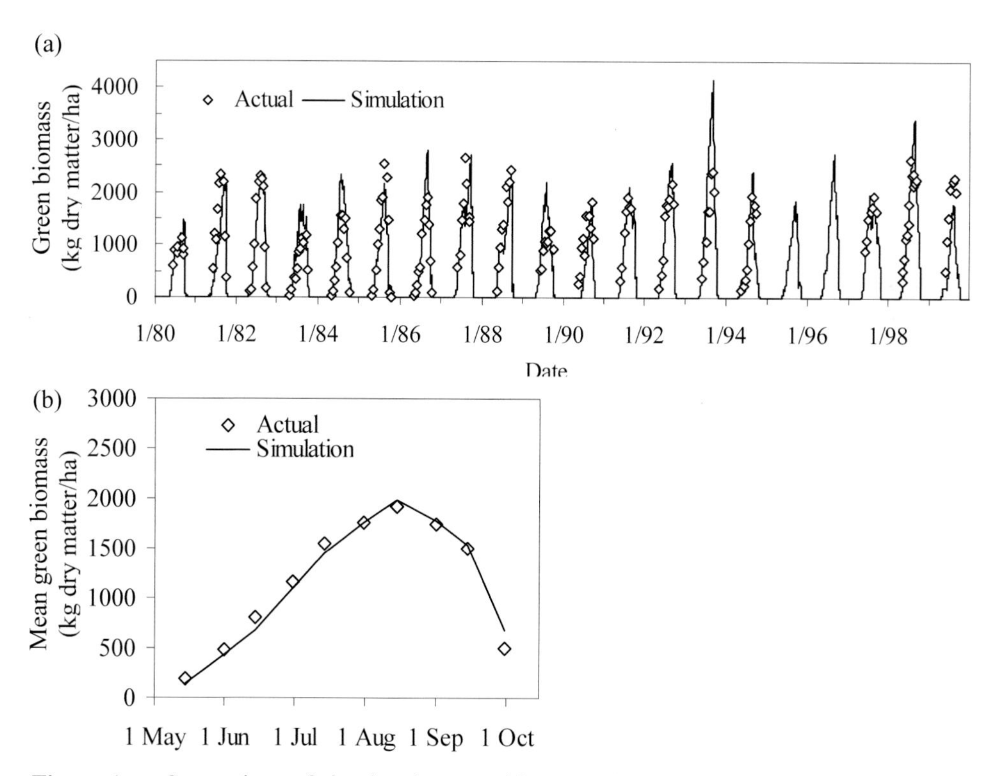

Figure 1 **a**. Comparison of simulated pasture biomass of 7 species common to the steppe of Mongolia and Inner Mongolia (line) with observed green pasture biomass (circles) at Xilingol, Inner Mongolia, 1980-1999. Measurements were not available for 1995-1996. **b**. Mean values for the same period (1980-1999).

Validation

Although estimation of the coefficients for equations is the most difficult problem in modelling, validation of the predictions of a model or DS tool is the most contentious. Validation generally refers to the accurate prediction of results observed in a particular experiment. It is contentious because of uncertainty about i) the accuracy of the experimental data, ii) the precise environmental conditions under which the data were collected, iii) possible differences in calibration of methods used for measurement, iv) the correct use of the model and v) the precision and accuracy of the model for the particular circumstances of the experiment. Conclusions about the validity of a model ultimately depend on its correct application, the accuracy of any data used for checking predictions and the accuracy and precision of its predictions.

Validation can be hampered by uncertainty about initialisation of the model (see below), particularly if initialisation values are based on a spatially limited sample and the data used for validation are measured at a larger scale. For example, soil core samples are expensive to collect, so minimal numbers of cores taken from a representative site are often used to initialise the soil profile for a whole paddock, or even a whole farm. In reality it is unknown if the sample is truly representative. If the initialisation values do not represent the actual properties of the predominant soil of the paddock or farm, then model predictions, for example, of herbage yields are unlikely to match actual observations. In this case, if predictions diverge from observations, the model is not necessarily in error and the initialisation could be inadequate.

Errors in model predictions may also be introduced by the time scales used for measurement of driving variables. If daily rainfall is used as a driver of plant growth, no account can usually made for rainfall intensity, which operates at a much finer scale. Infiltration and runoff values can differ markedly for the same daily rainfall measurement with very significant consequences for plant growth.

Given that most DS tools designed for use in grassland management are based on deterministic models, it is generally not feasible to calculate an error of prediction. However, the plausibility of output values and the sensitivity of the overall model can be checked by varying initialisation values for key variables. For example, inputs for soil depth can be varied over the measured range of values. This is usually a lengthy process generating large amounts of information, so additional tools or facilities are essential to help the user efficiently assess the response surface. Alternatively, the response of individual equations to changes in parameter values can be checked externally, for example, in a spreadsheet such as that available for the animal model used in GrazFeed (Freer *et al.*, 1997; www.pi.csiro.au/grazplan). The SGS pasture model developed recently in Australia (Johnson et al., 2003) was designed with an in-built facility to test the effect of parameter values on model responses.

All users of DS tools require accuracy and precision in the underlying models, but the level required depends on the intended application. Validation can be very difficult if small errors, when accumulated, lead to large errors at a higher level of the system. As McNamara (2004) points out, precise measures (1-5% error) of daily feed intake of even housed dairy cows, for example, are not available and may never be, but over a 300d lactation this small measurement error could potentially lead to a large cumulative error in predicted energy retention perhaps amounting to 30kg of body fat.

Despite these issues, a grassland farmer using a model primarily wants an affordable and useable tool with acceptable predictions, even though there may be few accurate records against which to test the predictions. In the absence of data, the predictions must at least approximate the user's expectations. Formal validation of the model is rarely possible on a farm and it may not matter that some processes are estimated by a relatively simple approximation. An example from GrassGro, which can model swards of mixed species, is the option to specify a fixed legume content for a pasture rather than to explicitly model the competition between grasses and legumes. Since the legume component of a pasture has a marked effect on animal food intake and production, the inclusion of a legume component is important. If species competition cannot be modeled with confidence, then under some circumstances specifying the legume content is a reasonable although crude approximation that ensures a more realistic simulation.

Initialisation

Before a DS tool can be used it must be initialised to represent the conditions on the farm or part of the farm that is under study. Initialisation can be time-consuming and expensive and where possible, the essential information should be automatically sourced from geo-referenced databases stored with the DS tool or accessed from the web. For example, local soil information and weather data should be accessible simply by specifying the geographic location of the paddock, enterprise or farm under study. Real progress towards achieving this is now possible by advances in computing and communications technology, although there may be IP issues that still require resolution to gain general access to web databases.

Spatial variability across paddocks or farms and temporal variability can cause non-uniform plant growth and development that requires integration for accurate initialisation (Ahuja & Ma, 2002). There are several approaches to simulating a highly variable site: averaging, sensitivity testing, integration of simulations of different spatial units (Beverley *et al.*, 2003) or stochastic modelling using probability distributions. Remote sensing holds promise as a method for measuring variability across landscapes. Initial values may also be obtained from a preliminary simulation run. As discussed above, the most common problems with estimating initial values are the cost and difficulty of measurement.

Technical support

The adoption of a DS tool is rarely a matter of selling a piece of software "off the shelf". Documentation, on-going technical support and some form of training for effective use are essential and must be funded. However, some tools are easy to use, require little training for familiarization and are suitable for distribution as software products. Other tools designed to set or evaluate strategic objectives, such as GrassGro, usually require significant training for effective use. An alternative to distributing software is the sale of comprehensive, regionally-specific analyses undertaken by an expert user working within a consultancy firm and perhaps with web-based delivery.

Web-based delivery of DS tools may also help contain the cost of technical support and ensure that only the latest versions of software are used for analyses. The University of New England has implemented technology to distribute slightly modified versions of the GRAZPLAN DS tools (Donnelly *et al.*, 2002) on a fee for service basis at educational institutions throughout Australia (Daily *et al.*, 2005).

Combining laboratory and software analyses is another approach to providing analyses using DS tools. The NUTBAL nutritional management system is operated by the Grazingland Animal Nutrition Laboratory at Texas A&M University (Stuth *et al.*, 2002). Clients collect fresh dung deposits from the field and submit the faecal sample to the laboratory, together with information describing the animals and grazing environment. The laboratory uses near infra-red (NIR) spectral analysis of faeces to estimate dry matter intake and diet quality. The estimates are used with the NUTBAL model to calculate the nutritional balance of the herd and advise modification to the animal's feed if necessary.

Funding

The scarcity of funds for production-based agricultural research and extension is now a major constraint to technological progress in farming and it is becoming more severe as resources are redirected to biotechnology, natural resource management and other important national priorities, such as the effects of climate change. It is ironic that endeavours in these disciplines could be enhanced with the use of appropriate DS tools. The problem is becoming more severe as the contribution of agriculture to employment and GDP of major economies is declining worldwide (Marsh, 2004; Keogh, 2004; Freshwater, 2004). Access to farming technology used to be free, primarily through public extension services, but these are being progressively withdrawn and advice is becoming more costly. The environment for research into model and DS tool development is particularly unfavorable, as it is largely funded by public outlays and the non-government sector provides only minor support for application of the technology to solve industry problems. A further major constraint is a worldwide shortage of biological scientists with training in applied mathematics and system analysis, who are willing to work in a field that is poorly resourced.

Support for grassland modelling, with or without agricultural applications, may have to come from non-agricultural sources. A partial solution to boost funding is to link DS tool development with well-funded public initiatives for landscape management or climate change and seasonal weather forecasting. The DS tools provide a powerful way to extend and evaluate these initiatives. The following example shows how application of a DS tool to a grazing system enables the value of a weather forecast to be assessed.

Case study: DS tools and evaluation of 3-monthly seasonal rainfall forecasts – how good does a seasonal rainfall forecast have to be to warrant action by a farmer?

Seasonal rainfall forecasts are routinely issued for many regions of the world and farmers can use them to guide weather-sensitive management decisions. However, the actual weather outcomes are subject to great uncertainty, so it is reasonable to question the reliability of decisions based on the forecasts. The question that must be answered is: what is the "break-even" probability for taking action in response to a forecast? This probability can be estimated by calculating the expected monetary value (EMV) of the alternative decisions (Vizard, 1994) and is equivalent to the cost:loss ratio (Wilks, 2001). If the monetary values of these management options are not known, then a DS tool like GrassGro, which is driven by the climatological record, can provide these as financial outcomes for a defined set of seasonal conditions; for example, the average gross margin of years with summers in the driest tercile. Combining the predictive capacity of GrassGro with calculation of the "break-even" probability (or cost:loss ratio) gives a unique and powerful way to determine whether a farmer should respond to a seasonal forecast for a particular enterprise. The following example demonstrates this approach for a bull-fattening enterprise in south-eastern Australia (Salmon *et al.*, 2003).

A beef producer at Branxholme, Victoria, (mean annual rainfall of 655mm) wanted to use surplus pasture at the end of the growing season in December 2002 by purchasing cheap bulls weighing 330kg and fattening them to 525kg for sale in the following November. Before making the decision to buy the bulls two questions needed answers. First, how many bulls should be purchased to make maximum profit? Second, what was the risk of the failure of autumn rains increasing the need to feed expensive grain to the bulls? The first step to answer these questions was to use GrassGro to simulate the perennial ryegrass, annual grass and subterranean clover pasture sown on the farm. The model was initialised to represent the yield and quality of the pasture and the condition of the bulls at the end of December 2002. Then, daily pasture production, intake of pasture and supplement and weight gain of the bulls were simulated for six stocking rates (1.5 to 4.0 bulls/ha) between December and the following November, using daily weather data from the climatological record from each of 46 years from 1957 to 2002. Annual gross margins were calculated for the bull fattening enterprise (Figure 2).

The simulation results indicate that on average the optimum number of bulls to buy would be 2.5/ha as this stocking rate achieved the highest mean gross margin without excessive risk of financial loss (Figure 2).

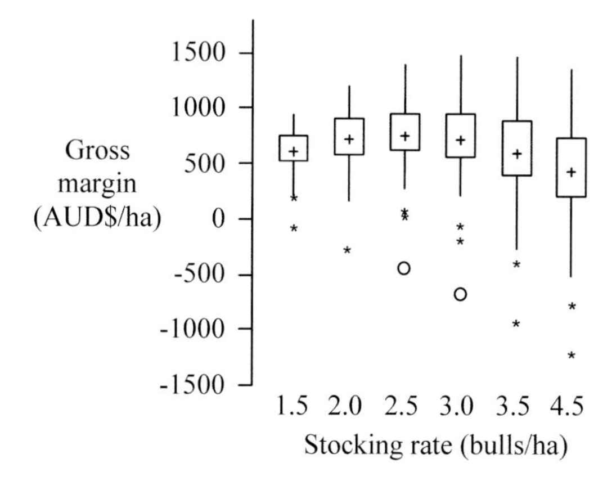

Figure 2 Boxplots showing the distribution of predicted annual gross margins at six stocking rates over the period 1957-2002 for a bull beef enterprise in Victoria, Australia. The mean gross margins are indicated by a cross symbol and outliers by an asterisk or open circles.

What about the producer's concerns of a delayed start to autumn growth with limited pasture feed? Should the number of bulls be reduced to say 1.5 rather than 2.5 bulls/ha, given that the seasonal rainfall forecast for the next 3 months (January to March) predicted a "dry" season (rainfall in the lowest tercile)? What is the required level of probability for the forecast being correct that would justify reducing the number of bulls purchased? This decision requires the calculation of the "break-even" probability or cost:loss ratio.

If p and (1-p) are the probabilities of a season being dry or not dry and v1 and v2 are the respective financial outcomes of purchasing the optimum number of bulls (2.5/ha), then

$$EMV_{2.5} = p*v1 + (1-p)*v2$$

Likewise, if v3 and v4 are the financial outcomes for reducing the number of bulls to 1.5/ha, then

$$EMV_{1.5} = p*v3 + (1-p)*v4$$

At the break-even probability $EMV_{2.5} = EMV_{1.5}$, that is

$$p*v1 + (1-p)*v2 = p*v3 + (1-p)*v4$$

So relative to v2 (not dry and buy 2.5 bulls/ha)

$$p = v4/(v1 + v4 - v3)$$

The ratio v4/(v1 + v4 – v3) is also known as the cost:loss ratio (Wilks, 2001).

The EMVs for calculating the break-even probability were obtained from the gross margins generated by GrassGro. For each stocking rate, gross margins for the 46 years simulated were ranked according to the amount of rain that fell in January to March of the same year. The mean gross margin was then calculated for the driest 33% of years and for the remaining 67% of years ("not dry"). The gross margins for each stocking rate (2.5 or 1.5 bulls/ha) under "dry" or "not dry" conditions were used to calculate the break-even probability (Table 1).

Table 1 Relative expected monetary values for calculation of the break-even probability for reductions in stocking rate from 2.5 to 2.0 or 1.5 bulls/ha.

	Effect on gross margin (AUD$/ha)	
Stocking rate (bulls/ha)	"Dry"	"Not Dry"
1.5	-395 (v3)	-192 (v4)
2.0	-333 (v3)	-57 (v4)
2.5	-360 (v1)	0 (v2)

Substituting the values for a stocking rate of 1.5 bulls/ha in Table 1 into the cost:loss ratio (-92/(-360 -192 + 395)) gives an illogical value of 1.22 for the break-even probability. This indicates that reducing the number of bulls purchased to 1.5/ha will always result in a financially worse outcome even if the forecast is correct and the January to March period turns out to be "dry". This is because (a) a stocking rate of 1.5 was less profitable than a stocking rate of 2.5 in all but one year (for the grain costs and beef prices used in the analysis), (b) summer rainfall did not greatly affect the distribution of pasture growth over the rest of the year in this environment and (c) the bulls were able to exhibit compensatory growth during spring. Reducing the number purchased to 2.0 bulls/ha gives a break-even probability for action of 0.68. This means that the probability of a seasonal rainfall forecast for January to March rainfall in the driest tercile must exceed 0.68 before the farmer should reduce the number of bulls purchased to 2.0/ha. In reality, the Australian Bureau of Meteorology rarely issues such forecasts that deviate from the underlying probability value of 0.33 (Bureau of Meteorology, 1997). The Canadian Meteorological Centre has concluded that the seasonal rainfall forecasts for that country have little or no value for decisions with the current forecasting system based on dynamical and empirical models (Gagnon & Verret, 2002).

Lags in biological processes mean that short-term outcomes are in large part determined by the current state of the biological system. A decision support tool like GrassGro is extremely powerful because it can describe difficult-to-measure attributes of the current system and capture the impact of important drivers of system response e.g. soil moisture, pasture mass and quality, root depths and plant growth stages, seed banks, livestock condition and reproductive status. Of equal importance is the way the GrassGro analysis of the tactical management decision takes into account the distribution of relevant weather events in the historical record. In this example the January to March forecast was not a good predictor of autumn growth. To be useful to a farmer, a seasonal weather forecast must: (a) predict the seasonal outcome for a period that is relevant to the particular grazing enterprise and (b) do so with a greater probability than can be obtained from analysis of the enterprise with a DS tool that accurately captures the current state of the grazing system and uses climatology. Perhaps the shift of resources away from agricultural research to climate forecasting in this case, falls within the umbrella of "failed themes in grassland science" referred to by L.R. Humphreys in his final address summing up the XIX IGC in Brazil in 2001 (Humphreys, 2001). There is certainly cause for concern.

Future developments

DS tools clearly have a valuable role in providing a highly structured and consistent framework of analysis for making informed decisions about the management of grasslands, but there is substantial room for improvement in their design and accuracy. Many of the limitations outlined above will be addressed by future technical developments and a new approach to model building.

Modular simulation frameworks and object oriented modelling

Multidisciplinary models are expensive to build, so eliminating duplication of effort to reduce costs is being attempted independently by several modelling groups throughout the world (Neil *et al.*, 1999; Moore *et al.*, 2001; David *et al.*, 2002). The objective is to reuse well-tested, component modules and supporting tools possibly developed by other scientific groups. This development is a significant technical advance in simulation capability and is based on module connectivity made possible by a software interface layer or "wrapper" that generates code to allow communication between modules (Figure 3). The technique is especially powerful because it permits connectivity between modules written in different computer languages, so even those modules do not have to be rewritten. Furthermore, relatively small development teams will be able to build comprehensive DS tools that use the best scientific modules developed by expert teams from other disciplines and perhaps even located in other countries. With internet facilities it should be possible to build, at will, tailor-made DS tools for specific applications drawing on worldwide expertise. The technology will provide the only likely cost-effective way to evaluate the integration of multiple enterprises on mixed farms over a range of seasons. Salmon et al. (2005) used this approach to provide a preliminary analysis of mixed farming for sheep meat and grain cropping in Australia using a model incorporating these features.

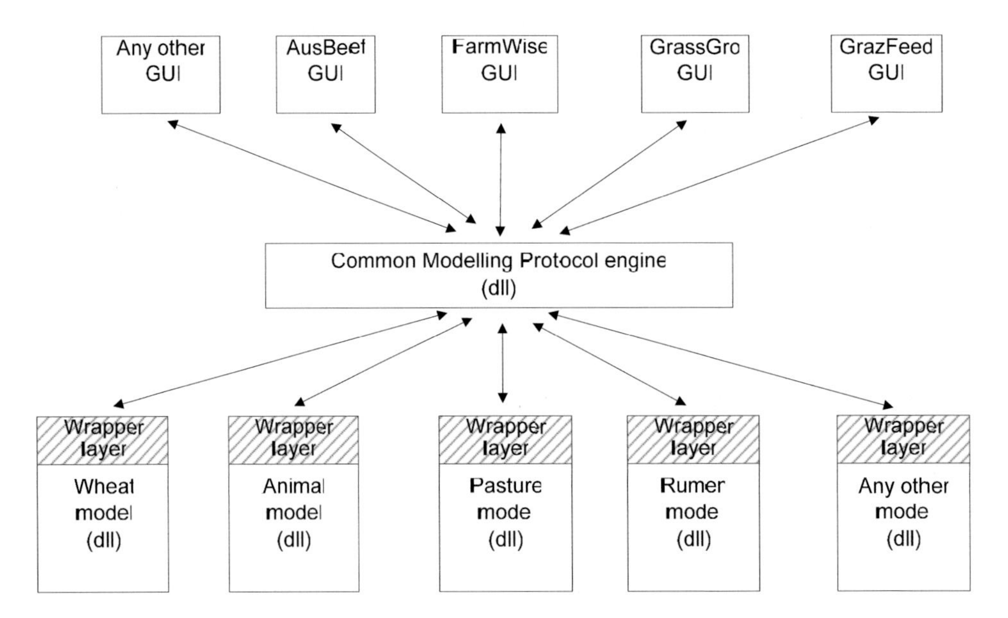

Figure 3 The CSIRO common modelling protocol is an example of a modular simulation framework that permits connectivity between models written in any language and complied as dynamic linked libraries (dll). The content of the models remains private at the cost of a relatively small code fragment or "wrapper layer" that allows relevant models to be linked together in a graphical user interface (GUI) through the Common Modelling Protocol engine to provide purpose-specific, tailor-made DS tools such as FarmWi$e or GrassGro.

Remote sensing

Linking models and DS tools with remote sensing techniques has potential to add value to farm decision making, especially where spatial scales can make model initialisation a difficult and expensive operation. The rapid development of remote sensing using spectral analysis to monitor forage resources, has potential to generate cheaper time series of plant data for validation and to provide digestibility estimates of forage quality which are essential inputs for DS tools. Comprehensive databases relating spectral analysis to herbage characteristics could be built for a wide range of pasture species at different stages of their growth cycle. Plant phenological stages predicted by a model could then be used to access estimates of herbage quality from the database.

Science

Many areas of grassland biology still require scientific resolution before a modelling approach can be considered totally adequate. Modelling grasslands is a complex multidisciplinary activity and requires substantial data. The need for comprehensive experiments conducted by skilled scientific teams was foreseen as an essential requirement more than 35 years ago by Morley (1968). His vision contrasts with current trends for trials that are more demonstration than research and located on farmer properties where careful monitoring is nearly always compromised. These studies are often intended to get the farming community involved in "research", which is desirable, but they contribute relatively little information of the type that is critically needed for model building and DS tool development. An exception is the commitment to long-term detailed observations at the whole-farm level combined with modelling at the "De Marke" experimental farm (Aarts, 2000). This should lead to the design of more sustainable nutrient management systems for dairy farming in the Netherlands.

For animal nutrition and production, some areas where information is insufficient for modelling include prediction of nutrient intake, the partitioning of absorbed and recycled nutrients, and the effect of nutrition on the quality or value of the product produced. For pasture plant production, the most urgent need is the parameterisation of additional important plant species sown or naturally occurring in pastures. The description of a plant genotype can be as broad as a plant functional type, for example, a generic annual grass that gets around the problem of describing all annual grasses. If warranted and if the information is available, the description can be more specific to represent a particular grassland species or cultivar. Since there are no standard procedures to guide this process, in practice a pragmatic choice is made about the level of information required for the intended use.

A successful analogue of this approach is used in the animal model in GrassGro and GrazFeed (Freer *et al.*, 1997) where breed names imply nothing more than a convenient reference to a functional animal type representing the genetic potential, which is scaled through the standard reference weight. This approach using functional types is very powerful as it leads to an ability to use the potential of generic models for predicting plant and animal production in any temperate region.

Second generation models

Market demands for timely delivery of quality products from grassland-based enterprises are increasing the need for DS tools with more accurate and flexible models to represent nutritional management. Estimating the voluntary food intake of grazing ruminants is still a challenge for the development of reliable DS tools. Models like that used in GrazFeed and which are suitable for use in advisory situations predict the intake and partitioning of dietary protein and energy but do not model the processes of tissue metabolism which control product quality. Considerable progress has been made with detailed mechanistic models for voluntary intake of grain diets operating through controls which act on metabolic functions, rumen fill and food breakdown rate (Nagorcka *et al.*, 2000). This opens the way to modelling fermentation in the rumen and the resulting concentration of individual volatile fatty acids and amino acids, which can then be used to predict the composition of weight, gain of lot-fed animals. The approach makes it possible to use models to explore opportunities for targeted feeding to meet precise carcase specifications, but it is not yet suitable for use in DS tools designed for grazing situations because initialisation of variables is too demanding. However, advances with NIR spectral analysis of feeds may provide a solution to this obstacle.

Conclusion

This paper has identified challenges and proposes actions essential for continued development and improvement of DS tools in a hostile funding environment. The actions include designing tools that are sensitive to the needs of users and model-developers and which do not compromise the integrity of the underlying biophysical models. This international Congress provides a unique opportunity to initiate this process and address the challenges.

Acknowledgements

Financial support from The Hermon Slade Foundation is gratefully acknowledged.

References

Aarts, H.F.M. (2000). Resource management in a "De Marke" dairy farming system. PhD Thesis, Wageningen University.

Ahuja, L.R. & L. Ma (2002). Parameterisation of agricultural system models: current approaches and future needs. In: L.R. Ahuja, L. Ma & T.A. Howell (eds.) Agricultural System Models in Field Research and Technology Transfer. Lewis Publishers, CRC Press, 273-316.

Beverley, C.R., A.L. Avery, A.M. Ridley & M. Littleboy (2003). Linking farm management with catchment response in a modelling framework. In: M.J. Unkovich & G.J. O'Leary (eds.) Solutions for a better environment. Proceedings of the 11th Australian Agronomy Conference, Geelong, Victoria, Australia, 2-6 February 2003. www.regional.org.au/au/asa/2003/i/5/beverly.htm

Bureau of Meteorology (1997). Seasonal Climate Outlook. National Climate Centre, Australian Government, Bureau of Meteorology, 1997-.

Cohen, R.D.H., J.P. Stevens, A.D. Moore & J.R. Donnelly (2003). Validating and using the GrassGro decision support tool for a mixed grass/alfalfa pasture in western Canada. *Canadian Journal of Animal Science*, 83, 171-182.

Daily, H.G., J.M. Scott & J.M. Reid (2005). Enhancing grasslands education with decision support tools. In: *Proceedings of the XX International Grasslands Congress,* Dublin, Ireland (In press).

David, O., S.L. Markstrom, K.W.Rojas, L.R. Ahuja & I.W. Schneider (2002). The object modeling system. In: L.R. Ahuja, L. Ma & T.A. Howell (eds.) Agricultural System Models in Field Research and Technology Transfer. Lewis Publishers, CRC Press. 317-330.

Donnelly, J.R., M. Freer, L. Salmon, A.D. Moore, J.R. Simpson, H. Dove & T.P. Bolger (2002). Evolution of the GRAZPLAN decision support tools and adoption by the grazing industry in temperate Australia. *Agricultural Systems*, 74, 115-139.

Freer, M., A.D. Moore & J.R. Donnelly (1997). GRAZPLAN: Decision support systems for Australian grazing enterprises. II. The animal biology model for feed intake, production and reproduction and the GrazFeed DSS. *Agricultural Systems*, 54, 77-126.

Freshwater, D. (2004). The new structure of US agriculture. *Farm Policy Journal*, 1, 38-46.

Gagnon, N. & R. Verret (2002). Relative economical value of CMC seasonal forecasts. Proceedings of the 16th American Meteorological Society Conference on Probability and Statistics in Atmospheric Sciences, Orlando, Florida, 26-30.

Hook, R.A. (1997). A directory of Australian modelling groups and models. CSIRO Publishing, Collingwood, Victoria, Australia, 312pp.

Humphreys, L.R. (2001). International grassland congress outlook – an historical review and future expectations. In: *Proceedings of the XIX International Grasslands Congress*, São Pedro, Brazil, 1085-1087.

Johnson, I.R., G.M. Lodge & R.E. White (2003). The Sustainable Grazing Systems Pasture Model: Description, philosophy and application to the SGS National Experiment. *Australian Journal of Experimental Agriculture*, 43, 711-728.

Keogh, M. (2004). Off the sheep's back and on to the tractor – the new reality of farming in Australia. *Farm Policy Journal*, 1, 4-16.

Marsh, J. (2004). The future of farming and farmers in the UK. Farm Policy Journal, 1, 18-28.

McNamara, J.P. (2004). Research, improvement and application of mechanistic, biochemical, dynamic models of metabolism in lactating dairy cattle. *Animal Feed Science and Technology*, 112, 155-176.

McCown, R.L. (2002). Changing systems for supporting farmers' decisions: problems, paradigms and prospects. *Agricultural Systems*, 74, 179-220.

Moore, A.D. (2005). Paying for our keep: grasslands decision support in more-developed countries. In: *Proceedings of the XX International Grasslands Congress*, Dublin, Ireland. (In press).

Moore, A.D., J.R. Donnelly & M. Freer (1997). GRAZPLAN: decision support systems for Australian grazing enterprises. III. Pasture growth and soil moisture submodels, and the GrassGro DSS. *Agricultural Systems*, 55, 535-582.

Moore, A.D., D.P. Holzworth, N.I. Herrmann, N.I. Huth, B.A. Keating & M.J. Robertson (2001). Specification of the CSIRO common modelling protocol. Report to Land and Water Australia, project CPI9, 58pp.

Morley, F.H.W. (1968). Computers and designs, calories and decisions. *Australian Journal of Science*, 30 (10), 405-9.

Nagorcka, B.N., G.L.R. Gordon & R.A. Dynes (2000). Towards a more accurate representation of fermentation in mathematical models of the rumen. In: J.P.McNamara, J. France & D.E. Beevers (eds.) Modelling Nutrient Utilisation in Farm Animals. CAB International, Wallingford, UK.

Neil, P.G., R.A. Sherlock & K.P. Bright (1999). Integration of legacy sub-system components into an object-oriented simulation model of a complete pastoral dairy farm. *Environmental Modelling and Software*, 14, 495-502.

Salmon, L., Frawley, C., Frawley, J., Lean, G. and Donnelly, J.R. (2003). Bull beef: production per head or per hectare? In: *Proceedings of the 1st Joint Conference of the Grassland Society of Victoria and the Grassland Society of NSW*, 75.

Salmon, L. Moore A.D. and Angus J.F. (2005) Simulation of pasture phase options for mixed livestock and cropping enterprises. In: *Proceedings XX International Grasslands Congress*, Dublin, Ireland. (In press).

Stuth, J.W., W.T. Hamilton & R. Connor (2002). Insights in development and deployment of the GLA and NUTBAL decision support systems for grazinglands. *Agricultural Systems*, 74, 99-114.

Thornley, J.H.M. (2001). Modelling grassland ecosystems. In: *Proceedings of the XIX International Grasslands Congress,* Saõ Pedro, Brazil, 1029-35.

Vizard, A. (1994). Finances and decisions. In: F.H.W. Morley (ed.) Merinos, money and management . A Post Graduate Foundation Publication, University of Sydney, 47-71.

Vizard, A. and Edwards, J. (1992). Technology transfer in the wool industry: challenges and opportunities. In: G.H. Smith (ed.) Effective technology transfer. *Australian Society of Animal Production*, 5-9.

Wilks D.S. (2001). A skill score based on economic value for probability forecasts. *Meteorology Applications* 8: 209-219.

Challenges and opportunities for animal production from temperate pastures

D.A. Clark
Dexcel, PB 3221, Hamilton, New Zealand, Email: Dave.clark@dexcel.co.nz

Key points

1. Temperate pastures offer a major opportunity to reduce the feed costs associated with ruminant production.
2. Pastures offer unique opportunities for producing high value components in feedstuffs that are beneficial to human health.
3. The increased use of pasture will not automatically lead to improved environmental outcomes – difficult challenges exist in reducing nitrogen and greenhouse gas pollution.
4. Grazed pasture systems offer advantages in animal welfare, provided management avoids the problems associated with climatic extremes, and the toxins associated with some pastures.
5. To remain competitive with total mixed ration systems, and environmentally acceptable, pastures with higher intake characteristics that allow a reduction in stock numbers per hectare and greater per animal productivity must be developed.

Keywords: pasture, animal performance, environment, animal welfare, greenhouse gases

Introduction

A distinction can be made between pasture and feedlot systems using total mixed rations (TMR) for dairy cattle, beef cattle and sheep on the basis of the feedstuff they use. But it is important to realize that economic forces have moved farmers to use pastures more like feedlots. Pastures are often supplemented with a vast range of feedstuffs making pasture-based systems more like feedlot systems. For several decades, researchers in W. Europe and Oceania have concentrated on grazing management to increase milksolids per ha; while those in North America have concentrated on increased production per cow from improved crops such as maize (*Zea mays* L.), lucerne (*Medicago sativa* L.) and by-products (Fick & Clark, 1998). Re-integration of grass into a TMR feeding regime will be challenging.

The concept of high utilisation through high stocking rates, coupled with management of calving and drying off or weaning, and sale and purchase of stock, to coincide with the natural grass growth cycle is still profitable. However, individual animal performance is compromised, leading to lost opportunities for cost reductions and improved product quality. There are increasing problems related to nitrogen (N) leaching, soil compaction, methane (CH_4) and nitrous oxide emissions, and animal welfare and survival, as new strains of high yielding cows are introduced. Where pasture is the major feed source, long-standing issues related to variable milk and meat quality, and the cost of plant capacity to handle high peak milk and meat processing associated with seasonal calving and lambing have not been solved (Clark, 2002).

In all ruminant enterprises feed costs are the major expense. For example, in a USA dairy enterprise feed costs typically represent 45-55% of total cash costs (Moore, 1998). He calculated the average annual cost of Missouri pasture land to be US$131/ha, including opportunity costs of land, labour and capital and concluded that although pasture was a cheap feed source, its cost was often not realised leading to poor utilisation. Hemme (2003) reported costs of milk production of US$10-20/100 kg milk for pasture-based systems in Australia and New Zealand compared with US$25-55/100 kg milk for EU countries where pasture contributes much less to total cow diet.

Cheeke (1993) listed potential advantages of large-scale, industrial-style livestock enterprises (hereafter referred to as feedlots), including improved animal welfare, improved nutrition, effective waste management, consistent high quality products and better worker benefits. But what is the reality? Can smaller-scale, pasture-based enterprises compete with feedlots? This paper compares ruminant performance under pasture-based and TMR systems and examines opportunities for pasture-based systems to more fully meet consumer demands. It identifies specific environmental and animal welfare challenges to pasture-based systems and suggests ways to answer these challenges. Finally, it outlines a blueprint for developing new pastoral systems that are profitable, produce healthy food and are morally defensible.

Animal performance from pastures

Dairy cows

Kolver & Muller (1998) showed that early lactation cows fed only high quality perennial ryegrass (*Lolium perenne*) pasture in spring ate 19.1 kg dry matter (DM)/cow per day or 3.4% of bodyweight and produced 29.6 kg milk/day. This was 4.5 kg/day less DM intake (DMI) and 14.5 kg less milk compared with TMR-fed cows. They showed that metabolisable energy (ME) was limiting milk yield rather than amino acids or metabolisable protein.

The non-fibre carbohydrate of cool-season grasses is 15-22%, compared with 35% for a TMR, the difference due to grain in TMR. Grass has high crude protein (CP) about 25% but is highly degradable in the rumen. Beever *et al.* (1986) identified high protein degradability, low non-fibre carbohydrate (NFC) and high protein intake as reasons for low efficiency of N utilisation in grazing cows. Loss of rumen degradable protein (RDP) can account for 1.5-3 kg of milk yield.

Data from Kolver *et al.* (2002) in Table 1 show that although grass and a TMR may both have high ME contents, the ability of the two diets to generate milk yield differs. This is due to the lower neutral detergent fibre (NDF) and higher NFC in the latter, and also physical differences like water content and rate of breakdown that determine intake (Waghorn, 2002). The higher DMI and starch content in the TMR are able to support much higher milk yield than a solely pasture diet.

Table 1 Mean annual nutrient composition of grass and total mixed ration (TMR) diets fed to dairy cows and associated milk and milksolids yield (from Kolver *et al.*, 2002).

Item	Grass	TMR
ME (MJ ME/kg DM)	11.7	11.8
Crude protein (%DM)	26.9	18.2
Degradable protein (%Protein)	57.8	56.2
Soluble protein (% protein)	30.8	33.1
Neutral detergent fibre (%DM)	42.5	34.3
Acid detergent fibre (%DM)	23.3	21.0
Non fibre carbohydrate (%DM)	16.1	36.3
Fat (%DM)	4.2	6.7
Milk (kg/cow)	5880	10100
Milksolids (kg/cow)	460	720

To achieve higher milksolids New Zealand farmers have utilised maize silage purchased off-farm and grazed all non-lactating dairy stock off-farm. These options have allowed extra milksolids to be produced from the farm. The full advantage from maize silage can only be obtained with increased stocking rate (Kolver *et al.*, 2001), which increases labour, machinery, feeding facility and inventory costs. The combination of perennial ryegrass - white clover pasture supplemented with maize silage does not lead to increased milk production per cow at peak but achieves a modest increase in whole lactation performance by increasing days in milk. Some farms are now stocked at 4-5 cows/ha producing 400-500 kg MS/cow per year. In these systems 75% of feed is from perennial ryegrass - white clover and 25% from maize silage.

Beef cattle

Keane & Allen (1998) compared different beef production systems, viz. Intensive – bulls finished on silage and concentrates slaughtered at 19 months, Conventional – steers finished on silage and concentrates and slaughtered at 24 months, and Extensive - steers finished off pasture and slaughtered at 29 months (Table 2).

Table 2 Inputs and outputs and economic returns for intensive, conventional and extensive beef finishing systems (from Keane & Allen, 1998).

Item	Intensive	Conventional	Extensive	Significance
Carcass weight (kg)	384	363	366	NS
Kill out proportion (g/kg)	568	541	535	***
Live weight gain (kg/day)	1.18	0.73	0.72	***
Stocking rate (animals/ha)	4.0	2.1	1.47	N/A[1]
Concentrates (kg/animal)	1705	1218	256	N/A
Carcass weight (kg/ha)	1536	756	538	N/A
Gross margin – interest (€/animal)	336	269	403	N/A

[1]Not applicable

The Intensive system had higher kill out proportions and daily live weight gains than the Extensive, and the total carcass output / ha was much higher under the Intensive system, because of the greater use of N fertiliser and concentrates. However, the extra carcass weight could not compensate for the extra feed costs so that the Extensive system had the higher gross margin / ha. These results demonstrate how to produce acceptable carcass weight and quality from extensive pasture-based systems.

Sheep

There is evidence that individual Suffolk ram lambs can grow at close to 600 g/day from weaning to slaughter (Held *et al.*, 1997) when they are well fed on a concentrate diet from weaning. However, Muir *et al.* (2003) reported a growth rate of 549 g/day for an East Friesian x Romney x Suffolk ram from birth to 52 kg live weight at 12 weeks of age on milk and a grass-clover pasture, and a group average of 437 g/day over the same period. However, lamb growth rates from pasture are often well below these figures presenting a challenge to achieve the growth potential of superior sheep genetics when pasture is the post-weaning feed.

Carpino *et al.* (2004) showed that Ragusano cheese made from milk from cows fed either a TMR or native Sicilian pastures differed in odor-active compounds with only 13 in the former compared with 27 (8 of them unique) in the latter. This work demonstrated how pasture feeding could produce unique cheeses with desired flavours. Of course the reverse is true, because plants such as chicory can impart undesirable flavours to milk, although these can be removed by plant breeding (Rumball *et al.*, 2003). Carotenoid pigments may be used as a biomarker for grass-fed sheep (Prache *et al.*, 2003) and presumably other species. Such a marker would identify carcasses as grass-finished and attract premiums in the some markets.

Cows grazing only pasture had 150 and 53% more conjugated linoleic acid (CLA) in their milk than cows grazing a diet with one-third and two-thirds pasture, respectively (Dhiman *et al.*, 1999). CLA is the only known antioxidant and anticarcinogen primarily associated with foods originating from animal sources, and chemical induced tumours in rat mammary gland and colon tissue were reduced when isomers of CLA were fed (Ip *et al.*, 1999). However, MacRae (2004) pointed out that CLA levels in butter, milk, lamb and beef of 4-7 mg/g total fatty acids (FA) were insufficient to reduce inflammation responses in humans. There is thus a major challenge to rumen microbiologists to substantially increase CLA production by rumen microbes.

It has been hypothesized that n-6 FA have a role in the inflammation processes associated with human coronary heart disease and cancer, and that n-3 FA can reduce inflammation by competing with the incorporation of n-6 metabolites into membrane cells (Gibney & Hunter, 1993). Present recommendations are to increase intake of n-3 FA towards the dietary optimum n-3: n-6 FA ratio of 0.4-0.5, assisted by increasing intake of linolenic acid derived from chloroplast lipids. Dewhurst *et al.* (2003) reported increased n-3: n-6 polyunsaturated fatty acid (PUFA) ratio in the lipid of animals fed fresh forages rather than concentrates.

High value proteins from cow's milk offer the opportunity of increased returns, but there is little evidence that pasture feeding will lead to quantitative or qualitative differences in these proteins. Some management systems such as once daily milking are likely to be practiced only where pasture is the major feed. Recent analysis of milk from cows milked once daily showed that higher absolute yields of lactoferrin are possible compared with milk from cows milked twice daily (Farr *et al.*, 2002). Lactoferrin is valued for its iron-binding, anti-bacterial properties. However, there is no guarantee that high value milk proteins will come from cows, since the human lactoferrin gene has been cloned and is produced by recombinant DNA technology to fortify infant milk powders. Cow lactoferrin will have to compete with this product.

Animal breeding, especially of dairy cattle, has developed under the assumption that diets based on harvested forages and concentrates that meet all nutrient requirements will be fed. Such cows will not perform the same on pasture, and cows bred for pasture systems should be used (Kolver *et al.*, 2004). Even cows bred for pasture feeding take longer to get back in calf and lose more condition after calving than cows bred thirty years ago. This indicates a need for continual focus on breeding for pasture-based systems if cows suited to such systems are to be available in the future. Failure to do so will erode the profitability, animal health and welfare advantages of such systems.

Long-term sustainability of pasture systems will depend on improved ability to convert forage protein into an animal product that will be bought by health-conscious consumers. Currently, 75% and 85% of protein fed to dairy and beef animals, respectively, ends up as waste N. Promising increases in milk production (Miller *et al.*, 2001) and live weight gain from grazing lambs (Lee *et al.*, 2001) have been reported from forages bred for higher levels of water-soluble carbohydrates (WSC). But these results may not be transferable to different environments. Recent work by Parsons *et al.* (2004) showed that perennial ryegrasses bred for higher levels of WSC had the same level of WSC as a control ryegrass when tested at warmer temperatures.

Pasture-based systems have failed to fully harness the beneficial effects of legumes. Research shows that legumes (e.g. white clover (*Trifolium repens*), red clover (*Trifolium pratense*), Lucerne and subterranean clover (*Trifolium subterraneum*)) are capable of increasing milk and meat production (e.g. Harris *et al.*, 1998; Waghorn & Clark, 2004) (Table 3). Those containing condensed tannins (e.g. birdsfoot trefoil (*Lotus corniculatus*)) may be particularly beneficial (Waghorn *et al.*, 1998), because they protect proteins from rumen degradation, and have anthelmintic and CH_4-reducing properties. Intensive grazing as a result of high stocking rates has led to decreased white clover content. Lower stocking rates will lead to less damage to clover stolons in spring and summer. Recent research suggests that legumes may have to be sown in strips within a grass-legume pasture to obtain high dietary levels of legumes (Marotti *et al.*, 2002), and to allow protection from grazing at critical stages.

Although the application of molecular biological technology to pasture improvement is not discussed in this paper it will lead to future opportunities. In particular, Cisgenics® (the movement of heritable material only within species) offers a method to improve plants without resorting to transgenic technology, which may be unacceptable to either producers and/or consumers (Elborough & Hanley, 2004).

Table 3 Performance of cows in mid-late lactation and growing lambs fed a range of diets relative to ryegrass. Comparisons were made with forages given as sole diets and as supplements with ryegrass for cows (from Waghorn & Clark, 2004).

Cows	Supplement (% DM intake)	Response (% of ryegrass)		Lambs	Daily gain (% of ryegrass)
		DMI	Milksolids		
Ryegrass vs.				Ryegrass vs.	
White Clover	-	-4	+9	White Clover	+74
Lotus corniculatus	-	+18	+68	*Lotus corniculatus*	+55
Lotus corniculatus	75	11	48	*Cichorium intybus*	+47
White Clover	25	8	22	*Hedysarum coronarium*	+72
White Clover	50	22	29	Lucerne	+41
White Clover	75	25	33	Red Clover	+15
Cichorium intybus	40	1	9		

Environment

Nitrogen

The management of N presents major challenges and opportunities for pasture-based agriculture. The challenge is to avoid environmental damage as N moves from the farm into the environment. The opportunities exist because N capture in high value human food products can substantially improve farm profitability. Intensive grazing not only degrades the environment (EC, 1991), but also has specific problems absent from cropping enterprises. A good example is the aggregation of N into urine patches and subsequent large leaching losses (Scholefield *et al.*, 1993).

Greenhouse gases

There are few comparisons of CH_4 emissions from grazing ruminants and those fed TMR diets. Robertson & Waghorn (2002) reported losses as a percentage of gross energy (GE) for pasture and TMR diets, respectively, of 4.9 and 5.0; 6.3 and 5.7; and 7.0 and 6.3 for cows in early, mid- and late lactation, with only the late lactation values being different ($P<0.05$). Dhiman *et al.* (2001) and Cushnahan *et al.* (1995) measured CH_4 losses of 8.2% and 7.8% of GE in lactating cows fed grass-dominant pasture. In a modelling exercise Benchaar *et al.* (2001) predicted that increasing intake and the proportion of concentrate in the diet would decrease CH_4 emissions by 7 and 40%, respectively. It may appear that increasing intake of grazing cows with concentrate supplementation is a simple way to reduce CH_4 emissions. However, van der Nagel *et al.* (2003) compared a TMR and pasture-based ration and showed that when the CO_2 emissions from cultivation, fuel and fertiliser were included in estimates of greenhouse gases (GHG) (CH_4 + CO_2) emissions, the TMR diet produced nearly twice as much GHG as pasture (1.53 vs. 0.84 kg CO_2 equivalents / kg milk). Currently, there are no animal–based interventions for reducing CH_4 emissions that fully meet requirements for product safety, welfare, cost and long term effectiveness. An obvious option of reducing stocking rate to decrease pasture consumption / ha and the proportion of feed used for maintenance is usually rejected on economic grounds. Carbon taxation, discussed later in the paper, could alter this perception.

Opportunities and challenges

Di & Cameron (2004) using a nitrification inhibitor (dicyandiamide (DCD)), marketed as eco-n™, showed that nitrate-N leaching was reduced from 85 to 21 kg N/ha per year for dairy cow urine applied at 1000 kg N / ha. This would reduce the annual average nitrate-N concentration under a urine patch from 25 to 7 mg N/litre, and reduce Ca and Mg leaching, but not ammonia volatilization. Annual pasture growth in the urine patch increased from 15.9 without eco-n™ to 18.3 t DM/ha with eco-n™. This work needs to be verified under a wide range of commercial farming systems, and care needs to be taken that the potential benefits are not lost, e.g. extra pasture should be used to increase per animal intake rather than to increase stocking rate.

However, these and other options are tactical approaches. Urgently needed are strategic approaches that address both the N input into a farm system and its capture into food products. During an annual cycle, more than 300 kg N/ha can be released into long-term temperate grassland (Clough *et al.*, 1998). If this release could be accurately predicted, artificial N inputs could be adjusted accordingly. These authors suggest the use of thermal units to predict N mineralisation. Lemaire & Meynard (1997) have developed a Nitrogen Nutrition Index to predict level and timing of N fertilisation. These tools coupled with decision support systems such as OVERSEER™ (Monaghan *et al.*, 2004) allow the inputs or processes that have the most impact on N pollution from farming to be identified and alternatives evaluated. Whitehead (1995) derived a general linear relationship between N intake and excretion in urine: thus 1.5% N in the diet leads to excreted N in urine at 45% of total N excretion, but 80% at 4% N in diet. Species with C_4 metabolism such as maize are usually more efficient at capturing carbon / unit of N than C_3 plants. The incorporation of maize silage with <1.5% N content into a pasture diet of 4% N will reduce N excretion at a given level of DMI. Maize is now used widely in dairying countries as either an integral part of TMR or as a supplement to grazed pasture. However, it does not necessarily follow that systems based on maize silage will decrease total N output from a system, especially where stocking rate, total herd DMI and total N consumed are increased. Peel *et al.* (1997) demonstrated the potential for dairy farm management of N outputs with a modelling exercise comparing three systems:

1. Conventional management following good practice and based on high output (economic optima for fertiliser, slurry broadcast, least cost minimum 18% CP in diet).
2. Reduced loss, high output system – tactical reduction in fertiliser, diluted broadcast slurry, incorporation of maize silage and no surplus RDP).
3. Minimal loss, reduced intensity – the same total output from a greater area – planned reduction in fertiliser, slurry injected, no surplus RDP.

All systems produced 6000 kg milk/cow. Nitrogen inputs were 472, 336, 266 kg/ha respectively, and N outputs (milk and meat) 72, 71, 58 kg/ ha, respectively, and total N losses were 175, 115, 57 kg/ha. Financial margins were reduced by 10% on the minimal loss system.

Table 4 shows that imposition of a carbon tax would make it economically feasible for New Zealand dairy farmers to reduce stocking rates by up to 30% (Clark & Lambert, 2002), but only if carbon taxation was greater than $NZ 44/tonne of CO_2. This would lead to a 25% reduction in CH_4 output / ha. It would also lead to important spillover benefits for nitrous oxide and nitrate leaching. However, individual farmers would lose $311/ha per year in economic farm surplus.

Table 4 The effect of stocking rate on milk yield, economic farm surplus (EFS) with and without a carbon tax and estimated CH_4 output (data from Macdonald *et al.* (2001) and CH_4 production calculated from Clark (2001)).

Stocking rate (cows/ha)	2.2	3.2	4.3
Milk (kg/ha/y)	12100	13800	14600
CH_4 (kg/ha/y)	233	309	385
EFS* ($/ha/y)	2661	2741	2325
EFS** ($/ha/y)	2430	2430	1940

*Economic farm surplus - details of calculation in LIC (2001).
**Net economic farm surplus - calculated as (EFS - cost of CO_2 tax (assumed at $43.70/tonne)) to give break even for optimum stocking rate for EFS (3.2 cows/ha) and the lowest stocking rate (2.2 cows/ha).

Animal welfare

Animals outside have freedom to move, and exposure to sunlight and fresh air – but this implies exposure to temperature extremes, wind, damaging ultraviolet radiation and high temperatures, unless adequate shelter is provided. In both indoor and outdoor environments there is exposure to deleterious organisms. Housing provides shelter from climatic extremes, but generates health issues of lameness, mastitis and physical constraint. It appears that a combination of the two approaches that uses low-cost shelter for certain climates, and times of the year, coupled with physical freedom at others would be beneficial.

Silanikove (2000) made the following recommendations to minimise heat stress in grazing animals. There should be shade/shelter available when ambient temperature exceeds 24°C and temperature-humidity index >70, water freely available, breed type limited and sufficient time given to acclimatise to hot conditions. Breed type is very important because the heat load on a cow producing 50 kg milk/day is much greater than on one producing 20 kg. In the former, it is possible to provide cooling regimes of fans and water sprinklers but these add cost to the system. Plant breeding offers options for reducing heat stress in hot environments by producing endophyte-free or endophyte modified tall fescue and perennial ryegrass (Woodfield & Matthew, 1999) with reduced levels of the vasoconstrictor, ergovaline.

Lacy-Hulbert *et al.* (2002) monitored cows of different genotypes on pasture and TMR diets for three years, and observed a higher incidence of coliform mastitis on the TMR fed cows. The TMR ration contained 23% starch and there is evidence that undigested starch in the large intestine encourages *Escherichia coli (E. coli)* growth (Huntington, 1997). Lacy-Hulbert *et al.* (2002) found coliform faecal count of cows on TMR to be 1000 fold higher compared with cows on pasture, consistent with research reviewed by Callaway *et al.* (2003).

A way ahead

If pastoral systems advocates believe they must produce at the same levels as feedlots then they will surely run into the same problems. The idea that genetic modification of plants and animals can deliver unlimited gains in productivity must founder on the rock of biological, chemical and physical limitations. There is a paradox here in that if we accept limits to production we may never discover the true potential of our systems, but if we push hard on genetic gains and management advances then we will create problems in pastoral systems at least as intractable as in feedlots. However, an acceptance that pastoral systems have limits, could mean that population growth slows more quickly and leaves more area for pasture-based as opposed to crop-based agriculture.

Currently, pastoral systems and feedlots compete on an even footing in the market place, with only organic produce commanding a premium. Pastoral systems based on sound scientific principles should be developed that allow them to capture at least the current premium paid by the consumer for organic product. The fact that a small but important group of consumers pays a premium for organic but not pastoral product suggests that they see no difference between pastoral and feedlot systems.

Pastoral agriculture can make an important contribution to a country's "clean and green" image, but it is difficult to quantify this in monetary terms. A New Zealand study (MfE, 2001) using contingent valuation methodology suggested that a "clean and green" image was worth several hundred million dollars to primary producers and the tourism industry. This estimate was based on interviews with overseas customers and tourists on their likely spending decisions should New Zealand's image be degraded. The effect of any improvement or degradation in environmental image will likely vary with country, depending on such factors as level of tourism and type and extent of exports. It is important not to overstate the case for pasture because all the 'problems' of feedlots can occur under pastoral systems if management concepts and practice are wrong. For example, environmental damage can occur under both systems and it is often a function of the absolute amount of the inputs coupled with spatial and temporal variation that exacerbates the effect of these inputs.

High intake pasture system

The challenge for pasture-based systems is to produce a year round diet that matches the energy, protein and mineral supply of a feedlot TMR, with equal intake potential, but under a low cost, predominantly grazing regime. Where year-round pasture growth occurs, total annual yield should be approximately 15 t DM / ha. A High Intake Pasture System diet (Clark, 2002) should be able to support 2 cows / ha at a live weight of 550-600kg with a production of 1.0 kg milksolids/kg live weight. The cow's annual DMI would be approximately 7.2 t DM (Table 5). Table 5 shows one scenario for a High Intake Pasture System that meets the above criteria. The scenario assumes a 50/50 split between autumn and spring calving to allow milk production for 365 days. In this scenario 25% of the farm area is sown in a tannin-containing legume (e.g. *Lotus corniculatus*) to obtain improved DMI and protein nutrition during the summer and autumn. Another 25% is sown in annual ryegrass for improved DMI and ME content in winter and early spring and this area is sown in maize for grain in spring to provide an energy dense feed throughout the year. Note that 50 % of the area is still sown in perennial ryegrass-white clover providing 53% of the cow's annual intake. This combination has the potential to provide feed for lactating cows throughout the year. However, it is still reliant on perennial ryegrass and white clover and is exposed to the deficiencies that this feed has for optimum dairy cow nutrition.

Table 5 An example of a High Intake Pasture System stocked at 2 cows/ha throughout the year

Forage	Proportion of area	Annual Feed Production adjusted for area (t DM/ha)	Proportion of cow's diet
Perennial ryegrass-white clover	0.5	7.7	0.53
Lotus corniculatus	0.25	2.4	0.17
Annual ryegrass	0.25*	1.8	0.13
Maize grain	0.25*	2.4	0.17
Total	1.00	14.3	1.00

*Annual ryegrass sown in autumn; maize sown in spring on the same area.

If establishment and management costs of High Intake Pasture Systems could be kept low then they would rival the conventional system for profitability with benefits for nitrate leaching and GHG emissions per unit milksolids, and animal welfare and survivability. The High Intake Pasture system offers advantages over a wholly ryegrass- white clover grazing system, decreasing the risk of dietary dependent disorders (e.g. facial eczema, bloat, endophyte toxicity, and hypomagnesaemia). It breaks the constraint set up by the natural growth cycle of perennial pastures so cows can be milked all year with longer lactations. The higher yields dilute the maintenance cost, and improve feed conversion efficiency from 13-15 kg DM/kg MS commonly achieved with pasture systems to 11.9 kg DM/kg MS, based on the above assumptions, closer to the 10-11 kg DM/kg MS for feedlot TMR (Kolver *et al.*, 2001). The grazing process still poses a formidable challenge to the High Intake Pasture system which must be solved if this system is to come close to matching the per cow productivity of feedlot TMR.

Conclusions

Pastures offer a low cost feedstuff that can be profitably used in dairy, beef and sheep enterprises in most temperate regions. Their increased use may solve some of the environmental and animal welfare problems associated with feedlots using large amounts of concentrate feed from cropping enterprises. However, pasture systems can generate a different set of problems, particularly when trying to maximize pasture utilisation through high stocking rates and year-round grazing. From an economic and environmental perspective pasture research should develop plants and farm systems that allow high individual animal intake. The consequent lower stocking rate will allow improvements in labour productivity, animal health and welfare and overall profitability. The challenge is enormous and so are the benefits.

References

Beever, D.E., M.S. Dhanoa, H.R. Losada, R.T. Evans, S.B. Cammell & J. France (1986). The effect of forage species and stage of harvest on the processes of digestion occurring in the rumen of cattle. *British Journal of Nutrition, 56,* 439-454.

Benchaar, C., C. Pomar & J. Chiquette (2001). Evaluation of dietary strategies to reduce methane production in ruminants: a modelling approach. *Canadian Journal of Animal Science, 81,* 563-574.

Callaway, T.R., R.O. Elder, J.E. Keen, R.C. Anderson & D.J Nisbet (2003). Forage feeding to reduce preharvest *Escherichia coli* populations in cattle, a review. *Journal of Dairy Science, 86*, 852-860.

Carpino, S., S. Mallia, S. La Terra, C. Melilli, G. Licitra, T.E. Acree, D.M. Barbano & P.J. van Soest (2004). Composition and aroma compounds of Ragusano cheese: native pasture and total mixed rations. *Journal of Dairy Science, 87*, 816-830.

Cheeke, P.R. (1993). Impacts of livestock production on Society, Diet/Health and the environment. Interstate Publishers, Danville, Illinois, 241pp.

Clark, D.A. (2002). Forage supply systems for pasture based dairying. *Proceedings of Facing the Challenges of the Supply Chain in 2010.* Shepparton, Victoria, Australia. The University of Melbourne: Institute of Land and Food Resources, Dookie College, 30-37.

Clark, D.A. & M.G. Lambert (2002). Environmental sustainability - costs and benefits for farmers. *Proceedings of the New Zealand Society of Animal Production,* 62, 219-224.

Clark, H. (2001). Ruminant methane emissions: a review of the methodology used for national inventory estimations. *Report prepared for the Ministry of Agriculture and Forestry.* June 2001.

Clough, T.J., S.C. Jarvis & D.J. Hatch (1998). Relationships between soil thermal units, nitrogen mineralization and dry matter production in pasture. *Soil Use and Management, 14*, 65-69.

Cushnahan, A., C.S. Mayne & E.F. Unsworth (1995). Effects of ensilage of grass on performance and nutrient utilization by dairy cattle. 2. Nutrient metabolism and rumen fermentation. *Animal Science, 60,* 347-359.

Dewhurst, R.J., N.D. Scollan, M.R.F. Lee, H.J. Ougham & M.O. Humphreys (2003). Forage breeding and management to increase the beneficial fatty acid content of ruminant products. *Proceedings of the Nutrition Society, 62*, 329-336.

Dhiman, T. R., G. R. Anand, L.D. Satter & M.W. Pariza (1999). Conjugated linoleic acid content of milk from cows fed different diets. *Journal of Dairy Science, 82*, 2146-2156.

Dhiman, T.R., K.C. Olsen, M.S. Zaman, I.S. MacQueen & R.L. Boman (2001). Methane emissions from lactating dairy cows fed diets based on conserved forage and grain or pasture. *Journal of Animal Science*, 79, (Supp. 1), 289.

Di, H.J. & K.C. Cameron (2004). Treating grazed pasture soil with a nitrification inhibitor, eco-n™, to decrease nitrate leaching in a deep sandy soil under spray irrigation – a lysimeter study. *New Zealand Journal of Agricultural Research, 47*, 351-361.

EC (1991). Council directive of 12 December concerning the protection of waters against pollution caused by nitrates from agricultural sources. 91/676 EEC, Legislation 1375/1-1375/8, European Community, Brussels.

Elborough, K.M. & Z. Hanley (2004). Pasture biotechnology – not as you know it. *Proceedings of New Zealand Society of Animal Production, 64,*101-104.

Farr, V.C., C.G. Prosser, D.A. Clark, M. Broadbent, C.V. Cooper, D. Willix-Payne & S.R. Davis (2002). Lactoferrin concentration is increased in milk from cows milked once-daily. *Proceedings of the New Zealand Society of Animal Production,* 62, 225-226.

Fick, G.W. & E.A. Clark (1998). The future of grass for dairy cattle. In: Grass for Dairy Cattle. J.H. Cherney & D.J.R. Cherney (eds). CAB International, University Press, Cambridge, 1-22.
Gibney, M.J. & B. Hunter (1993). The effects of short- and long- term supplementation with fish oil on incorporation of N-3 polyunsaturated fatty acids into cells of the immune system in healthy volunteers. *European Journal of Clinical Nutrition, 47*, 255-265.
Harris, S.L., M.J. Auldist, D.A. Clark & E.B.L. Jansen (1998). Effects of white clover content in the diet on herbage intake, milk production and milk components of New Zealand dairy cows housed indoors. *Journal of Dairy Research, 65*, 389-400.
Held, J., A.L. Slyter, B. Read & B. Long (1997). The effect of sire selection on lamb growth and carcass traits. http://ars.sdstate.edu/sheepext/sheepday97/97-5.htm
Hemme, T. (2003). (R)evolution in dairy farming. Paper presented at the IDF World Dairy Summit in Brugge, September 2003. http://www.ifcndairy.org
Huntington (1997). Starch utilization by ruminants: from basics to the bunk. *Journal of Animal Science,* 75, 852-857.
Ip, M.M., P.A. Masso-Welch, S.F. Shoemaker, W.K. Shea-Eaton & C. Ip (1999). Conjugated linoleic acid inhibits proliferation and induces apoptosis of normal rat mammary epithelial cells in primary culture. *Experimental Cell Research, 250*, 22-34.
Keane, M.G. & P. Allen (1998). Effects of production system intensity on performance, carcass composition and meat quality of beef cattle. *Livestock Production Science, 56*, 203-214.
Kolver, E. S. & L. D. Muller (1998). Comparative performance and nutrient intake of high
producing Holstein cows grazing or fed a total mixed ration. *Journal of Dairy Science,* 81, 1403-1411.
Kolver, E.S., J.R. Roche, M.J. de Veth, P.L. Thorne & A.R. Napper (2002). Total mixed rations versus pasture diets: evidence for a genotype x diet interaction in dairy cow performance. *Proceedings of the New Zealand Society of Animal Production, 62*, 246-225.
Kolver, E.S., J.R. Roche, D. Miller & R. Densley (2001). Maize silage for dairy cows. *Proceedings of the New Zealand Grassland Association, 63*, 195-201.
Kolver, E.S., B. Thorrold, K. Macdonald, C. Glassey & J. Roche (2004). Black and white answers on the modern dairy cow. *SIDE Proceedings, pp.*165-185.
Lacy-Hulbert, S.J., E.S. Kolver, J.H. Williamson & A.R. Napper (2002). Incidence of mastitis among cows of different genotypes differing in nutritional environments. *Proceedings of the New Zealand Society of Animal Production 62*, 24-29.
Lee, M.R.F., E.L. Jones, J.M. Moorby, M.O. Humphreys M.K. Theodorou, J.C. MacRae & N.D. Scollan (2001). Production responses from lambs raised on *Lolium perenne* selected for an elevated water-soluble carbohydrate concentration. *Animal Research, 50*, 441-449.
Lemaire, G. & J.M. Meynard (1997). Use of the Nitrogen Nutrition Index for analysis of the agronomical data. In: G. Lemaire (ed.) Diagnosis on the nitrogen status in crops. Heidelberg, Springer-Verlag, 45-55.
LIC (2001). Dairy Statistics 2000/01. Livestock Improvement Corporation, Hamilton, New Zealand.
Macdonald, K.A., J.W. Penno, P.K. Nicholas, J.A. Lile, M. Coulter, & J.A.S Lancaster (2001). Farm systems – impact of stocking rates on farm efficiency. *Proceedings of the New Zealand Grassland Association, 63,* 223-227.
MacRae, J.C. (2004). Nutritional opportunities for longer-term, sustainable, ruminant production. *Proceedings of the New Zealand Society of Animal Production,* 64, 77-83.
Marotti, D.M., D.F. Chapman, G.P. Cosgrove, A.J. Parsons & A.R. Egan (2002). New opportunities to improve forage based production systems. *Proceedings of the New Zealand Society of Animal Production, 62*, 273-277.
MfE (2001). Valuing New Zealand's clean green image. http://www.mfe.govt.nz/publications/sus-dev/clean-green-image-value-aug01/
Miller, L.A., J.M. Moorby, D.R. Davies, M.O. Humphreys, N.D. Scollan, J.C., MacRae & M.K. Theodorou (2001). Increased concentration of water-soluble carbohydrate in perennial ryegrass (*Lolium perenne* L.): Milk production from late-lactation dairy cows. *Grass and Forage Science, 56*, 383-394.
Monaghan, R.M., D. Smeaton, M.G. Hyslop, D.R. Stevens, C.A.M. de Klein, L.C Smith, J.J. Drewry & B.S. Thorrold (2004). A desktop evaluation of the environmental and economic performance of model dairy farming systems within four New Zealand catchments. *Proceedings of the New Zealand Grassland Association, 66,* 57-67.
Moore, K.C. (1998). Economics of grass for dairy cattle. In: Grass for Dairy Cattle. J.H. Cherney & D.J.R. Cherney (eds). CAB International, University Press, Cambridge, 373-391.
Muir, P.D., N.B. Smith & C. Lane (2003). Maximising lamb growth rates – just what is possible in a high performance system. *Proceedings of the New Zealand Grassland Association, 65,* 61-63.
Parsons, A.J., S. Rasmussen, H. Xue, J.A. Newman, C.B. Anderson & G.P Cosgrove (2004). Some 'high sugar grasses' don't like it hot. *Proceedings of the New Zealand Grassland Association, 66,* 265-271.

Peel, S., B.J. Chambers, R. Harrison & S.C. Jarvis (1997). Reducing nitrogen emissions for a complete dairy system. In: S.C. Jarvis & B.F. Pain (eds). Gaseous nitrogen emissions from grasslands. CAB International, Wallingford, 383-390.

Prache, S., A. Priolo & P. Grolier (2003). Persistence of carotenoid pigments in the blood of concentrate–finished grazing sheep: its significance for the traceability of grass feeding. *Journal of Animal Science,* 81, 360-367.

Robertson, L.J. & G.C. Waghorn (2002). Dairy industry perspectives on methane emissions and production from cattle fed pasture or total mixed rations in New Zealand. *Proceedings of the New Zealand Society of Animal Production, 62,* 213-218.

Rumball, W., R.G. Keogh, J.E. Miller & R.B. Claydon (2003). Cultivar release 'Choice' forage chicory (*Cichorium intybus* L.). *New Zealand Journal of Agricultural Research, 46,* 49-51.

Scholefield, D., K.C. Tyson, E.A. Garwood, A.C. Armstrong, J. Hawkins & A.C. Stone (1993). Nitrate leaching from grazed grass lysimeters: effects of fertilizer input, field drainage, age of sward, and patterns of weather. *Journal of Soil Science,* 44, 601-613.

Silanikove, N. (2000). Effects of heat stress on the welfare of extensively managed domestic ruminants. *Livestock Production Science, 67,* 1-18.

Van der Nagel, L.S., G.C. Waghorn & V.E. Forgie (2003). Methane and carbon emissions from conventional pasture and grain-based total mixed rations for dairying. *Proceedings of the New Zealand Society of Animal Production,* 63,128-132.

Waghorn, G.C. (2002). Can forages match concentrate diets for dairy production? *Proceedings of the New Zealand Society of Animal Production,* 62, 261-266.

Waghorn, G.C. & D.A. Clark (2004). Feeding value of pastures for ruminants. *New Zealand Veterinary Journal,* 52, 320-331.

Waghorn, G.C., G.B. Douglas, J.H. Niezen, W.C. McNabb & A.G. Foote (1998). Forages with condensed tannins – their management and nutritive value for ruminants. *Proceedings of the New Zealand Grassland Association, 60,* 89-98.

Whitehead, D.C. 1995. Grassland Nitrogen. CAB International, Wallingford, 397 pp.

Woodfield, D.R. & C. Matthew (1999). Ryegrass endophyte: an essential New Zealand symbiosis. *Grassland Research and Practice Series 7,*168 pp.

Optimising financial return from grazing in temperate pastures

P. Dillon[1], J.R. Roche[2], L. Shalloo[1] and B. Horan[1]

[1]*Dairy Production Department, Teagasc, Dairy Production Research Centre, Moorepark, Fermoy Co. Cork, Ireland*

[2]*Dexcel Ltd., Hamilton, New Zealand*

Key Points

1. The increased interest in pasture-based systems of milk production in recent years has been largely generated through lower product prices and rising costs of production.
2. Pasture based systems of milk production decrease unit production costs, through lower feed and labour expenses, and reduced capital investment.
3. Systems utilising grazed pasture will be optimised in regions where pasture production potential is high, variability in seasonal pasture supply and quality is low, manufacturing milk accounts for a large proportion of total production, and where large areas of land are available at relatively low cost.
4. Pasture based systems may allow greater global sustainability (through reduced use of fuel, herbicides and pesticides), increased product quality, improved animal welfare and increased labour efficiency.

Keywords: pasture based-system, confinement TMR system, cash cost, grazed grass

Introduction

The recently rejuvenated interest in grazing systems of milk production for dairy cows in many temperate and subtropical regions of the world (especially Europe and USA) is a result of lower inflation-adjusted milk prices, the proposed removal of some subsidies and tariffs, rising labour, machinery and housing costs, and perceived environment and animal welfare concerns associated with intensive dairying.

The decision to change from intensive feeding systems to grazing requires a thorough evaluation of the whole farm system (Parker *et al.*, 1992). Clark and Jans (1995) outlined the advantages and disadvantages associated with grazed forage versus concentrate-based feeding systems (Table 1).

Studies by Hanson *et al.* (1998), Kriegl (2001) and the CIAS (2001) found that pasture-based dairy systems were more profitable than confinement systems of similar size. Pasture-based systems are capable of high milk output per ha at low cost (Penno *et al.*, 1996). In contrast, confinement production systems have higher costs but are able to support higher milk output per cow than pasture based systems.

A key issue associated with grazing is the lack of control over feed quality and availability associated with feed intake, while the infrastructure associated with intensive dairies and cropping systems designed to supply concentrate feeds to these dairies, places a significant financial burden on intensive dairy operations. Grazing systems were associated with reduced feed and waste management costs when compared with confinement systems in the U.S. Capital investment changed from depreciating assets (machinery and buildings) to static or appreciating assets (cows and land) according to Hamilton *et al.* (2002).

Table 1 Comparison of grazed grass and confinement total mixed ration systems

System of production	Grazed grass based	Confinement/TMR
Feed costs	Low	High
Feed quantity	Variable	High
Stocking rate	Critical	Ignored
Milk supply profile	Seasonal	Constant
Labour requirements	Seasonal	Constant
Decision support	Rudimentary	Sophisticated
Automation opportunity	Low	High
Effluent management	Low	High
Agrochemical usage	Low	High
Energy usage	Low	High
Capital investment	Low	High

Agricultural policy has major implications for the type of production system that develops in different countries. The EU Common Agricultural Policy (CAP) set up in 1957 aimed to guarantee food security at stable and reasonable prices to producers, by maximising production and protecting domestic agriculture from foreign competitors (Whetstone, 1999). The continuing reform of the CAP with the desire to make production more market focused suggests more unstable and unpredictable milk prices in the future.

Similar challenges exist in many other regions. Several factors have caused structural changes in the US dairy industry in the last decade. A large number of traditional dairy operations were small. In 1994, 21% of the dairy farms had fewer than 30 milking cows in Wisconsin, and only 9 percent greater than 100 cows. Many of the dairy farmers with fewer than 100 cows had obsolete facilities and were approaching retirement age. In the US in 1993, farms with 100 cows or more making up 14% of the dairy operations, had over 50% of the cows and produced 55% of total milk production. By 2000, farms with 100 or more cows accounted for 20% of operations, had 66% of the cows and produced more than 70% of total milk production (NASS, February 2004). Furthermore, inflation adjusted milk price has been steadily declining at approximately 2.5% per year and profitability in dairying has been under pressure as a result of rising costs. Like Europe, a reduction in government involvement in price support is likely in the near future.

Some farmers in the U.S. have compensated for declining profitability by increasing scale and/or movement of dairying from traditional dairy regions (the North-East or mid-West) to drier climates (West) with accessible grain and forage supplements. The potential for dairy farmers in Europe to relocate or expand to compensate for increasing production and social costs is limited. EU policy of continuing milk quotas until at least 2015, while allowing milk prices to decline will require dairy farmers to become more efficient. Tighter operating margins globally are leading many dairy farmers to examine alternative methods of farming to cut costs and protect future investments and livelihoods.

The deregulation of the Australian industry began in 1999 and has meant the discontinuation of regulated sourcing and pricing of milk for liquid consumption. The overall impact of deregulation was to hasten the decline in the number of dairy farms, which has fallen from 22,000 in 1980 to fewer than 10,000 in 2004. Australian dairy farmers, like their New Zealand neighbours, now operate in a completely deregulated industry environment, where international prices are the major factors determining the price received for their milk. As a result of this, making greater use of grazed pasture as a base feed has gained more interest in the last decade.

Another factor driving the rejuvenation of interest in grazing systems is the consumer perception that intensive agriculture is not animal friendly. While animal productivity often increases, perceived environmental, ethical (e.g. animal welfare, genetic manipulation), and food safety issues (e.g. residues) associated with larger dairy operations (factory farming) are now receiving increased attention by consumers and policy makers. Compliance with legislation on the environment, food safety and animal welfare already pose problems for dairy farming systems and are likely to be of greater concern in the future. Future farming systems need to be seen to be sustainable in terms of the environment and animal welfare. Grazing dairy systems are perceived to be more animal friendly than total confinement but increased grazing intensity has been shown to have negative effects on aspects of the environment such as river nutrient loading.

Effect of location and climate on milk production systems

As the liberalisation of world trade continues and the international competition for dairy markets accelerates, farmers must consider the competitiveness of their production systems. A concerted effort must be made to understand the strengths and weaknesses of various systems of production by making comparisons of the performance of the component pieces of the business to international standards or benchmarks. For this reason, an analysis of the long term economic viability of various systems of milk production was undertaken to determine how competitive livestock systems in various regions might be in a world market.

Table 2 shows a comparison of the meteorological data for Southern Ireland (Moorepark Research Centre, Co. Cork Ireland, 52.1°N 8.3°W, approximately 60m above sea level; 1960-2003); North Island of New Zealand (Ruakura, Research Centre, Hamilton, New Zealand, 37.3°S 175.1°E, approximately 40m above sea level; 1998-2001); South Island of New Zealand (Lincoln University, Canterbury, New Zealand, 43.5°S 172.5°E; approximately 40m above sea level; 2001-2004); Victoria, South-Eastern Australia, (Department of Primary Industries Ellinbank, Warragul, Victoria, Australia, 38.2°S 145.9°E; approximately 120m above sea level; 1995 to 1998) and Arlington University Farm, Columbia County, Wisconsin, USA (43.3°N 89.4°W; approximately 330m above sea level; 1961-1990).

Latitudinal positioning, continental and ocean current influences and other factors result in vastly different climates in the reported regions. For pasture-based systems of milk production approximately 1,000 mm of rainfall evenly distributed through the year is a requirement. In all five regions rainfall is evenly distributed throughout the year, but in the South Island of New Zealand total rainfall is less than adequate for optimum grass production. Lack of water is compensated for by irrigation for approximately 6 months.

Below average soil temperatures of 8 °C pre-vernalization (early-Winter) and 5.5 °C in late-winter/Spring grass growth is minimal. This occurs for one to two months in Southern Ireland and five months in Wisconsin, USA, while in Victoria, Australia and New Zealand's North and South Island mean air temperature rarely declines below 5 °C, suggesting that herbage growth rate seldom ceases. The high mid-summer temperatures (in particular minimum temperature) in Victoria Australia, Wisconsin USA and to a lesser extent the North Island of New Zealand can result in lower herbage growth (depending on rainfall) and poorer forage quality during summer, resulting in lower cow performance and/or a supplementation requirement to maintain milk yield.

Table 3 shows the seasonality of grass production and composition for Southern Ireland (Moorepark Research Centre; 300 kg N/ha/year), North Island of New Zealand (Ruakura, Research Centre; 200 kg N/ha/year), South Island of New Zealand (Lincoln University; 200 kg N/ha/year) and Victoria Australia (Ellinbank, Warragul; 300 kg N/ha/year). The highest grass production was obtained in both the North and South Islands of New Zealand, the lowest in Victoria Australia while Southern Ireland was intermediate. The greatest seasonality of grass production was obtained in Southern Ireland (26:1), the lowest in the North Island of New Zealand (4:1), while both South Island New Zealand (5:1) and Victoria Australia (7:1) were intermediate.

In Ireland the lowest herbage growth rates were the result of low winter temperatures while in both the North and South Islands in New Zealand and Victoria Australia they were the result of high summer temperatures and/or inadequate summer rainfall. Victoria Australia has lower winter growth rates than New Zealand's North Island, even though both regions are at similar latitude. The lower winter herbage production in Victoria is probably due to lower temperatures during the winter period, possibly a result of the continental influence. This continental influence appears equivalent to 5° latitude, as growth rates in Victoria are similar to those measured in Canterbury, New Zealand.

Winter herbage growth rates in the North Island of New Zealand allow cows to graze for the entire year on the home farm. Milk to supplement price differential means very little supplement is purchased. Low winter herbage growth rates in the South Island of New Zealand result in stock being moved onto other farms, where crops (usually brassica) are grown and fed *in situ* with cereal and grass silage, and straw. In southern Ireland cows are normally housed for two to three months and supplemented with conserved grass silage. In Victoria (Australia) the winters are mild and herbage growth rates of 15 to 30 kg DM/ha per day allow cows to graze year round. Growth rates rarely exceeded 75 kg DM/ha per day in spring and declined to less than 10 kg/ha per day during the summers of 1997 and 1998. Crushed barley is fed strategically at milking throughout the year.

Farming system

Table 4 shows a comparison of the physical characteristics of selected groups of dairy farmers in New Zealand (LIC, 2003), Victoria Australia (Dairy Australia 2003), Ireland (Fingleton, 2003) and USA (both confinement and pasture based systems) (IFCN, 2003). Dairy farming in New Zealand is characterized by large herd size, modest milk production per cow, low level of concentrate supplementation, high stocking rate (2.7 to 2.8 cows/ha) and a high number of cows per labour unit. Dairy farming in Victoria Australia is fairly similar to New Zealand with larger farm size and similar herd size and milk production level. Stocking rate is variable and dependent on concentrate supplementation level, while pasture as a proportion of total feed consumed is lower. Dairying in Ireland is characterized by smaller farm size, small herd size, lower milk production per ha than New Zealand and Australia, lower stocking rates, higher level of concentrate supplementation than New Zealand and lower number of cows per labour unit. In contrast both confinement and pasture based systems of milk production in Wisconsin, USA are characterised by medium herd size, high levels of milk production per cow, low stocking rates, high levels of concentrate supplementation and a low cow number per labour unit.

Table 2 Average rainfall, and maximum, minimum and mean temperature in southern Ireland, North Island New Zealand, South Island New Zealand, Gippsland (Victoria) Australia, and Columbia County, Wisconsin, U.S.A.

Month*		1	2	3	4	5	6	7	8	9	10	11	12	Annual
Ireland	Rainfall, mm/month	109	92	81	66	61	68	54	92	77	114	101	109	1024
	Mean Max Temp, °C	8.9	9.1	10.5	12.5	15.6	17.8	20.0	19.6	17.2	13.9	10.7	9.3	13.8
	Mean Min Temp, °C	2.3	2.2	3.2	3.8	6.3	9.2	11.4	10.9	8.8	6.7	3.9	2.7	6.0
	Mean Ave. Temp, °C	5.2	5.6	7.1	8.2	11	13.6	15.7	15.2	12.9	10.2	7.3	6	9.8
NZ North Island	Rainfall, mm/month	156	104	100	75	90	86	51	75	83	96	102	99	1117
	Mean Max Temp, °C	15.1	14.5	16.5	18.2	20.2	21.4	23.9	25.1	24.0	21.3	18.5	15.1	19.5
	Mean Min Temp, °C	6.1	3.7	6.0	8.0	9.9	11.4	12.4	13.6	12.3	10.4	8.3	5.4	9.0
	Mean Ave. Temp, °C	10.6	9.1	11.2	13.1	15.1	16.4	18.1	19.3	18.1	15.9	13.4	10.3	14.2
NZ South Island	Rainfall, mm/month	39	38	48	47	64	22	59	40	33	83	25	55	553
	Mean Max Temp, °C	10.8	12.6	15.6	16.4	18.0	21.2	21.8	20.4	20.5	15.6	15.6	13.5	16.8
	Mean Min Temp, °C	0.8	2.8	4.7	5.7	7.7	10.1	11.9	10.8	9.7	6.6	4.2	2.6	6.5
	Mean Ave. Temp, °C	8.0	5.5	7.8	10.1	10.9	12.7	15.5	16.6	15.3	14.7	11.0	9.7	11.5
Australia	Rainfall, mm/month	133	118	101	88	85	63	65	64	50	90	91	81	1029
	Mean Max Temp, °C	11.4	13.6	15.1	17.7	19.3	20.8	25.2	25.6	22.2	17.6	15.2	13.5	18.1
	Mean Min Temp, °C	4.2	5.5	5.7	7.7	8.8	9.3	12.8	12.8	10.7	8.1	7.6	5.6	8.2
	Mean Ave. Temp, °C	7.8	9.6	10.4	12.7	14.1	15.1	19.0	19.2	16.5	12.9	11.4	9.6	13.2
U.S.A.	Rainfall, mm/month	24	21	49	74	86	102	89	103	102	60	52	37	799
	Mean Max Temp, °C	-4.3	-1.5	5.1	14.1	21.1	26.1	28.2	26.8	22.3	15.7	6.7	-1.5	13.2
	Mean Min Temp, °C	-13.8	-11.3	-4.7	1.7	7.5	12.6	15.3	14.0	9.7	3.9	-2.4	-10.3	1.9
	Mean Ave. Temp, °C	-9.1	-6.3	0.2	7.9	14.3	19.3	21.8	20.5	16.0	9.8	2.1	-5.9	7.6

*Month 1 = January in the northern hemisphere and July in the southern hemisphere

Table 3 Pasture growth and quality in southern Ireland, North Island New Zealand, South Island New Zealand and Gippsland (Victoria) Australia

Month*		1	2	3	4	5	6	7	8	9	10	11	12	Total
Ireland	Pasture growth, kg/ha per day	3	7	29	53	78	70	59	56	41	18	6	4	12,726
	CP, % DM		19.0	18.5	19.3	17.7	15.6	16.0	17.2	15.3	16.2	20.4	19.0	
	NDF, % DM		40.2	40.5	41.5	41.5	44.6	46.1	47.9	45.2	46.0	45.5	46.0	
	ME, MJ/kg DM		11.4	12.4	12.2	12.4	11.0	10.8	10.6	11.0	10.7	11.2	10.5	
NZ North Island	Pasture growth, kg/ha per day	42	41	78	82	89	80	64	43	27	22	35	23	19,035
	CP, % DM	19.8	22.6	24.9	22.7	22.5	19.8	20.5	21.6	17.1	22.0	25.9	22.5	
	NDF, % DM	38.5	39.6	42.9	43.6	45.3	43.7	43.5	43.8	55.4	44.6	40.7	42.2	
	ME, MJ/kg DM	12.4	12.4	12.2	12.1	11.4	11.8	10.6	10.0	8.3	10.4	11.5	12.5	
NZ South Island	Pasture growth, kg/ha per day	18	24	43	59	70	80	78	67	63	40	33	23	18,182
	NDF, % DM	35.4	40.0	35.7	37.5	42.9	41.5	45.3	41.8	40.9	37.9	38.5	45.2	
	ME, MJ/kg DM	11.9	12.5	12.3	12.3	11.6	11.9	11.2	11.5	11.4	11.7	10.9	11.2	
Australia	Pasture growth, kg/ha per day	14	28	45	57	55	38	32	8	12	17	26	24	10,870
	CP, % DM	27.7	27.2	27.5	23.5	21.4	18.3	14.9	15.5	16.3	21.6	27.5	24.5	
	NDF, % DM	43.3	46.2	47.3	46.5	47.2	51.9	57.8	64.1	57.4	52.1	46.8	43.5	
	ME, MJ/kg DM	12.2	11.9	11.8	11.7	11.1	10.8	10.2	10.1	9.8	10.7	11.6	11.3	

*Month 1 = January in the northern hemisphere and July in the southern hemisphere.

Table 4 Physical characteristics of dairy farming systems in various countries

Variable	New Zealand	Australia	Ireland	US Grazing	US Confined
Farm size (ha)	103	229	24	80	168
Number of cows	271	312	45	64	115
Milk yield/ cow (l)	3,678	4,800	4,588	7,779	10,243
Fat + protein/ cow (kg)	323	350	343	544	832
Replacement rate (%)	18	15	19	-	33
Concentrate (kg/cow)	150	400	750	-	4,500
Stocking rate (cows/ha)	2.67	2.48	1.88	0.80	0.68
Cows/ labour Unit	97	80	44	-	40

Table 5 shows the milk production and live-weight profile for research herds in Ruakura, Research Centre New Zealand (3.3 cows/ha; no purchased supplement; Macdonald, 1999), Ellinbank, Victoria Australia (4.7 cows/ha; 1 tonne DM crushed barley; approximately 0.2 ha turnips/cow during summer; Grainger, 1998), and Moorepark Research Centre in Ireland (2.5 cows/ha; 350 kg purchased supplement (Horan *et al.*, 2004). Milk production profile is highly seasonal in all three countries. The relative milk production per cow in the herds of the selected groups of dairy farmers were 79%, 85% and 67% compared to that in the research herds in New Zealand, Australia and Ireland respectively. This indicates a greater potential milk production from pasture in all three countries especially Ireland. The mechanism used to match feed demand to feed supply is to have calving concentrated in spring due to the seasonal nature of grass production and some climatic constraints. Concentrating calving in early spring allows maximum use of grazed grass during lactation. In Ireland calving is concentrated in the months of February and March, while in the North Island of New Zealand and Victoria, Australia it is concentrated in July and August. There is also a growing shift towards split calving in Victoria, Australia.

The system of milk production in the United States is primarily based on total confinement, with cows fed a high concentrate total mixed ration (TMR). Such dairy farms can be characterized as capital, labour, and management-intensive businesses. For the purposes of comparison, two distinct types of dairy enterprises have been characterised within the US (Wisconsin). A confinement system of milk production in the current study is where cows are housed and milked in adjoining barns. In such herds, cows are calved and milked all year round and are fed on mechanically harvested feed. The second US system of production is a rotational grazing system, where grazed grass constitutes more than 30% of the total forage eaten by the dairy cow. Similar to the confinement system, these herds are not seasonal, although calving may be 'bunched' into two periods of the year. Winter forages are grown on the farm, and grain/concentrate is likely to be fed in near conventional amounts, however unlike the confinement system, this grain is unlikely to be produced on farm.

Table 5 Milk production in southern Ireland, North Island New Zealand and Gippsland (Victoria) Australia

Month*		1	2	3	4	5	6	7	8	9	10	11	12	Total
Ireland	Milk yield, kg/d 6,832		17.4	25.6	29.6	28.6	27.4	26.2	23.2	19.7	16.4	10.8	10.0	
	Fat, %		4.14	3.85	3.75	3.59	4.08	3.69	3.91	4.17	4.26	4.66	4.00	4.01
	Protein, %		3.45	3.31	3.35	3.28	3.37	3.44	3.53	3.64	3.85	4.01	3.66	3.54
	Live weight		515	498	495	498	508	519	529	541	553	565	578	527
NZ North Island	Milk yield, kg/d 4,680		19.8	22.6	21.2	18.2	16.3	15.6	12.9	10.8	8.9	7.4		
	Fat, %		5.0	4.4	4.4	4.5	4.7	4.8	4.9	5.1	5.4	5.7		4.89
	Protein, %		3.4	3.3	3.4	3.5	3.6	3.5	3.4	3.6	3.9	4.1		3.57
	Live weight		448	450	456	470	477	477	476	469	479	492		469
Australia	Milk yield, kg/d 5,617	17.3	21.0	22.7	24.4	23.1	20.3	16.9	15.1	12.8	11.0			
	Fat, %	4.5	4.5	4.4	4.3	4.3	4.3	4.4	4.6	4.9	5.3			4.55
	Protein, %	3.5	3.2	3.1	3.1	3.1	3.1	3.1	3.2	3.4	3.5			3.23
	Live weight	525	484	465	490	496	512	522	513	519	529			506

*Month 1 = January in the northern hemisphere and July in the southern hemisphere.

All EU member states have experienced very significant structural changes in relation to milk production since the introduction of milk quotas in April 1984. Figure 1 shows the change in milk output per farm from 1983 to 2003 for EU countries as well as New Zealand and Australia. At the introduction of the milk quota scheme in the EU there were 32,700 and 62,010 milk producing farms in Denmark and Ireland respectively, while in 2004 this number has reduced to 6,600 (20%) and 26,500 (43%), respectively. Over this same period average milk quota per farm has increased from 152,000 and 73,790 kg to 675,000 and 215,000 kg for Denmark and Ireland, respectively. However in New Zealand the number of dairy farmers has only reduced from 15,881 to 12,751, while nationally the quantity of milk processed has doubled (6,956 to 14,599 million l). Over the same period cow numbers have increased from 2,280,273 to 3,851,302 and average herd size has increased from 114 to 302 cows (LIC, 2003).

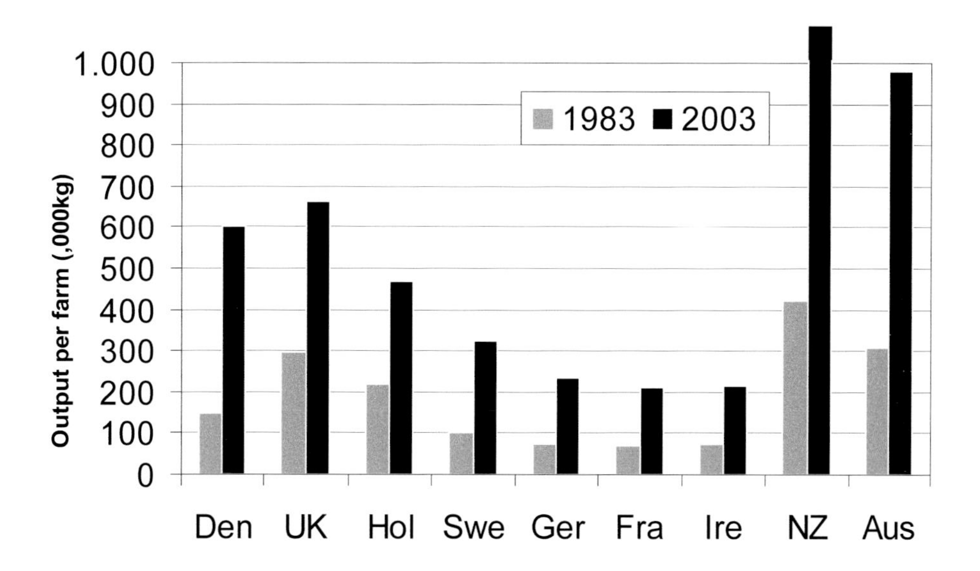

Figure 1 Development in milk output per farm over the period 1983 to 2003 in selected countries

A comparison of the profitability of various milk production systems

Table 6 shows a comparison of the receipts, cash costs, total costs and net margin of milk production for New Zealand (Dexcel, 2004), Ireland (Fingleton, 2003), Victoria Australia (Australian Dairy Industry, 2004) and Wisconsin, USA (both confinement and pasture-based) (Kriegl, personal communication) for three years (2001, 2002 and 2003). Cash costs included all costs directly incurred in the production of milk, for example feedstuffs, fertilizer etc. as well as external costs through to confinement systems within the US. Averaged over the three years the lowest milk price was obtained in New Zealand and Australia (19 and 18 € cent/l respectively), the highest in the US (33 and 32 € cent/l grazing and confinement systems respectively), while the Irish milk price (30 € cent/l) was similar to the US. The lowest cash costs per l were achieved in New Zealand and Australia (13 and 15 € cent/l respectively), the highest in the US (25 and 28 € cent/l grazing and confinement systems respectively), while the Irish costs were intermediate (22 € cent/l). The highest margin per l was obtained in the

Irish system (12 € cent/l), followed by US Grazing (10 € cent/l), similar for US Confinement and New Zealand (6 € cent/l) and lowest for Australia (4 € cent/l).

In general terms among the countries chosen, higher milk prices and higher costs of production are associated with systems incorporating more conserved feeds and higher concentrate supplementation while in more intensive pasture-based countries the costs of production are lower. Both the New Zealand and Australian dairy industries operate in a deregulated environment and consequently receive the prevailing world market price for their milk. In contrast, farm gate milk prices for Irish and US producers tend to be higher as a result of various price support mechanisms in operation in these regions. The between-year variability in milk price was greatest in the US, while milk price in Ireland under the CAP was the most stable of the countries studied. Total receipts incorporate both the revenue from the sale of livestock and surplus feeds as well as the revenue achieved from milk sales. The higher values for other income in both Ireland and the US reflect the greater value of livestock sales from the farm, again largely as a result of protectionist policies in both the EU and US in their respective beef systems.

The cash costs associated with milk production show large variation between systems. Costs associated with concentrate supplementation (feed, machinery repairs and maintenance) increase with system intensification, while in lower input pastoral systems, forage costs (fertiliser, contractor charges, etc.) will be more important. Overall cash costs tend to be lower on the pasture-based systems due to the reduction in feed and labour costs, and total cost/l tended to increase with increasing intensification. The highest costs of production were observed in both US systems of production, resulting mainly from higher feed, hired labour and repair and maintenance costs. New Zealand farmers had the lowest costs of production. Relative to both New Zealand and Australia, Irish farmers had higher costs associated with feed, repairs and maintenance as well as higher land rental charges.

Intensive pasture-based systems produce less milk/cow than confinement systems due to lower dry matter intakes. This can be seen in the US where the confinement farms produce an extra 2,464 l/cow compared to the grazing farms. However, the reduced cost of 4 € cent/l associated with the reduced requirement for facilities and equipment, the feeding of stored feeds, manure storage and disposal costs, and labour on the US grazing system results in a higher margin of 4 € cent/l of milk produced in the latter.

In non-cash costs, depreciation charges associated with the US and to a lesser extent Ireland were higher than those associated with Australia and New Zealand. These higher depreciation charges are a function of greater capital investment requirements. The cash and total costs of production observed in the current analysis compare very favourably with the costs observed for other EU systems of production displayed in Table 7 (Boyle *et al.*, 2002). This study found that the costs of milk production in Ireland were among the lowest in the EU. Similar to the current study, it also showed that within the EU, systems of production based on conserved forages, high levels of concentrate supplementation and low pasture utilization (Denmark and Holland) had higher costs, which were attributed to higher feed costs and high annual interest and depreciation charges.

Table 6 International comparison of total receipts, cash costs and profitability per litre

Variable € cent/l	New Zealand			Australia			Ireland			US (Grazing)			US (Confinement)		
Year	00/01	01/02	02/03	00/01	01/02	02/03	00/01	01/02	02/03	00/01	01/02	02/03	00/01	01/02	02/03
Cash Receipts	19.0	21.2	16.0	18.4	20.7	15.0	29.5	31.3	27.8	30.8	37.8	30.6	30.0	36.0	30.5
Other Income	1.8	2.1	1.6	2.6	2.3	3.0	4.0	4.0	4.0	6.4	5.6	8.5	8.5	6.3	7.5
Total receipts	20.8	23.3	17.6	21.0	23.0	18.0	33.5	35.3	31.8	37.2	43.4	39.1	38.5	42.3	38.0
Feed	2.0	2.3	2.4	3.2	4.2	5.4	3.0	3.4	3.6	7.0	8.2	8.3	6.7	7.9	7.4
Hired labour	1.3	1.5	1.6	0.6	0.8	0.7	0.5	0.6	0.6	1.3	1.3	1.4	2.6	2.8	3.1
Fertilizer	1.8	1.9	1.6	1.1	1.0	1.0	1.5	1.7	2.1	1.1	1.0	0.6	0.8	0.9	0.7
Medicine/ Vet/ AI	1.0	1.1	1.0	2.0	2.0	2.0	1.2	1.1	1.3	1.2	1.4	1.5	1.5	1.7	1.7
Repairs/maintenance	1.1	1.2	1.0	1.1	1.3	1.1	1.7	1.5	1.6	2.0	2.5	1.8	2.0	2.3	2.0
Interest	2.4	2.5	2.9	1.4	1.3	1.4	0.8	1.1	0.8	2.0	2.1	2.0	2.6	2.5	2.1
Water charges/rates	1.0	1.0	1.0	0.6	0.6	0.8	0.0	0.0	0.0	0.0	0.0	0.0	0.0	0.0	0.0
Rental charges	0.0	0.0	0.0	0.3	0.3	0.3	2.4	2.3	2.0	1.8	2.0	2.2	3.0	3.3	3.2
Other	1.9	2.0	1.7	3.7	4.5	3.3	8.4	7.4	8.0	7.4	7.6	8.5	8.1	8.4	8.9
Total cash costs	12.5	13.5	13.2	14.0	16.0	16.0	19.5	19.1	20.0	23.8	26.1	26.3	27.3	29.8	29.1
Depreciation	1.1	0.5	1.3	1.5	1.3	1.5	2.3	1.8	2.2	4.6	4.6	3.9	4.6	4.7	4.6
Total costs	13.6	14.0	14.5	15.5	17.3	17.6	21.8	20.9	22.2	28.4	30.7	30.2	31.9	34.5	33.7
Net margin	7.2	9.3	3.1	5.6	5.7	0.4	11.7	14.4	9.6	8.9	12.7	8.9	6.6	7.8	4.3

Table 7 EU milk sector competitiveness: costs (€ cent/l) 1998/1999

	Ireland	Belgium	France	Germany	Holland	Denmark
Feed	5.2	4.3	4.5	4.8	5.6	7.6
Hired labour	2.2	0.1	0.2	0.4	0.3	2.2
Fertilizer	2.1	0.9	1.4	0.8	0.8	0.6
Medicine/ Vet/ AI	2.3	1.4	1.5	2.1	1.9	2.1
Repairs/maintenance	2.3	1.5	1.9	2.8	2.2	3.0
Interest	0.8	1.5	0.9	1.0	3.3	6.1
Rental charges	1.3	1.2	1.7	2.1	1.8	0.9
Other	2.2	3.7	8.1	6.2	6.6	5.8
Total cash costs	18.2	14.8	20.3	20.2	22.4	28.2
Depreciation	1.5	4.9	4.8	5.6	6.9	3.9
Total costs	19.8	19.8	25.1	25.8	29.3	32.1

Boyle *et al.* (2002) also showed that Ireland's cost advantage over the other EU countries had fallen gradually over the period 1989 to 1999, probably as a result of an introduced maize subsidy in Europe and the reduction in world grain prices, both of which favoured non-pasture based production. However the same study showed that in terms of economic costs (cash costs plus imputed costs for family labour, capital and owned land) Ireland has fared less well within the EU, due to relatively low scale and low land and labour productivity.

Figure 2 shows the relationship between costs of milk production per l and the proportion of grazed grass in the dairy cow diet (derived from data from Boyle *et al.*, 2002; Clark and Jans, 1995). This relationship suggests that for a 10% increase in grazed grass in the feeding system the cost of milk produced will be reduced by 2.5 € cent/l. Consequently one strategy to reduce the impact of a lower milk price is to increase the grazed grass proportion of the diet.

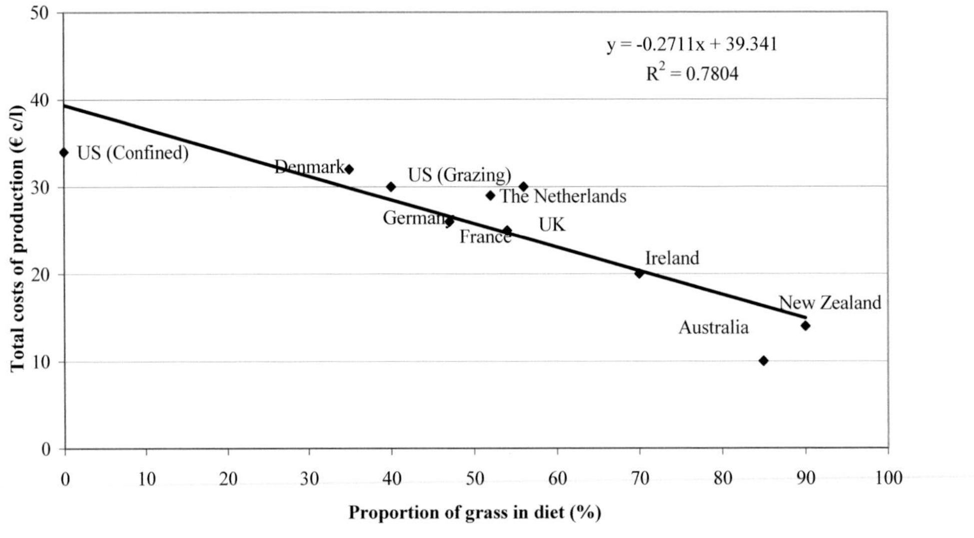

Figure 2 Relationship between total costs of production and proportion of grass in cows diet

Considerations in choosing a system of milk production

A spectrum of milk production systems exists from pasture-only to total confinement. Identifying the optimum system of milk production and the key drivers of profitability for any particular location is therefore extremely important. There are a number of prevailing economic factors that determine which production system is most suitable and these include milk price, supplement cost, land quality and availability, labour cost and availability, environment issues and consumer demand.

Milk price

The New Zealand dairy industry operates in a deregulated environment, which has resulted in a relatively low milk price, with large fluctuations from year to year. The deregulation of the Australian industry began in 1999 and has meant the discontinuation of regulated sourcing and pricing of milk for liquid consumption. This has resulted in a milk price similar to New Zealand. Milk price in Australia and New Zealand is also influenced by revaluation of domestic currency against the US$ dollar. In the EU the decoupling of direct payments from production as a result of CAP reform will result in a reduced milk price. Milk quota restrictions will require Irish producers to target lower costs of production, as increased scale to compensate for declining profitability is not an option in the short term. Like Europe, a reduction in government involvement in price support in the US has resulted in reduced milk price. In all countries the overall impact of reduced milk price has been a large reduction in dairy farm numbers, with a corresponding increase in scale of operation. However economies of scale will only be effective as long as unit cost of production is less than milk price. In scenarios where scale economies do not allow the unit cost of production to be less than milk price, then a lower cost milk production system will be required. These results indicate that grass-based systems have the potential to have the lowest cost per unit of production. A limiting feature of all grass-based systems is the lack of flexibility to rapidly increase milk production if milk price increases. However, increased production costs per l will be inevitable when farmers move to maximise profitability during periods of elevated milk price and /or depressed concentrate price.

Land value

The ability to avail of the increased profitability of pasture-based systems may be curtailed by land costs (both rental and purchase). Access to land at economically feasible prices is crucial to the future success of pasture based dairy systems. High land prices reduce the potential return on investment and lower the milk price (or raise the supplement price) at which intensification becomes more profitable than expansion.

High land costs are found in small farms in the US and large Irish farms while price of land is low in Australia. Land productivity ranges from 1,000 to 25,000 kg milk per ha, reflecting international variation in stocking rate of 0.3 to 2.5 cows/ha (IFCN, 2003). The differences in productivity result from differences in the production system, which differ from country to country as well as within country. In regions of high livestock density land prices reach very high levels due to the competition for land among producers. In New Zealand, high land prices are a result of strong demand due to scarcity of suitable land and the relatively small amounts traded annually. Similarly, in Ireland land ownership is very high among farmers (approximately 90%) and also has a high status value. Only 1% of the total land is traded each year, which increases competition in the market. Recent economic prosperity has resulted in much of the land traded being used for development purposes.

Cost of alternative feeds

The cost at which alternative feeds can be grown or purchased will also determine the suitability of confinement systems. It is envisaged that the cost of conserved forages will continue to increase due mainly to increases in contractor charges associated with inflation in labour, energy and machinery costs. There has been a movement in all countries regardless of the level of grass growth towards grazing management strategies that can increase the proportion of grazed grass and reduce the dependency on high cost conserved feeds. The profitability of concentrate supplementation is determined by the milk to concentrate price ratio and the level of additional milk production achieved in response to supplementation. If the market value of the additional milk achieved outweighs the costs of concentrate inclusion and pasture utilisation is not compromised, higher supplementation levels will yield greater farm profit. However, if milk price continues to decline, the economic feasibility of concentrate use within the dairy feed budget declines as the marginal benefit of increased milk output is outweighed by the cost of the additional supplementation.

Labour availability

With the cost of labour and the proportion of the cost of production represented by labour increasing annually, labour efficiency becomes an increasingly important performance measure to evaluate and monitor on dairy farms. Labour efficiency is related to the level of mechanisation, which may therefore translate into higher capital investments. Dairy farms with higher levels of labour and capital efficiency tend to have higher levels of labour and management income per operator. The efficiencies (capital and labour) of many labour saving technologies (e.g. milking parlours) are only captured with increasing herd size. Successful dairy farms will be those that can adapt to changing economic conditions and evaluate and adopt cost-effective labour saving new technologies. Using international comparisons of labour efficiency data from IFCN (2002), the labour efficiency (hours per milking cow) of a Waikato New Zealand 229 cow herd, South Island New Zealand 447 cow herd, Wisconsin US 70 cow herd and Wisconsin 600 cow herd were 20, 19, 106 and 61, respectively (comparable to 210, 247, 93 and 156 kg milk per hour, respectively). These results indicate that higher levels of labour efficiency can be achieved on pasture-based systems.

The industry demography shows a clear trend toward fewer, but larger and more productive herds. In the 40-year period form 1955 to 1995, the number of farms in the United States decreased from 2.7 million to 137,000, while the average herd size increased from 8 to 69 cows. Milk production per cow increased from less than 2,724 kg per cow per year to more than 7,265 kg. Tremendous development, adoption and management of new production-enhancing technologies over the past few decades has led to rapid increases in herd size and milk production levels. A recent Cornell University study reported that if current trends continue, in 2020, total milk production will be 17-20% greater, cow numbers will be 15% lower and the number of farms will decline by 85%.

Maintaining herd sizes large enough to obtain the major economies of scale and using capital resources efficiently is important across all farm systems. We can expect to see opportunities through an increase in the number of farm partnership operations. These operations will provide an opportunity to improve capital, labour efficiency, profitability, and quality of life, given an appropriate management structure. Facility and equipment replacement decisions will also be very important to the longer-term success of these as well as single-family operations. The economic advantages of pasture-based systems are greater for producers starting a dairy operation. Start up costs for grazing systems are only 40 to 50% of the initial costs associated with the traditional confined dairy facilities.

Environment

Nutrient management presents major challenges and opportunities for pasture-based agriculture. The challenge is to avoid environmental damage while increased nutrient efficiency could substantially improve farm profitability. Under current consideration in the EU is the Nitrates Directive (91/676/EEC) which states that 'the amount of livestock manure applied to land each year, including by the animals themselves, shall not exceed 170kg organic N per hectare', (OJEC, 1991, 91/L375/EEC; 7). Such legislation is likely to reduce production efficiency and potential profitability and may lead to higher input systems incorporating crops such as maize in the dairy diet. The cultivation of maize requires the ploughing of permanent grassland containing very large reserves of N (7 t N /ha approx.) in the plough layer (Gardiner and Radford, 1980; McGrath and Zhang, 2003). This N, which is otherwise stable, becomes exposed and therefore available when the soil is ploughed. Whitmore *et al.* (1992) suggested that cultivation of long-term grassland could result in the release of up to 500 kg of N/ha in the first year of cultivation. These quantities are beyond the uptake capacity of maize and hence lead to substantial losses of N to the environment. Similarly, there are few comparisons of greenhouse gases emissions from grazing ruminants and those fed TMR diets. However van der Nagel *et al.* (2003) states that total greenhouse gas emissions from TMR diets were nearly twice as much as from pasture (1.53 vs. 0.84 kg CO_2 equivalent/kg of milk). Such a shift to higher input systems may therefore be more damaging to the environment than the traditional grass based system.

The implementation of such legislation in grazing systems could reduce profitability and limit the potential for future expansion. Similar public concerns are echoed in New Zealand, Australia and US with different nutrients causing problems at different sites. Strategic approaches that increase the efficiency of N in the farming system, while at the same time maintaining animal performance are urgently needed.

Consumer preference and product quality

The primary drivers for changing consumption patterns in dairy products is consumer affluence and perceived health benefits. Lomborg (2001) reports that we are three times wealthier now than we were in 1950. Greater retailer awareness of food constituents and demand for healthier foods will be of significant importance in the coming years. Milk fat produced on pasture has higher levels of unsaturated fatty acids (42%) and particularly monounsaturated fatty acids (MUFA)(33%) compared to that produced in confinement (34 and 27% respectively). MUFA are superior to polyunsaturated fatty acids (PUFA) as they lower blood cholesterol through the reduction of detrimental low-density lipoprotein (LDL) cholesterol (Gibney, 1993). The second relative benefit of pasture derived milk products relates to the concentration of conjugated linoleic acid (CLA). CLA has been shown to be protective against heart disease (Lee *et al.*, 1994), certain types of cancer (Schultz *et al.*, 1992) and obesity (Pariza *et al.*, 1996) and acts as an antioxidant (Ha *et al.*, 1990) and growth promoter (Chin *et al.*, 1994). More CLA is produced by cows on pasture than by cows on indoor diets with milk from cows on pasture containing approximately 16 mg CLA/g fat compared to approximately 5 mg/g fat indoors (Stanton *et al.*, 1997; Murphy, 2000). With increasing health awareness among consumers, the requirement for healthier products will increase the demand for pasture derived milk products in future years.

Acknowledgements

The authors graciously acknowledge the data provision by C. Grainger, Department of Primary Industries, Ellinbank, Victoria, T. Kriegl, University of Wisconsin, U.S.A., C Glassey and A. van Bysterveldt, Dexcel, New Zealand. The technical assistance of J. Lee in compiling the data is also graciously acknowledged.

References

Australian Dairy Industry (2004). Australian Bureau of Agricultural and Resource Economics. Internet: www.dairyaustralia.com.au

Boyle G.E. (2002). The competitiveness of Irish Agriculture. Published by the Irish Farmers Journal Dublin, 41-58.

Chin, S.F., J.M. Storkson, K.J. Albright, M.E. Cook & M.W. Pariza (1994). Conjugated linoleic acid. A powerful anticarcinogen from animal fat sources. *Journal of Nutrition*, 124, 2344-2349.

CIAS (2001). Grazing project combines strength of on-farm research station studies. Res. Brief 12, Centre for Integrated Agricultural Systems. University of Wisconsin. 2001. Internet site: http://www.wisc.edu/cias/pubs/brief/012.html

Clark D.A. & F. Jans (1995). High forage use in sustainable dairy systems. In: *Proceedings of the IVth International Symposium on the Nutrition of Herbivores,* p497-522, Clermont-Ferrand (France), September, 11-15.

Dexcel (2004). Economic Survey of New Zealand Dairy Farmers. Dexcel Limited, Hamilton, New Zealand. Internet site: http://www.dexcel.co.nz

Fingleton W. (2003). The situation and outlook for dairying 2002/2003. In: *Proceedings Teagasc Situation and Outlook in Irish Agriculture* 10th December 2003, Teagasc, Dublin, 1-11.

Gardiner, M.J. & T. Radford (1980). Soil associations of Ireland and their land use potential. Soil Survey Bulletin No. 36. Teagasc, 19 Sandymount Avenue, Dublin 4, 142 pages.

Gibney, M. J. (1993). Record of Advancement in Animal Nutrition in Australia. Published by the Department of Biochemistry, Microbiology and Nutrition, University of New England, ISBN 1 86389 054 8 p145-152.

Grainger, C. (1998). ABC farms final Report. Dept. of Natural Resources and Environment, 1998. pp65.

Ha, Y., P. Storkson & J. M. Pariza (1990). Inhibition of benzo(a)pyrene-induced mouse forestomach neoplasia by conjugated dienoic derivatives of linoleic acid. *Cancer Research*, 50, 1097-1101.

Hamilton S.A., G.L. Bishop-Hurley & R. Young (2002). Economics of a pasture-based system. In: Dairy Grazing Manual, eds. Bishop-Hurley G.L., Kallenbach R., Roberts C., Hamilton S.A. pp137-141. MU Extension, University Missouri-Columbia.

Hanson G.D., S. Cunningham, A. Ford, L.D. Muller & R.L. Parsons (1998). Increasing intensity of pasture use with dairy cattle: An economic analysis. *Journal of Production Agriculture,* 11, 175-179.

Horan B., P. Dillon, P. Faverdin, L. Delaby, F. Buckley & M. Rath (2004). Strain of Holstein-Friesian by pasture-based feed system interaction for milk production, bodyweight and body condition score. *Journal of Dairy Science* (in press).

IFCN, (2002) Dairy Report 2002. International Farm Comparison Network. ISSN 1610-434X. Internet: http//: www:ifcndairy.org.

IFCN, (2003) Dairy Report 2003. International Farm Comparison Network. ISSN 1610-434X. Internet: http//: www.ifcndairy.org.

Kriegl, T. (2001). Wisconsin Grazing Dairy Profitability Analysis- A Preliminary five year comparison of the cost of production of selected conventional and grazing Wisconsin dairy farms. University of Wisconsin Centre of Dairy Profitability.

Lee, K.N., D. Kritchevsky & M.W. Pariza (1994). Conjugated linoleic acid and atherosclerosis in rabbits. *Atherosclerosis,* 108, 19-25.

LIC. (2003). Dairy statistics 2002 – 2003. Livestock Improvement Corporation, Hamilton, New Zealand.

Lomborg, B. (2001). The skeptical environmentalist: measuring the real state of the world. Cambridge University Press.

Macdonald, K. (1999). Determining how to make inputs increase your economic farm surplus. *Proceedings of the 51st Ruakura Farmers Conference,* 51, 78-87.

McGrath, D. & C. Zhang (2003). Spatial distribution of soil organic carbon concentrations in grassland of Ireland. *Applied Geochemistry,* 18, 1629-1639.

Murphy, J.J. (2000). Synthesis of milk fat and opportunities for manipulation. In: *Proceedings of the British Society of Animal Science* Occasional Publication (eds. R.E. Agnew, K.W. Agnew, & A.M. Fearson), 25, 201-222.

NASS, (2004). U.S. Department of Agriculture, National Agricultural Statistics Service. Internet: http://www.usda.gov/nass/

Parker, W.J., L.D. Muller & D.R. Buckmaster (1992). Management and economic implications of intensive grazing on dairy farms in the North eastern States. *Journal of Dairy Science,* 75, 2587-2597.

Pariza, M., Y. Park, M. Cook, K. Albright, & W. Liu (1996). Conjugated Linoleic Acid reduces body fat mass in overweight and obese humans. *FASEB Journal*, 10, A3227.

Penno J.W., K.A. Macdonald & A.M. Bryant (1996). The economics of the No. 2 Dairy systems. In: *Preoceedings of the Ruakara Farmers Conference,* 48, 11-19.

Schultz, T.D., B.P. Chew, W.R. Seaman & L.O. Luedecke (1992). Inhibitory effect of conjugated dienoic derivatives of linoleic acid and b-carotene on the in vivo growth of human cancer cells. *Cancer Letters*, 63, 125-133

Stanton, C., F. Lawless, J. Murphy & B. Connolly, (1997). In Animal Fats- BSE and After, Society of Chemical Industries/ P.J. Barnes and Associates/Lipid Technology, p 27-41.

Van der Nagel, L.S., G.C. Waghorn & V.E. Forgie (2003). Methane and carbon emissions from conventional pasture and grain-based total mixed rations for dairying. *Proceedings of the New Zealand Society of Animal Production*, 63, 128-132.

Whetstone, L. (1999). The perversity of agricultural subsidies, In: J. Morris, and Roger Bate, R., (eds.) Fearing Food, Risk, Health and Environment, Butterworth- Heinemann, London, UK., 123pp.

Whitmore, A.P., N.J. Bradbury, & P.A. Johnston (1992). The potential contribution of ploughed grassland to nitrate leaching. *Agriculture, Ecosystems and the Environment*, 39, 221-233.

Section 1

Appropriate plants for grazing

Intake characteristics of diploid and tetraploid perennial ryegrass varieties when grazed by Simmental x Holstein yearling heifers under rotational stocking management

R.J. Orr, J.E. Cook, K.L. Young, R.A. Champion and A.J. Rook
IGER North Wyke, Okehampton, Devon EX20 2SB U.K., Email: robert.orr@bbsrc.ac.uk

Keywords: perennial ryegrass, varieties, intake rate, rotational stocking, cattle

Introduction Orr *et al.* (2003) measured large differences in dry matter (DM) intake rate between 15 intermediate-heading perennial ryegrass varieties when they were continuously stocked with sheep and subsequently explored the extent to which, for 5 of these varieties, these differences could be explained by chemical and morphological traits (Orr *et al.*, 2004a) which could be targeted in grass breeding programmes. Here, four of the 15 varieties, which within ploidy had low or high intake characteristics when grazed by sheep, were rotationally stocked with cattle and intake and sward factors were measured.

Materials and methods Four intermediate-heading perennial ryegrass varieties (see Table 1) were rotationally stocked with yearling Simmental x Holstein beef heifers (195 ± 3.0 kg) from April to September in 2002 and 2003. The 4 varieties were each sown in 3 replicate blocks to create 12 (0.6 ha) areas and each area was further subdivided into 30 paddocks. Twelve groups of 4 heifers grazed 1-day paddocks and moved to a new paddock at approximately 15.00 h. Herbage intake rate was measured in September 2002 and July and August 2003, using a weighing technique described by Orr *et al.* (2004b), during the first hour on the paddocks. Herbage samples were collected from the grazed horizon for assessment of DM, N, WSC and DOMD content. Eating times were recorded over 24 h using IGER behaviour recorders and compressed sward height was measured at the start (CSH_{IN}) and end (CSH_{OUT}) using an Ashgrove platemeter. The partition of herbage mass and the number of tillers per m^2 were measured within circular quadrats, with the same diameter as the platemeter, in which the grass was cut to ground level using scissors.

Results Mean CSH_{IN} was 10 cm and CSH_{OUT} was 6 cm. Intake rate was significantly higher in August 2003 for heifers grazing Glen than those grazing Belramo (27.5 vs. 20.6 g DM/min; *F* prob. = 0.019), but there were no significant differences between varieties in the two other measurement periods. For sward factors (Table 1), N concentration was significantly higher for AberExcel than Rosalin and there were significant differences in leaf mass within the diploid and tetraploid varieties. Mean daily liveweight gains over the two grazing seasons by the heifers were not significantly different.

Table 1	Belramo *(diploid)*	Glen *(diploid)*	Rosalin *(tetraploid)*	AberExcel *(tetraploid hybrid)*	s.e.d	*F* prob. *Belramo v Glen*	*Rosalin v AberExcel*
Intake rate (g DM min^{-1})	23.8	27.0	26.7	26.9	2.13	0.190	0.933
Eating time (min $24h^{-1}$)	526	538	513	531	23.6	0.627	0.469
DOMD (g DOM kg^{-1} DM)	694	714	695	703	1.28	0.177	0.573
N (g kg^{-1} DM)	35	37	34	38	1.6	0.309	0.047
WSC (g kg^{-1} DM)	138	141	140	135	4.5	0.548	0.283
Leaf mass (kg DM ha^{-1})	1053	1381	1266	945	113.2	0.027	0.030
Tillers (m^{-2})	9,431	10,640	8,496	6,615	776.7	0.171	0.052
Daily liveweight gain (kg)	0.98	1.07	1.00	0.94	0.050	0.103	0.232

Conclusions Mean intake rates, eating times and daily liveweight gains were not significantly different between the varieties when rotationally-stocked with cattle. It is evident from these and previous results with sheep (Orr *et al.*, 2003), when intake rates were significantly higher for Glen than Belramo and for AberExcel than Rosalin, that not only does the grass plant display considerable plasticity under different management but that there are interactions with genotype. In order to understand further these interactions grass variety evaluations using continuously stocked cattle swards are required in order to separate the effects of defoliation interval from those of grazing style of the different animal species. Then it will be possible to develop new varieties for grazing use that are matched with appropriate grazing management recommendations.

References

Orr, R.J., J.E. Cook, R.A. Champion, P.D. Penning & A.J. Rook (2003) Intake characteristics of perennial ryegrass varieties when grazed by sheep under continuous stocking management. *Euphytica,* 134, 247-260.

Orr, R.J., J.E. Cook, R.A. Champion & A.J. Rook (2004a) Relationships between morphological and chemical characteristics of perennial ryegrass varieties and intake by sheep under continuous stocking management. *Grass and Forage Science*, 59, 389-398.

Orr, R.J., S.M. Rutter, N.H. Yarrow, R.A. Champion & A.J. Rook (2004b) Changes in ingestive behaviour of yearling dairy heifers due to changes in sward state during grazing down of rotationally-stocked ryegrass or white clover pastures. *Applied Animal Behaviour Science*, 87, 205-222.

The effect of early and delayed spring grazing on the milk production, grazing management and grass intake of dairy cows

E. Kennedy[1,2], M. O'Donovan[1], J.P. Murphy[1], L. Delaby[3] and F.P. O'Mara[2]

[1]*Teagasc, Dairy Production Research Centre, Moorepark, Fermoy, Co. Cork, Ireland, Email: ekennedy@moorepark.teagasc.ie,* [2]*Faculty of Agri-Food and Environment, NUI Dublin, Belfield, Dublin 4, Ireland,* [3]*INRA, UMR Production du Lait 35590 St. Gilles, France*

Keywords: dairy cows, grazing date, stocking rate

Earlier access to pasture can increase the overall proportion of grazed grass in the diet of the spring calving dairy cow. Further benefits can also be achieved from early turnout, including improved animal production, increased sward utilisation and enhanced sward quality (O'Donovan *et al.*, 2004). The objective of this study was to compare the effect of initial spring grazing date and stocking rate on the performance of spring calving dairy cows.

Materials and methods Sixty-four spring calving dairy cows, 32 primiparous and 32 multiparous, were randomly assigned to one of four grazing treatments (n=16/treatment) which were balanced for lactation number (1.9 ± 1.47), days in milk (58 ± 9.0), milk yield (28.7 ± 5.47kg), milk fat content (39.6 ± 4.26 g/kg), milk protein content (32.1 ± 2.23 g/kg), bodyweight (511 ± 52.0 kg) and body condition score (2.8 ± 0.35). Eight 'filler cows' were used to achieve the required stocking rate. Two swards, early grazed (E) and late grazed (L), were created, and two stocking rates (SR), high (H) and medium (M), were applied across them. Half of the area was grazed between 16 February and 4 April creating the early grazed sward. The rest of the area (late grazed sward) remained ungrazed. The study was completed over four 21-day rotations from 16 April to 3 July 2004. Twenty-one paddocks, previously measured and permanently fenced, were allocated to each treatment, each with a residency time of 24 hours. The SR imposed were 5.5 cows/ha (EH), 4.5 cows/ha (EM), 6.5 cows/ha (LH) and 5.5 cows/ha (LM). The SR on the LH treatment was reduced to 5.9 cows/ha for the final rotation. Subsequent to each grazing 60 kg N/ha was applied therefore the early grazed sward received an extra 60 kg N/ha. Milk yield was recorded daily and milk composition was determined weekly. Live weight was recorded weekly and body condition score was measured every three weeks. Milk production carryover effects of each of the feeding regimes were measured for 8 weeks subsequent to this study. During this period all cows were managed as a single herd and offered a DHA (daily herbage allowance) of 22 kg DM/cow per day.

Results There was a significant difference between treatments for milk yield (P<0.001), SCM yield (P<0.001), milk protein concentration (P<0.001), milk fat yield (P<0.01) and milk protein yield (P<0.001). The EM treatment recorded superior performance levels for these parameters when compared to all other treatments. The LH treatment recorded the lowest production performance while the production performance of the EH and LM treatments were between those of the EM and LH treatments. No significant differences in liveweight or BCS were recorded. During the carryover period there were no significant differences between treatments for milk yield, milk composition, bodyweight or BCS.

Table 1 Effect of initial grazing date and stocking rate on milk production performance bodyweight and body condition score of spring calving dairy cows

	EH	EM	LH	LM	Rse	Sig.
Milk yield (kg/day)	22.7^{a}	24.5^{b}	20.9^{c}	22.4^{a}	2.05	***
Fat (g/kg)	38.9^{a}	37.8^{a}	40.0^{a}	37.8^{a}	3.42	NS
Protein (g/kg)	32.9^{a}	34.1^{b}	32.1^{c}	32.7^{a}	0.91	***
SCM yield (kg/day)	20.9^{a}	22.5^{b}	19.4^{c}	20.4^{ac}	1.71	***
Fat yield (g/cow per day)	871.8^{a}	917.8^{b}	829.5^{a}	845.9^{a}	77.69	**
Protein yield (g/cow per day)	744.4^{a}	831.4^{b}	670.2^{c}	733.2^{a}	65.07	***
Liveweight (kg)	503^{a}	514^{a}	509^{a}	516^{a}	28.4	NS
BCS	2.76^{a}	2.75^{a}	2.71^{a}	2.73^{a}	0.159	NS

NS=Non-significant, **=P≤0.01, ***=P≤0.001. ^{abc}values in the same row not sharing a common superscript are significantly different.

Conclusions The results suggest that increased milk production is achievable with early grazed swards. If a medium SR is imposed across such swards a high production performance can be obtained. The date of initial spring grazing and the subsequent stocking rate imposed have direct effects on milk production performance.

References

O'Donovan, M., Delaby, L. and Peyraud J.L. (2004) Effect of time of initial grazing date and subsequent stocking rate on pasture production and dairy cow performance. *Animal Research 53,* 489-502

Performance of meat goats grazing winter annual grasses in the Piedmont of the southeastern USA

J-M. Luginbuhl and J.P. Mueller
Campus Box 7620, North Carolina State University, Raleigh NC 27695-7620, USA,. Email: *jean-marie_luginbuhl@ncsu.edu*

Keywords: goats, grazing, performance

Introduction In the Southeastern United States, meat goats (*Capra hircus hircus*) are becoming increasingly important contributors to the income of many small producers. Meat goats perform well in grazing situations if grazing management practices match their grazing behavior. Nevertheless, little research data are available from the region specifically directed toward forage feeding programs for goats reared for meat production. Hart *et al.* (1993) reported that growing Alpine, Angora and Nubian kids grazed on high quality *Triticum aestivum* forage gained 50 g/d, whereas Kiesling *et al.* (1994) reported gains ranging from 65 to 141 g/d in growing Angora goats grazing *Secale cereale*. This 3-year (YR) grazing study was designed to evaluate the performance of replacement does and wethers grazed on *Secale cereale* (SC, var. Elbon), *Lolium multiflorum* (LM, var. Marshall) and *Triticum secale* (TS, var. Resource Seeds 102).

Materials and methods The experimental area was divided into 9 plots of 0.19 ha each in a randomized complete block design with 3 replications. Forage species were sod-drilled in fall and fertilized with ammonium nitrate (56 kg N/ha) each November and February. Each year, 54 yearling goats (purebred Boer, ¾ Boer and ¼ Landrace, and ½ Boer and ½ Landrace; initial body weight (BW): 29 kg) were stratified by BW, placed in 6 blocks of 9 animals with similar BW, assigned randomly to one of nine plots, and managed using controlled rotational grazing with Premier® electronetting. In YR 1, all 6-tester goats were females, whereas in YR 2 and YR 3, 36 females and 18 castrates were used. Each goat was treated for elimination of gastrointestinal parasites (Ivermectin) at the start of grazing. Goats were moved to a fresh strip of grass 3 to 4 times per week depending on forage availability, and immediately back fenced. Additional goats (2 to 14 goats/plot) were used as put-and-take animals to control forage growth. Goats had free-choice access to a mineral mixture, water and movable shelters and goats were weighed on two consecutive days at the beginning and end of the experiment. In YR 2 and YR 3, two of the 6-tester goats/plot were castrates. In YR 3, blood and ruminal fluid samples were collected from castrates which were then harvested at a commercial facility.

Results YR 1: grazing periods ranged from 25 February – 14 April for SC, 28 February – 19 May for LM, and 28 February – 21 April for TS, with 6-tester goats/plot. YR 2: 3-tester goats/plot were grazed on each forage species from 22 January - 3 March, and then with 6-tester goats/plot until 8 April (SC), 4 May (LM) and 23 April (TS). YR 3: grazing started with 6-tester goats/plot for each forage species from 9 to 28 December. All goats were removed from the experimental plots on 28 December due to lack of forage, with the exception of LM plots, on which 3 goats/plot were left grazing until 18 January. For TS, 3 goats/plot were grazed from 11 to 20 January, at which date 51 cm snow fell in 24 hours. Grazing resumed with 6-tester goats/plot on each plot on 24 February. Grazing ended on 31 March for SC, 10 May for LM, and 20 March for TS. Crude protein values of forage samples hand-plucked periodically from experimental pastures averaged 21.5, 23.3 and 23.0% for LM, SC and TS, respectively. Forage species had no effect on ADG in YR 1, 2 or 3 (avg: 136, 151, 142 g/d, for LM, SC and TS, respectively), but castrates gained more weight than does ($p<0.01$) in YR 2 (139 vs 94 g/d) and YR 3 (201 vs 137 g/d). Gain per ha was greater ($p<0.05$) for LM than SC and TS (YR 1: 504, 235, 293 kg; YR 2: 288, 195, 234 kg; YR 3: 532, 251, 137 kg). In YR 3, the pH of ruminal fluid, ruminal ammonia and carcass yield from castrates grazing LM, SC and TS averaged 6.67, 25.7 mg/dL and 51.3%, respectively. Plasma urea N (16.4, 21.9, 24.1 mg/dL), ruminal acetate (62.0, 60.7, 57.7 mM/100mM), propionate (22.0, 25.2, 27.0 mM/100mM) and acetate:propionate (2.83, 2.43, 2.22) differed between forage species ($p<0.05$).

Conclusions Results indicated that these winter annual grasses were of excellent quality and exceeded the nutritional requirements of growing replacement stock. Growing goats achieved satisfactory weight gains when fed only on these forages under controlled rotational grazing management, but LM resulted in superior per hectare live weight gains.

References

Hart, S.P., T. Sahlu & J.M. Fernandez (1993). Efficiency of utilization of high and low quality forage by three goat breeds. Small Ruminant Research, 10, 293-301.

Kielsing, H.E.,C.E. Barnes, T.T. Ross, J. Libbin, M. Ortiz, R.L. Byford & K.W. Duncan (1994). Grazing alfalfa and winter annual cereals with Angora goats. Agricultural Experimental Station Research Report 684. New Mexico State University Las Cruces, NM.

The importance of patch size in estimating steady-state bite rate in grazing cattle

E.D. Ungar, N. Ravid, T. Zada, E. Ben-Moshe, R. Yonatan, S. Brenner, H. Baram and A. Genizi
Department of Agronomy and Natural Resources, Institute of Field and Garden Crops, the Volcani Center, POB 6 Bet Dagan, Israel 50250, Email: *eugene@volcani.agri.gov.il*

Keywords: intake, jaw movements, chew-bites, acoustic monitoring

Introduction Since the pioneering work of Black and Kenney (1984), various intake studies have been conducted at the spatial scale of a single feeding station ("patch") to elucidate the processes that determine instantaneous intake rate (e.g. Laca *et al.*, 1994). While these are well-suited for patch depletion studies, it is less clear how well they represent non-patchy and relatively homogeneous environments (Ungar & Griffiths, 2002). Clearly, grazing should be restricted to the upper grazing horizon (i.e. layer of bites), but sample duration may be insufficient to characterize steady-state behaviour, especially when grazing commences on an empty mouth. We examined the impact of feeding station size on bite rate and jaw movement allocation between bites and chews.

Materials and methods Six Israeli-Holstein dairy heifers were used. Treatments were three sizes of feeding station (0.16, 0.30 and 0.53 m^2) and two initial sward heights (10 and 20 cm; mean herbage mass per unit area: 105 g/m^2 and 149 g/m^2, respectively), in a full factorial design. Animals were allowed to deplete the patch without interference. Grazing sessions were recorded on video with acoustic monitoring, using a microphone on the forehead of the animal. Analysis was from the first bite until depletion of the upper grazing horizon of the patch, or until the animal first paused, raised its head or moved its head away from the patch. The acoustic signal was sequenced aurally; each sound burst produced by a jaw movement was classified as a pure chew, pure bite or chew-bite. Factors in the ANOVA were animal, patch size, sward height and the size x height interaction.

Results The allocation of jaw movements and bite rate responded strongly to the duration of the grazing session, expressed here as the total number of jaw movements (Figure 1). The allocation of jaw movements was initially to pure bites, which we attribute to the mouth being empty. On the smallest patches, this initial run of pure bites depleted much of the upper grazing horizon of the patch and hence constituted a high proportion of the total number of events in the sequence. As the size of the feeding station increased, more jaw movements were performed ($P < 0.001$) and the proportion of jaw movements allocated to pure chews increased ($P = 0.01$). Bite rate was higher on the shorter sward (62.9 vs. 57.4/min; $P = 0.02$), as was the rate of jaw movement (77.8 vs. 74.2/min; $P = 0.03$) and the proportion of pure bites (0.47 vs. 0.33/min; $P = 0.02$), however the interaction between patch size and initial sward height was not significant in any analysis. It is possible that representative values for the sward type were not reached at even the largest patch size. For comparison, mean bite rate of a group of 9 heifers (including 5 from above) grazing continuous expanses of the same herbage was 47.5/min.

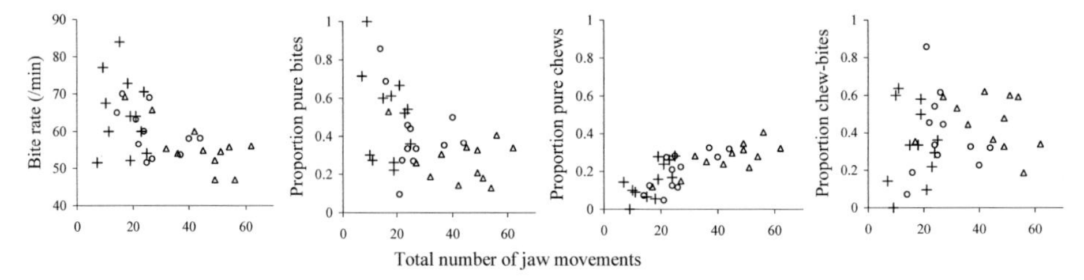

Figure 1 Response of bite rate and the allocation of jaw movements between the three types to total number of jaw movements in the sequence (Patch size: + = small, O = medium, Δ = large).

Conclusions Small patches do not permit the estimation of steady-state jaw allocation, and hence bite rate, on homogeneous, continuous swards. For such purposes, patch size should be closer to the potential area that can be reached by a stationary animal.

References

Black, J.L. & P.A. Kenney (1984). Factors affecting diet selection by sheep. II Height and density of pasture. *Australian Journal of Agricultural Research* 35, 565-578.

Laca, E.A., E.D. Ungar & M.W. Demment (1994). Mechanisms of handling time and intake rate of a large mammalian grazer. *Applied Animal Behaviour Science*, 39, 3-19.

Ungar, E.D. & Griffiths, W.M. (2002) The imprints created by cattle grazing short sequences of bites on continuous alfalfa swards. *Applied Animal Behaviour Science,* 77:1-12.

Spring calving suckler beef systems: influence of grassland management system on herbage availability, utilisation, quality and cow and calf performance to weaning

M.J. Drennan, M. McGee, S. Kyne and B. O'Neill

Teagasc, Grange Research Centre, Dunsany, Co. Meath, Ireland, Email: mdrennan@grange.teagasc.ie

Keywords: grassland management systems, suckler cow

Introduction Suckler beef systems in Ireland are primarily based on grass. Suckler systems vary in intensity but many operate low input systems and participate in REPS (Rural Environmental Protection Scheme). As there is a considerable cost associated with second-cut silage this research compared a two-cut system with a simplified low input one-cut system.

Materials and methods Data were collected over three consecutive years from two, rotationally grazed (mid-April to Oct./Nov.) systems using a total of 188 spring-calving Limousin × Friesian and Simmental × (Limousin × Friesian) cows and their progeny to weaning. The systems were (i) High (H); stocking rate (SR) of 0.77 ha/cow unit, 206 kg/ha nitrogen (N), two silage cuts and (ii) Low (L); SR of 0.95 ha/cow unit, 102 kg/ha N and one silage cut. Pre- and post-grazing sward heights and mass were measured using a rising plate meter and cutting (4 cm stubble height) and weighing strips (0.54 m x 4.5 to 5 m) of grass, respectively. Herbage yield and grass crude protein (CP) and dry matter digestibility (DMD) were determined in years 1 and 3.

Results There was no significant effect of grazing system on cow liveweight or body condition score changes or calf liveweight gains at pasture over the entire grazing season in any of the three years (Table 1). Pre-grazing heights were similar for both systems in the three years, but post-grazing heights (and yield) were lower ($P<0.05$) for H than L in year 1. There was no significant difference between systems in herbage DMD either pre- or post-grazing. In year 1 herbage CP was lower pre-grazing (n.s.) and post-grazing ($P<0.01$) and, in year 3 lower ($P<0.001$) both pre- and post-grazing for L than H (Figure 1).

Table 1 Cow liveweight and body condition score changes, calf liveweight gains and, herbage availability and *in vitro* digestibility for the High (H) and Low (L) grazing systems over three years

	Year 1			Year 2			Year 3		
System	H	L	s.e.	H	L	s.e.	H	L	s.e.
Cow weight gain (kg)									
Turnout – June	26.8	33.4	9.6	58[a]	45[b]	4.3	46	58	5.4
June – housing	31.0	31.5	8.7	26[a]	41[b]	4.2	37	36	3.4
Turnout – housing	57.8	65.4	11.4	84	86	6.0	83	94	5.4
Cow body condition score change									
Turnout - housing (units)	-0.02	-0.19	0.14	0.44	0.47	0.12	0.41	0.59	0.09
Calf weight gain (kg)									
Turnout - housing	252	256	6.8	237	234	3.6	238	241	3.8
Grazing heights (cm) Pre	12.1	12.6	0.48	11.4	11.4	0.22	11.6	10.9	0.24
Post	5.7[a]	6.3[b]	0.17	5.6	5.8	0.10	6.3	6.2	0.11
Grazing mass (kg) Pre	2022	2369	163.0	-	-	-	2325	2541	140.0
Post	424[a]	555[b]	33.0	-	-	-	1005	1003	72.7
In vitro dry matter digestibility (g/kg)									
Pre	750	764	10.4	-	-	-	761	747	5.7
Post	674	655	14.0	-	-	-	640	641	7.3

* Columns, within year with different superscripts are significantly different, $P<0.05$

Conclusions Cow and calf performance at pasture was similar between the management systems. Grass DMD did not differ between the systems but CP levels were lower for L than H reflecting the lower N fertiliser application.

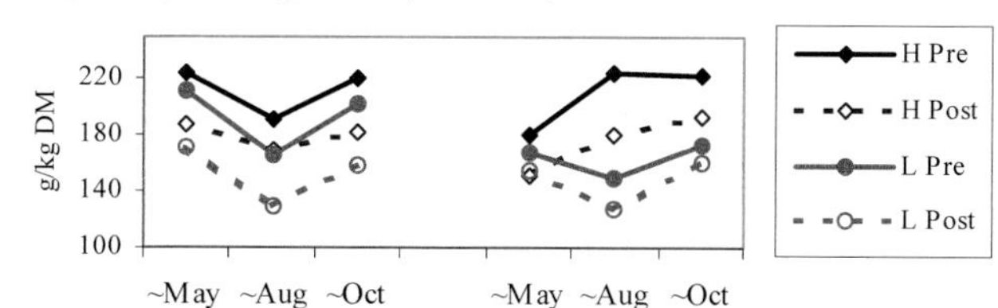

Figure 1 Pre- and post-grazing herbage crude protein for systems H and L in years 1 and 3

Production and plant density of Sulla grazed by sheep at three growth stages

H. Krishna and P.D. Kemp
Institute of Natural Resources, Massey University, Palmerston North, New Zealand, Email: p.kemp@massey.ac.nz

Keywords: *Hedysarum coronarium*, persistence, grazing frequency, herbage mass

Introduction Sulla is one of the few temperate forage legumes that contain enough condensed tannins to improve the efficiency with which livestock use protein (Marshall *et al.*, 1979). However, it usually is productive only for approximately 14 months in New Zealand, and little is known of its response to grazing. This paper reports on the production and persistence of Sulla cv. Necton, when using growth stage as the criterion for time of grazing by sheep in a maritime, temperate environment.

Materials and methods Sulla cv. Necton was sown (15 kg/ha of inoculated seed) in October in medium fertility, Typic Fragiaqualf soil (pH 6, Olsen P 12 µg/g) near Palmerston North, New Zealand. A randomised complete block design with 3 treatments (LV (late vegetative), MSE (mid-stem elongation), and EF (early flowering) growth stages) and 4 blocks was used. Individual treatment plots were 29 X 6.6 m. Each time it reached the set growth stages, Sulla was grazed with mature Romney ewes for 365 days after sowing (DAS). The first grazing of LV, MSE, and EF was 83, 90, and 111 DAS, respectively. Grazing intensity was set at approximately 70% of herbage removed, including most leaves. Herbage mass was measured pre- and post-grazing using ground level cutting in three 0.3 m^2 quadrats/plot. Leaf and stem were dissected in the pre-grazing samples before drying.

Results Sulla was highly productive in all treatments and its EF stage had the highest 365-day herbage mass accumulated for three grazings (Table 1). The mean post-grazing herbage masses for the EF, MSE, and EF treatments were 1,616, 1,465, and 1,972 kg DM/ha, respectively. The residual herbage consisted almost entirely of stem. The ratio of leaf mass : stem mass was lower at EF than at LV or MSE stages (Table 1). In February, the grazed herbage in the EF treatment had 2.3% N and a DM digestibility of 72%. Plant density was greater in the EF treatments after 365 days (Table 1), but all treatments failed to persist >14 months. The grazing of LV and MSE treatments in winter was the main cause of their lower plant density than the EF treatments (Table 2).

Table 1 Net herbage mass accumulation, leaf: stem ratio and plant density of Sulla cv. Necton over 365 days from sowing under infrequent, hard grazing with sheep

Growth stage	No. of grazings	Herbage mass kg DM/ha	Leaf: stem ratio	Plants/m^2 0 DAS	Plants/m^2 365 DAS
LV	4	21,780	2.0	67	15
MSE	4	22,020	2.1	62	16
EF	3	24,700	1.3	45	32
LSD 5%		1990	0.2	NS	6

Table 2 Plant density of Sulla cv. Necton in spring (September) after being grazed or not grazed in winter (June/July) at the late vegetative (LV) and the mid-stem elongation (MSE) growth stages, by sheep

Treatment	Plant/m^2	
	LV	MSE
Grazed in winter	14	16
Ungrazed in winter	34	39
LSD 5%	4	4

Conclusions The productivity of Sulla cv. Necton was confirmed with >20 t DM/ha in all treatments when grazed hard (3-4 times) during the 365 day period post sowing. Winter grazing damaged plant density and it was difficult to use winter growth. Grazing at early flowering avoided winter grazing and increased plant survival, but decreased feed quality through increased stem. Necton Sulla is highly productive under infrequent grazing but its intolerance of winter grazing and short lifespan limits its usefulness.

References

Marshall, D.R., P. Broue, J. Munday (1979). Tannins in pasture legumes. *Australian Journal of Experimental Agriculture and Animal Husbandry*, 19, 192-197.

Management of pasture quality for sheep on New Zealand hill country

D.I. Gray, J.I. Reid, P.D. Kemp, I.M. Brookes, D. Horne, P.R. Kenyon, C. Matthew, S.T. Morris and I. Valentine
College of Sciences, Massey University, Private Bag 11 222, Palmerston North, New Zealand Email: D.I.Gray@massey.ac.nz

Keywords: decision-making, farmer knowledge, feed budgeting, planning, tactical management

Introduction The control of pasture quality over spring is central to the achievement of high levels of sheep performance on hill country. Despite this, with the exception of the work of Lambert *et al.* (2000), little is known about how farmers actually manage pasture quality. The purpose of this research was to describe how a high performing hill country farmer manages pasture quality on their sheep area over spring and from this develop a framework that will assist other farmers improve their pasture management.

Method and materials The case study farmer (647 ha, 7,770 s.u.) was selected because of his high levels of performance for the district and expertise in tactical feed management. Data collection was primarily through monthly semi-structured interviews supported by field observations. Interview data were transcribed verbatim and analysed using qualitative techniques to develop a model of the farmer's decision-making processes.

Results and discussion The control of sheep pasture quality requires farmers to make important strategic and tactical decisions (Figure 1). Strategic decisions aim to match feed supply with pasture growth over the spring and maintains grazing pressure so that average pasture cover (APC) levels do not exceed 1200 kg DM/ha. Key decisions in this area include lambing date, stocking rate, sheep performance levels, pasture cover at set-stocking, stock purchase and sale dates, shearing policy and weaning date. Equally important are the tactical decisions to minimise within- and between-block variation in pasture cover levels (≈ 1200 kg DM/ha) during mid- to late-spring. Key tactical decision areas include: (1) ensuring the correct distribution of pasture cover at set-stocking, (2) setting stocking rate and pasture cover levels at set-stocking for the different sheep mobs that best match feed demand to pasture growth, (3) integrating cattle to help control the steeper contour sheep paddocks and (4) using fortnightly monitoring and micro-budgeting to match feed demand with feed supply.

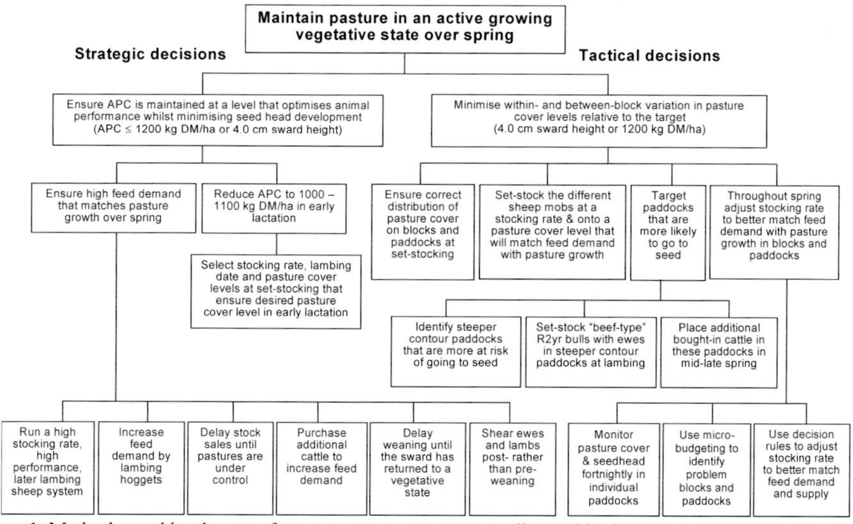

Figure 1 Methods used by the case farmer to manage pasture quality on his sheep area

Conclusions This study highlights that the control of pasture quality on hill country is complex, requiring farmers to make a range of important strategic and tactical decisions. The model presented in this paper provides a framework that other farmers can use to improve their management of pasture quality on hill country.

References

Lambert, M.G., M.S. Paine, G.W. Sheath, R.W. Webby, A.J. Litherland T.J. Fraser, & D.R. Stevens (2000). How do sheep and beef farmers manage pasture quality. In: *Proceedings of the New Zealand Grassland Association*, 62, 117 - 121.

Perennial ryegrass variety differences in nutritive value characteristics

T.J. Gilliland[1], R.E. Agnew[2], A.M. Fearon[3] and F.E.A. Wilson[1]

[1]*Department of Agriculture and Rural Development for Northern Ireland, [1]Plant Testing Station, Crossnacreevy Belfast BT6 9SH, Email: trevor.gilliland@dardni.gov.uk, [2]ARINI, Hillsborough, BT26 6DR, [3]Agriculture and Food Science Centre, Newforge Lane, Belfast BT9 5PX, UK.*

Keywords: perennial ryegrass, varieties, herbage quality

Introduction Animal grazing performance at grass is predominately determined by herbage intake rates, with high yielding dairy cows requiring up to 20 kg/d DM within a limited grazing time (Gibb, 1998). Grass nutritional factors such as seasonal patterns in digestibility and water-soluble carbohydrate levels have been linked to animal productivity (Davies *et al.*, 1991), while sward surface height, herbage mass, bulk density and green leaf mass have been shown to promote high grazing intake (Barrett *et al.*, 2001). Furthermore, fatty acid profiles have been shown to improve the unsaturated fatty acid composition of milk, with potential human health benefits (Parodi, 1997). Recent CAP funding changes are expected to intensify the drive to optimise margin over costs. Given that grazed grass is the cheapest ruminant feed, it is expected that nutritive value characteristics of grass varieties will become increasingly important relative to total productivity, both as a breeding objective and as an evaluation criteria by variety testers and by farmers. This study examined the genetic diversity in such parameters among a wide range of perennial ryegrass (*Lolium perenne* L.) varieties, as an indicator of the heterogeneity among current varieties and the prospects for improvement by selective breeding.

Materials and methods A range of UK recommended perennial ryegrass diploid and tetraploid varieties of widely differing maturities were managed over four growing seasons, under a nine-cut simulated grazing system. In addition to yield parameters, at each cut, water-soluble carbohydrate (WSC) concentration was assessed on herbage dried within one hour of cutting (60ºC). Further 350g dry samples were subjected to modified acid detergent fibre (MADF) analysis and 350g fresh herbage samples were stored at -20ºC and then tested for lipid composition, including the linoleic acid (C18:2, *cis*-9, *cis*-12) and α-linolenic acid (C18:3, *cis*-9, *cis*-12, *cis*-15) fractions of total fatty acid content. Sward surface height (SSH), extended tiller height (ETH), bulk density ($BD_{>6}$) and herbage mass ($HM_{>6}$) characteristics of sward geometry were measured on a subset of varieties.

Results In overview, this study showed that existing registered perennial ryegrass varieties differ significantly in a range of nutritive value and sward structural parameters and would, therefore, differ in the output performances they could support in a grazing herd. Water-soluble carbohydrate concentration differed significantly ($P<0.001$) between varieties with tetraploids higher than all diploid varieties, except those selectively bred for this trait. Significant differences were also recorded between varieties of different maturity type and stage of physiological development ($P<0.001$). Similarly, differences in herbage digestibility were observed between varieties ($P<0.001$) and maturities ($P<0.05$), but only between ploidies when the very high digestibility, high WSC diploids were excluded. Significant varietal differences were also recorded in the proportion of linoleic acid ($P<0.05$) and of α-linolenic acid ($P<0.05$). These differences were not associated with either ploidy or maturity classes and were generally low in varieties with highest digestibility/WSC contents. Comparison of canopy structure characteristics among the intermediate varieties revealed significant differences ($P<0.001$), overlaid by temporal patterns of variation associated with maturity and season.

Conclusion The presented studies showed that overall, differences in animal value parameters were poorly associated with variety yield potential, but were strongly influenced by the physiological stage of development of the grass when harvested. It was concluded that the observed genetic diversity, indicated a good potential for achieving selective breeding improvements from within existing genepools. However, achieving such changes without a linked yield penalty may be difficult in certain cases and precise sward management to maintain grass in its optimum physiological condition, may be vital in fully exploiting 'animal value' factors on-farm.

References

Gibb, M.J. (1998). Animal grazing/intake terminology and definitions. In: M.G. Keane & E.G. O'Riordan (eds.) Pasture Ecology and Animal Intake. Proceedings of a Workshop held in Dublin, September 1996. Occasional Publication No. 3, 21-37. Dublin: Teagasc.

Barrett, P.D., A.S. Laidlaw, C.S. Mayne & H. Christie (2001). Pattern of herbage intake rate and bite dimensions of rotationally grazed dairy cows as sward height declines. *Grass and Forage Science,* 56, 362-373.

Davies, D.A., M. Fothergill & D. Jones (1991). Assessment of contrasting perennial ryegrasses with and without white clover, under continuous sheep stocking in the uplands. 3.Herbage production, quality and intake. *Grass and Forage Science,* 46, 39-49.

Parodi, P.W. (1997). Cow's milk fat components as potential anticarcinogenic agents. *Journal of Nutrition*, 127, 1055-1060.

A survey of European regional adaptation in Italian ryegrass varieties

T.J. Gilliland[1] and A.J.P. van Wijk[2]
[1]*Department of Agriculture for Northern Ireland, Plant Testing Station, Crossnacreevy, Belfast BT6 9SH, UK, Email: trevor.gilliland@dardni.gov.uk* [2]*Centre for Genetic Resources the Netherlands, P.P. Box 16, 6700 AA Wageningen, The Netherlands*

Keywords: plant testing, registration, agro-climatic zones, environmental response

Introduction Ryegrass is widely adapted to cool temperate eco-zones and breeders often submit individual varieties for testing in a number of EU countries. National testing programmes often combine data from several trial sites that may differ climatically, but not from sites in other member states, despite the possibility of high ecological similarity. Given increasing interest in 'animal value' characters (soluble sugars, lipids, sward geometry), additional testing for these would be valuable but is prohibited by capped or declining funding. Data sharing between EU national authorities could be advantageous but is inhibited by the lack of statistically valid data on the sensitivity of each performance parameter to agro-climatic conditions across the EU. This paper, reports the preliminary stages of the 'EuroVCU' (herbage) desktop study of ryegrass variety performances across an extensive range of EU national test centres. Analysis of the resulting data sets quantifies the genotype by environment responses of current varieties and could provide a validated protocol for future data sharing.

Materials and methods Due to its wide use and similarity in management, Italian ryegrass (*Lolium multiflorum* LAM.) was chosen as the model grass species for this pilot study. In 2003, the scientists responsible for conducting official grass variety tests in each of the current member states of the EU were contacted. They were invited to join the EuroVCU (herbage) consortium and asked to compile information on the testing history, test decisions and availability of data for each of the 169 varieties listed on the 22nd EU Common Catalogue. This constituted Phase I of the study and the data were summarised and processes with the aim of deriving a core set of varieties that would provide 10 pair-wise variety-links between each of the contributing member states. Phase II of the study involved gathering performance, management and metrological data for the core varieties from each site. This constituted the full data set from which analyses of the consistency/inconsistency of individual varieties across management and climatic variation could be measured, of stability/instability of individual performance characteristics across varieties and protocols and climatic variants, and finally the similarity/dissimilarity of test centres for common variety performance results on a character-by-character basis.

Results Twenty EU regions responded to the Phase I call (Austria, Belgium, Croatia, England & Wales, Estonia, Finland, France, Germany, Hungry, Italy, Lativa, Netherlands, N. Ireland, Norway, Republic Ireland, Scotland, Slovenia, Slovakia, Sweden, Switzerland), which generated a total of 679 variety x country reports. The pattern of variety testing appeared disjointed and impossible to interpret prior to Phase II data being gathered. For example, of the 169 EU registered varieties, only three were tested in half or more of the countries surveyed (Ajax, Danergo and Mondora). Conversely, 22 varieties were tested in only a single country and a further 18 were not tested by any contributors (AM1, Bella Bionda, Califa, Classic, EF486 Dasas, Kitil, Locobelo, Marvel, Menichetti, Multisolc AX9, Primadonna, Ralino, Rouky, Sultano, Tauro, Teanna, 110DE, 111DE). The average number of varieties tested per country was 35, but there were big differences, with some having tested more than 70 varieties, while others having only tested a few. This probably indicates the different importance of Italian ryegrass in each region, which may be linked to climatic differences. In total, Phase I generated over 2,600 data entries on variety test history, current status and results availability, from which 44 varieties were selected to provide the pair-wise comparisons in Phase II (Abercomo, Adin, Ajax, Atalja, Atos, Baresi, Barextra, Barmultra, Bartolini, Bartoluchi, Bofur, Danergo, Exalta, Fabio, Fenil, Fredrik, Gemini, Gisel, Gordo, Lemtal, Ligrande, Lolita, Lubina, Malmi, Meroa, Minaret, Mondora, Montblanc, Multimo, Rio, Sabalan, Sikem, Sultan, Taurus, Tetraflorum, Tonic, Total, Tribune, Tur, Turgo Pajbjerg, Urbana, Zarastro, Zenith, Zorro). Phase II was sent to all contributors plus the authorities in the remaining countries (Bulgaria, Cyprus, Czech Republic, Denmark, Greece, Luxembourg, Poland, Romania and Spain). It is anticipated that the participant numbers in Phase II may increase by the October 2004 deadline, which has been set to allow for final report completion by Spring 2005.

Conclusions The concept of this study is to use common varieties as replicates of each national site. This strategy makes it possible to compare how each performance characteristic varies between different locations and to identify common agri-climatic zones on a character-by-character basis. The EuroVCU (herbage) study, therefore, provides the necessary knowledge to validate future data sharing between variety testing authorities.

Acknowledgements The members of the EuroVCU consortium are thanked for their essential contribution to this study.

Effect of perennial ryegrass cultivars on the fatty acid composition in milk of stall-fed cows

A. Elgersma[1,4], H.J. Smit[1], G. Ellen[2] and S. Tamminga[3]
[1]*Plant Sciences Group, Crop and Weed Ecology, Wageningen University, Haarweg 333, 6709 RZ Wageningen, The Netherlands, Email: anjo.elgersma@wur.nl,* [2]*NIZO food research, PO Box 20, 6710 BA Ede, The Netherlands,* [3]*Animal Sciences Group, Animal Nutrition, Wageningen University, Marijkeweg 40, 6709 PG Wageningen, The Netherlands,* [4]*University of Ghent, Belgium*

Keywords: fatty acids, conjugated linoleic acid, water-soluble carbohydrates, volatile fatty acids, herbage

Introduction Herbage provides bulk feed for ruminants and plant lipids, especially C18:3, are a major source of benefical fatty acids (FA) in milk. There are very few direct comparisons allowing a precise evaluation of the effects of the basal forage diet on milk FA composition. Grass quality differences can affect rumen metabolism and there could be opportunities to change the composition of ruminant products through choice of grass cultivar. To test this hypothesis, six cultivars were fed to dairy cows in a stall-feeding trial with fresh grass to evaluate the effect of grass cultivar on rumen VFA and milk FA composition during the growing season.

Materials and methods Twelve Holstein Friesian dairy cows were used in a stall-feeding trial with fresh grass to evaluate the effect of grass cultivar on milk fatty acid (FA) composition during the growing season. Six diploid perennial ryegrass (*Lolium perenne* L.) cultivars were used: Abergold, Respect and Agri (intermediate heading) and Herbie, Barezane and Barnhem (late heading). They were cut daily during three 14-d periods between July and August at the same target yield. The experiments consisted of two 3x3 Latin square trials, in each of which three cultivars were fed to two groups of three cows. Half of the cows had a rumen fistula. Dry matter intake (DMI), milk production (MP) and milk composition (MC) were recorded daily in individual cows. Rumen liquid samples were taken from the fistulated cows and analysed for volatile fatty acid (VFA) composition. Levels of individual FA in grass and milk were determined by gas chromatography.

Results The dry matter (DM) yield during the three harvest periods was on average 2433 kg/ha in early July and 2090 kg/ha thereafter. The leaf blade proportion of DM increased from 0.67 to 0.87 to 0.91 during the season. The biggest range among cultivars was found in early June (Table 1); in later harvests leaf blade proportions varied from 0.84 to 0.90 and from 0.90 to 0.93, respectively. The six cultivars were rather variable in their chemical characteristics. Barnhem and Abergold had the highest ($P < 0.001$) WSC and the lowest ($P < 0.01$) NDF concentrations. Barezane and Respect had a lower WSC concentration ($P < 0.001$) than the other cultivars. Barnhem had the lowest and Barezane the highest CP concentration ($P < 0.001$). However, there were no significant differences among cultivars in FA concentration (22.1 g/kg DM) or proportions of FA. Average proportions of the major FA C18:3, C16:0 and C18:2 were 0.74, 0.13 and 0.10, respectively.
Despite the variation in quality parameters among the cultivars, their DMI (16.6 kg DM/d) did not differ, and MP (27.4 kg/d) and MC (34 g/kg protein, 40 g/kg fat and 45 g/kg lactose) were similar (Smit *et al.*, 2005). Rumen VFA concentrations did not differ among cultivars. No variation in milk FA composition was found. The mean proportions of individual FA C16:0, C14:0, C18:0, C18:1 *cis*-9, vaccenic acid and rumenic acid were 263, 121, 108, 178, 33 and 14 g/kg FA, respectively.

Conclusions Despite variation in morphological and chemical characteristics of the herbage, no variation in DMI, VFA concentrations, milk production or milk FA composition was found. The latter may be due to the lack of variation in grass FA concentration and composition in the cultivars studied.

Table 1 Chemical characteristics (g/kg DM) of six perennial ryegrass cultivars, averaged over 3 cuts taken early and late July and late August, and leaf blade proportion of DM in early July

Cv	CP	NDF	WSC	Leaf
Abergold	160	399	192	0.66
Respect	159	429	152	0.59
Agri	157	423	170	0.57
Herbie	156	412	172	0.64
Barezane	166	414	158	0.73
Barnhem	150	400	195	0.79
Mean	158	413	173	0.67
s.e.d.	2	6	5	0.03
Sign.	***	**	***	***

Sign.: ** : $P < 0.01$; *** : $P < 0.001$

References

Elgersma A., G. Ellen, H. van der Horst, B.G. Muuse, H. Boer & S. Tamminga (2003). Influence of cultivar and cutting date on the fatty acid composition of perennial ryegrass. *Grass and Forage Science,* 58, 323-331 and Erratum in ibid. 59, 104.

Smit H.J., B.M. Tas, H.Z. Taweel, J. Dijkstra, A. Elgersma & S. Tamminga (2005). Effects of perennial ryegrass cultivars on traits for improved animal performance. *Proceedings of the Twentieth International Grassland Congress (this publication).*

Survey of tetraploid and diploid perennial pastures in the Waikato for number of spores produced by the fungus *Pithomyces chartarum*

J.P.J. Eerens[1], W.W. Nichol[2], J. Waller[1], J.M. Mellsop[1], M.R. Trolove[1] and M.G. Norriss[2]
[1]*AgResearch Ruakura, Private Bag 3123, Hamilton, New Zealand* [2]*Wrightson Research, PO Box 939, Christchurch, New Zealand. Email: Han.eerens@agresearch.co.nz*

Keywords: diploid, facial eczema, perennial ryegrass, *Pithomyces chartarum*, tetraploid

Introduction Facial eczema (FE) is a disease of livestock, caused by a toxin released into the bloodstream after digestion of spores of *Pithomyces chartarum,* a fungus residing in necrotic plant material in the base of pastures (di Menna & Bailey, 1973). Spore numbers tend to be highest in warm, humid conditions, where high post grazing residuals have lead to a build up of necrotic plant material. Tetraploid perennial ryegrass pastures tend to be more palatable, and with lower post gazing residuals, than equivalent diploid pastures; thus we hypothesised that spore numbers would be lower in tetraploid pastures. A survey of tetraploid and diploid pastures was carried out to investigate the relationship between FE spore numbers, and perennial ryegrass ploidy levels.

Material and methods Fifty pairs of diploid and tetraploid perennial ryegrass based pastures from 37 farms in the greater Waikato area were sampled for FE spores during March, April and May 2004. Eight cultivars were represented in the diploid pastures, while cv. Quartet (4 weeks later flowering than most of the diploid pastures) was the sole ryegrass in 84% of the tetraploid pastures, with 16% based on cv. Banquet. Both paddock and farm information was collected for sowing date, sown cultivars, soil type (silt/ash, sandy loam, clay, peat), soil fertility status (Olsen-P either <20 or $\geq$ 20) and stocking rate. Before each assessment, pasture cover (measured using a capacitance probe) and days since last grazing were recorded, and a grass sample was collected. FE spore numbers were assessed in a sub-sample (20-25 g) while a second (100 g) sub-sample was oven dried for 24 hours at 80°C to determine dry matter (DM) percentage. Data analyses used the Generalised Linear Mixed Model procedure of Genstat 7. Back transformed means are presented in this paper.

Results Average spore numbers in 2004 were below the long-term average, but higher spore counts were observed each month in diploid perennial ryegrass pastures relative to tetraploid pastures (Table 1). The largest difference was in April ($P<0.05$). Similarly pastures sown on clay soils contained consistently more FE spores than pastures on the other three soil types, the difference being significant compared to the other three soil types in March, but only compared to ash/silt soil in April ($P<0.05$). In April, pastures on peat soils contained more ($P<0.05$) spores than those on ash/silt soil. Spore counts were higher when Olsen-P levels were lower (<20) than with higher levels (>20), the difference was significant in March ($P<0.05$). No other factor had a significant effect on spore counts at any time. Farms and paddocks had some (NS) effect on FE spore counts that may have been due to varying weather conditions, and/or differences in farm management strategies.

Table 1 *Pithomyces chartarum* spore numbers over three consecutive months on pastures with different ploidy levels, soil types and soil fertility status (spores/g DM pasture)

Ploidy	March	April	May	Total
Diploid[1]	127,000	105,000	64,000	339,000
Tetraploid[1]	99,000	66,000	38,000	240,000
LSR (5%)[2]	1.62	1.6	1.9	1.5
Soil type				
Ash/Silt	78,000	44,000	45,000	189,000
(Sandy) Loam	107,000	75,000	40,000	272,000
Clay	326,000	121,000	66,000	511,000
Peat	59,000	119,000	51,000	250,000
LSR (5%)	2.6	2.2	3.9	2.1
OlsenP status				
Olsen-P <20	245,000	127,000	101,000	508,000
Olsen-P $\geq$20	51,000	54,000	24,000	160,000
LSR (5%)	3.5	2.8	5.2	2.4

[1] Diploid, 8 cv's; Tetraploids Quartet (84%), Banquet (16%)
[2] LSR (5%) stands for least significant ratio

Conclusions Pastures that mostly consisted of the late flowering cv. Quartet showed lower spore numbers relative to diploid pastures. Further work is required to confirm whether this is a ploidy effect and not flowering date or some other cultivar factor. The survey also indicates that spore numbers are likely to be higher when Olsen-P levels are low (<20) and on clay soils. Higher spore numbers on clay soils could be a consequence of high moisture retention capacity, or differences in pasture growth rate and grazing pressure.

References

Di Menna, M.E. & J. R. Bailey (1973). *Pithomyces chartarum* spore counts in pasture. *New Zealand Journal of Agricultural Research,* 16, 343-351.

Diverse forage mixtures effect on herbage yield, sward composition, and dairy cattle performance

M.A. Sanderson[1], K. Soder[1], N. Brzezinski[2], S. Goslee[1], H. Skinner[1], M. Wachendorf[2], F. Taube[2] and L. Muller[3]
[1] *USDA-ARS, Pasture Systems and Watershed Management Research Unit, Building 3702, Curtin Road, University Park, PA, 16802-3702 USA, Email:mas44@psu.edu,* [2] *University of Kiel, Hermann-Rodewald Str. 9, D 24118 Kiel, Germany,* [3] *Dept. of Dairy and Animal Science, The Pennsylvania State University, University Park, PA 16802 USA*

Keywords: pastures, forage mixtures, biodiversity

Introduction Managing complex mixtures of plants to take advantage of spatial and temporal variability in land and climate may be one ecological approach to increase productivity of pastures. We tested the hypothesis that complex mixtures of forage species would yield more herbage and reduce weed competition compared with a simple grass-legume mixture in grazed pastures.

Materials and methods Four mixtures were established in replicated 1-ha pastures at University Park, Pennsylvania, USA in the autumn of 2001: 1) orchardgrass (*Dactylis glomerata* L.)-white clover (*Trifolium repens* L.); 2) orchardgrass, white clover, chicory (*Cichorium intybus* L.); 3) orchardgrass, tall fescue (*Festuca arundinacea* Schreb.), perennial ryegrass (*Lolium perenne* L.), red clover (*Trifolium pratense* L.), birdsfoot trefoil (*Lotus corniculatus* L.), and chicory; and 4) six species mix plus white clover, alfalfa (*Medicago sativa* L.), and bluegrass (*Poa pratensis* L.). The experimental design was a randomized complete block with two replicates. The pastures were subdivided into smaller paddocks and stocked rotationally with lactating Holstein cows during April to August in 2002 and 2003. Four cows grazed each treatment. Herbage allowance was 25 kg/cow per day of dry matter. Cows were fed a 13% crude protein corn-based concentrate (1 kg/4 kg milk) in two equal feedings after milking. Cows were moved to a fresh paddock after morning and afternoon milking. Herbage intake was estimated by the chromic oxide technique during May, June, July, and August in each year. Lactating cows were not available after August 1 of each year, therefore, pastures were mob grazed with 21 dry cows for one day in mid August (2003) and early September (2002 and 2003) to complete the grazing season. Animal performance was not measured on the dry cows. Pre-grazing and post-grazing herbage mass was measured twice each week during the grazing season with a calibrated rising plate meter. The botanical composition of each pasture was measured during each grazing cycle by hand separating clipped samples before and after grazing. Data were analyzed as a randomized complete block design. Treatments (mixtures) were fixed effects and blocks were random effects.

Results In 2002, herbage yield was lower ($P<0.05$) on the orchardgrass-white clover mixture compared with the more complex forage mixtures (Table 1). In 2003, with much greater rainfall, there were no significant differences in herbage yield among mixtures. In 2002 the yield increase with increased seeded species richness resulted from adding a highly productive species (chicory), an example of the 'sampling effect' mechanism for explaining plant species diversity effects. Weed proportions were similar ($P>0.05$) for the two- and three-species mixtures, whereas the six- and nine-species mixtures had lower ($P<0.05$) weed populations than the simple mixture. Sown species composition of the pastures changed greatly during the experiment, with the complex mixtures simplifying to fewer species.

Table 1 Herbage yield, weed components, and dairy cow productivity. Weed proportion, milk yield, and herbage intake are means of two years

Mixture	Herbage yield 2002	Herbage yield 2003	Weed proportion	Milk yield	Herbage DMI
	kg DM/ha		% of DM	kg/cow per day	
2-species	4800	9000	18	34.1	12.1
3-species	7400	9900	15	35.3	12.1
6-species	7900	11300	11	34.4	12.1
9-species	7400	9000	4	34.3	11.4
SEM	280	766	2.1	1.1	0.49

Conclusions Complex forage mixtures were more productive than a simple grass-legume mixture during drought and also had reduced weed pressure. Individual animal performance was similar among simple and complex mixtures. Increasing plant species diversity on pastures may be a short-term way to increase forage productivity and reduce weed competition. Stability of species composition in complex mixtures, however, may be a problem in the long term.

Potential yield of cocksfoot (*Dactylis glomerata*) monocultures in response to irrigation and nitrogen

A. Mills[1], D.J. Moot[1], R.L. Lucas[1], P.D. Jamieson[2] and B.A. McKenzie[1]
[1]*Agriculture and Life Sciences Division, PO Box 84, Lincoln University, Canterbury, New Zealand, Email: millsa@lincoln.ac.nz,* [2]*Crop & Food Research, Private Bag 4704, Christchurch, New Zealand*

Keywords: *Dactylis glomerata*, orchardgrass, irrigation, nitrogen, yield

Introduction Cocksfoot is a widely sown grass in temperate pastures. However, while potential yield of cocksfoot can exceed 28 t DM/ha per year, it is often restricted by water, temperature and nitrogen (N). Of these, Peri *et al.* (2002) showed that N was severely limiting in all seasons. The aim of this study was to confirm the potential yield of cocksfoot and quantify the extent of yield reductions due to environmental constraints.

Materials and methods The experiment was a split plot design with three replicates established in October 2003 onto an 8-year old 'Grasslands Wana' cocksfoot monoculture. Irrigation was the mainplot (irrigated (I) or dryland (D)) and N (0 (-N) or 800 kg/ha/y (+N)) the subplot. Irrigation maintained the soil moisture deficit above 50 mm in the top 0.5 m. Nitrogen was applied in eight split applications of 100 kg N/ha at the beginning of ~28-30 d regrowth periods during active growth. Dry matter production was determined from a 0.2 m^2 quadrat cut at the end of each regrowth period. The site was mown to 30 mm and herbage removed from the site.

Results Accumulated DM production was 22.6 t DM/ha per year for I+N (Figure 1a). In comparison, yield was 10.5, 15.1 and 7.5 t DM/ha/y for I+N, D+N and D-N treatments, respectively. The seasonal effect of different temperatures was accounted for by estimating yield against thermal time (Tt) (Figure 1b). Yield increased by 6.5 kg DM/°Cd above a base temperature (T_b) of 2.5°C under non-limiting N and moisture conditions. In the absence of N, yield consistently increased by 2.9 kg DM/°Cd whereas the D-N treatment produced 2.1 kg DM/°Cd. A broken stick model was used to determine slope for the D+N treatment. The first period was water stressed and D+N pastures grew at a similar rate (1.5 kg DM/°Cd) as the D-N treatments. In the second period water stress had been alleviated by rainfall and D+N treatments produced a similar rate (7.1 kg DM/°Cd) to the I+N treatment. The point of inflexion was 2015 Tt units (17/3/04) indicating an eight week lag phase after rainfall began. The broken stick model increased the R^2 from 88.6 to 99.3%. The LSD of the slope was 0.83.

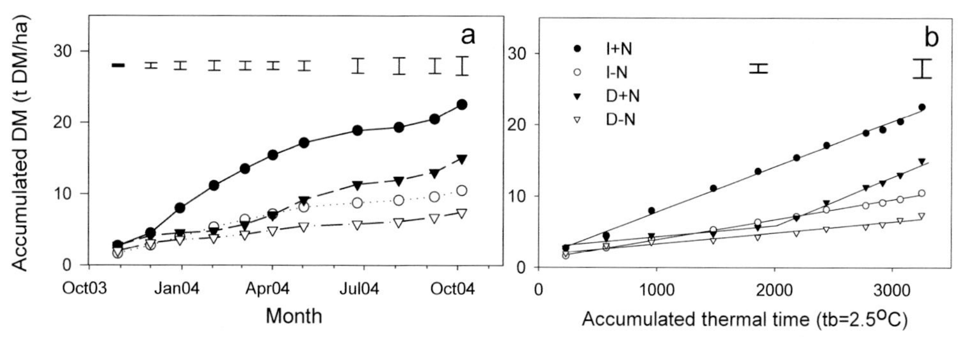

Figure 1 Accumulated DM production of 'Grasslands Wana' cocksfoot, at Lincoln University, Canterbury, New Zealand, against time (a) and thermal time (b) Error bars are LSD ($p \leq 0.05$) for the I*N interactions

Conclusions Cocksfoot pastures grown in Canterbury have the potential to produce 23-28 t DM/ha per year. However, without irrigation or N, yield was 15.1 t DM/ha per year lower. The addition of N through legume fixation or strategic fertiliser use could double the current yield, particularly in periods of feed deficit in the autumn and spring. However, during drought, water stress may limit N uptake and yield will be proportional to total accumulated shoot N. Linear relationships between DM production and Tt provide a repeatable basis for extrapolating results for comparison of cocksfoot monoculture growth in other environments and interpreting seasonal temperature effects.

Acknowledgements A. Mills acknowledges AGMARDT, Lincoln University and the Sinclair Cummings Trust for financial assistance during her PhD study.

References

Peri, P. L., D. J. Moot, & R. J. Lucas (2002). Urine patches indicate yield potential of cocksfoot. *Proceedings of the New Zealand Grassland Association,* 64, 73-80.

Intake and milk production of lactating dairy cows grazing diverse forage mixtures over two grazing seasons

K.J. Soder[1], M.A. Sanderson[1], J.L. Stack[2], L.D. Muller[2]
[1]*USDA-ARS, Pasture Systems & Watershed Management Research Unit, Building 3702, Curtin Road, University Park, PA 16802-3702 USA, Email: ksoder@psu.edu,* [2]*Department of Dairy and Animal Science, The Pennsylvania State University, University Park, PA 16802, USA*

Keywords: pastures, forage mixtures, diversity, intake, milk production

Introduction Voluntary intake and stocking rate are key determinants of animal performance on pasture. Greater plant diversity in grassland plant communities has been linked to increased primary production, greater stability in response to disturbance, and reduced weed pressure. Thus, increasing plant diversity may be one approach to improving animal productivity. An experiment was conducted to determine the effects of forage diversity on intake and milk production of lactating dairy cows over two grazing seasons.

Materials and methods Four diverse forage mixtures were established in replicated 1-ha pastures at University Park, Pennsylvania, USA in the autumn of 2001: 1) orchardgrass (*Dactylis glomerata* L.)-white clover (*Trifolium repens* L.); 2) orchardgrass, white clover, chicory (*Cichorium intybus* L.); 3) orchardgrass, tall fescue (*Festuca arundinacea* Schreb.), perennial ryegrass (*Lolium perenne* L.), red clover (*Trifolium pratense* L.), birdsfoot trefoil (*Lotus corniculatus* L.), and chicory; and 4) six species plus white clover, alfalfa (*Medicago sativa* L.), and bluegrass (*Poa pratensis* L.). The experimental design was a randomized complete block with two pasture replicates. Pastures were subdivided and rotationally grazed with lactating Holstein cows during April to August in 2002 and 2003. Four 3-wk periods were conducted during each of two grazing seasons, so that by the end of each grazing season, all cows had been on all mixtures. Herbage allowance was 25 kg DM/cow per day. Cows were fed a 13% crude protein corn-based concentrate (1 kg/4 kg milk daily) in two equal feedings after milking. Cows were moved to a fresh paddock after each milking (2x/d). Herbage intake was estimated by the chromic oxide technique during each period. Daily milk yield and weekly milk composition samples were collected.

Results Pasture quality was within the range summarized by other regional studies (Table 1). A significant year effect was noted for most nutrients and yields with greater differences noted during 2002 (a drought year). Forage yield was affected by pasture mixture in 2002. Pasture DM intake was not affected by pasture mixture; however, there was a significant year effect (Table 2). Milk yield, milk fat, and milk protein were not affected by pasture mixture or year. Milk urea nitrogen was significantly affected by year.

Table 1 Nutrient composition of pasture diversity treatments

	Crude Protein		NDF		IVDMD[1]		Forage Yield	
	2002	2003	2002	2003	2002	2003	2002	2003
Mixture	% of DM		% of DM		% of DM		kg DM ha^{-1}	
2-species	21.8^{a}	23.2^{a}	36.6^{a}	40.8^{a}	66.6^{a}	66.7^{ac}	4800^{a}	9000
3-species	20.0^{b}	24.6^{ab}	31.6^{b}	31.8^{b}	70.4^{bc}	69.2^{a}	7400^{b}	9900
6-species	21.2^{ab}	24.1^{a}	29.3^{b}	35.4^{b}	67.1^{ac}	60.2^{b}	7900^{b}	11300
9-species	22.5^{a}	25.9^{b}	24.5^{c}	28.2^{b}	70.9^{b}	64.4^{c}	7400^{b}	9000
SEM	0.5	0.5	1.5	1.5	1.2	1.2		

[1]IVDMD=in vitro DM digestibility, ^{a,b,c}means in the same column with different superscripts differ ($P < 0.05$)

Table 2 Pasture intake, milk production, and milk composition

	Pasture Intake		Milk Yield		Milk Fat		Milk Protein		Milk Urea N	
	2002	2003	2002	2003	2002	2003	2002	2003	2002	2003
Mixture	kg DM/d		kg/d		kg/d		kg/d		mg/dl	
2-species	13.7	11.1	33.7	34.5	1.20	1.11	0.90	0.85	12.4	14.5
3-species	13.7	10.5	34.6	36.2	1.19	1.14	0.93	0.92	11.1	15.2
6-species	13.6	10.5	34.2	34.6	1.18	1.16	0.92	0.87	12.3	13.4
9-species	12.9	10.2	33.4	35.2	1.19	1.10	0.89	0.88	12.5	15.0
SEM	0.5	0.5	1.1	1.1	0.07	0.07	0.04	0.04	0.30	0.30

Conclusions Pasture mixture did not affect intake or productivity of dairy cows. Managing for a moderately complex mixture of forages on pasture may result in greater carrying capacity of the pastures due to increased forage productivity, particularly in drought years, while maintaining high levels of animal productivity.

A comparison of perennial ryegrass cultivars differing in heading date and grass ploidy for grazing dairy cows at two different stocking rates

M. O' Donovan[1], G. Hurley[1], L. Delaby[2] and G. Stakelum[1]
[1]*Teagasc, Moorepark Production Research Centre, Fermoy, Co. Cork, Ireland, Email : ghurley@moorepark.teagasc.ie , [2]INRA, UMR Production du Lait 35590 St Gilles, Rennes, France.*

Keywords: grass varieties, dairy cows, stocking rate

Introduction Animal productivity is the ultimate performance indicator of any new grass cultivar. Direct assessment is complex and expensive. Recent research has identified a number of important sward factors that influence intake and production. The most important factor appears to be green leaf mass. The objective of this two year study was to investigate the effects on milk yield, composition and grass intake of grass cultivars with contrasting heading dates (HD) and grass ploidies (PL) when grazed at different stocking rates.

Materials and methods The design of this study was a 2 × 2 × 2 factorial arrangement of treatments. Four cultivars consisting of two HD (intermediate - I and late - L) and two grass ploidies (diploid – D and tetraploid - T) were compared at two grazing stocking rates (HSR and LSR). The grass cultivar treatments were: (i) LT, (ii) LD, (iii) IT, (iv) ID. Prior to experiment, eighty Holstein-Friesian cows were blocked on lactation number (3.6 ± 0.7), days in milk (43.9, ± 8.6 days), previous 3-week milk yield (34.2, ± 3.3) and liveweight (546, ± 38.9). They were randomly assigned to one of the eight grazing treatments and managed as individual herds of ten. The study began in early April and finished in late September, lasting a total of 26 weeks. The high (HSR) and low (LSR) stocking rates were 4.8 and 4.3 cows/ha up to early June and 3.9 and 3.5 cows/ha thereafter. Mean concentrate input was 139kg /cow over the two years. Individual animal dry matter intake (DMI) was measured on six occasions during the experiment using n-alkanes. Milk yield was recorded daily. The concentrations of fat, protein and lactose were determined in one successive morning and evening milk sample per week. Daily milk yield, milk constituent yield, milk composition and body weight (BW) were analysed using covariate analysis in SAS.

Results Cows grazing the LHD cultivars had higher milk yield (+0.8kg/cow/day, $P<0.01$), lactose concentration (+0.61g/kg, $P<0.001$) and SCM (+0.5kg, $P<0.09$) compared to cows grazing IHD cultivars. Cows grazing at the LSR had significantly higher ($P<0.001$) milk yield (+1.9kg/cow per day), SCM yield (+1.44kg/cow per day), fat yield (+61.9g/cow per day), protein yield (+66.5g/cow per day), lactose yield (+59.7g/cow per day) and lactose concentration (+0.61g/kg) than cows grazing at the HSR. Heading date ($P<0.06$) and grass ploidy ($P<0.09$) approached significance for protein yield; cows grazing the LHD (+22.2g/kg DM) and T (+19.9g/kg DM) cultivars had higher milk protein yield than their comparative herds. Cows grazing at the LSR had higher BW (+13.6kg, $P<0.001$) than at the HSR. Cows grazing LHD cultivars had higher GDMI (+1.0kg, $P< 0.001$) than cows grazing IHD cultivars. Cows at the LSR had higher GDMI (+1.2kg, $P<0.001$) than the cows grazing at the HSR. The interaction between HD × SR for GDMI approached significance. The interaction between HD × SR was due to the difference in GDMI between cultivar groupings at their respective stocking rates. At the HSR cows on the LHD cultivars had a GDMI of 1.6 kg higher than those on the IHD cultivars, while at the LSR cows grazing the LHD cultivars had a GDMI of 0.72kg higher than the cows grazing the IHD cultivars.

Table 1 Effect of heading date, grass ploidy and stocking rate on the milk production performance, body weight and body condition score of spring calving dairy cows

Heading date (HD)	Intermediate				Late							
Grass ploidy (PL)	Diploid		Tetraploid		Diploid		Tetraploid					
Stocking rate (SR)	HSR	LSR	HSR	LSR	HSR	LSR	HSR	LSR	Rse	HD	PL	SR
Milk yield (kg/d)	23.2	24.8	24.0	25.2	24.1	25.7	23.7	26.8	1.94	**	NS	***
SCM (kg/d)	17.4	18.7	18.1	19.0	17.9	19.2	17.9	20.1	1.80	+	NS	***
Fat conc (g/kg)	39.1	37.6	39.4	39.2	38.9	38.5	39.1	37.4	3.97	NS	NS	NS
Protein conc (g/kg)	33.6	33.6	33.7	34.9	33.5	34.1	33.7	33.7	1.56	NS	NS	NS
Lactose conc (g/kg)	45.7	45.9	45.6	46.1	45.7	46.3	45.3	46.4	1.03	NS	NS	***
Body weight (kg)	568	580	560	575	566	581	571	584	20.7	NS	NS	***
GDMI (kg cow/day)	16.5	18.0	16.4	18.0	17.5	17.9	18.1	19.1	1.53	***	NS	***

Conclusions Adopting a strategy of increasing SR to graze IHD cultivars did not increase milk production performance. This suggests that the milk production performance achieved by LHD cultivars was equally effective at high and low stocking rates.

In situ rumen degradability of perennial ryegrass cultivars differing in ploidy and heading date in Ireland

V. Olsson[1,2], J.J. Murphy[1], F.P. O'Mara[3], M. O'Donovan[1] and F.J. Mulligan[2]
[1]*Teagasc, Dairy Production Research Centre, Moorepark, Fermoy, Co Cork, Ireland, Email: volsson@moorepark.teagasc.ie,* [2]*Department of Animal Husbandry and Production, Faculty of Veterinary Medicine and* [3]*Department of Animal Science and Production, Faculty of Agri-Food and the Environment, University College Dublin, D4, Ireland*

Keywords: perennial ryegrass, cattle, rumen degradability, heading date, ploidy

Introduction Grazed grass is the predominant feed in Irish dairy and beef cattle production systems. Knowledge of the degradability characteristics of protein in Irish forages is necessary for the establishment of protein values (PDIE and PDIN values) for these. This knowledge would also facilitate more accurate formulation of supplements for grass diets with the potential to reduce nitrogen (N) excretion.

Materials and methods Perennial ryegrass (*Lolium perenne*) swards of 4 cultivars were established as monocultures at Moorepark Research Centre in 1998 and 2000. The cultivars differed in heading date and ploidy: Spelga (S) – intermediate (I) diploid (D), Napoleon (Nn) – I tetraploid (T), Portstewart (P) – late (L) D and Millenium (M) – L T. Manual grass cuts, representing herbage selected by grazing dairy cows, were taken daily 21-26 May and 9-14 July in 2001 and 13-18 May, 1-6 July, 12-17 August and 9-14 September in 2002. Grass samples were frozen, chopped, freeze-dried, milled through a 1 mm screen and pooled by sampling time. Two g of each sample (n=24) were incubated in nylon bags (5×10 cm; 50µm pore size) in each of 3 rumen cannulated Holstein Friesian steers, offered a diet of 75% grass silage and 25% concentrate. All samples were incubated together and 1 bag per sample was removed at 0, 2, 4, 8, 12, 24 and 48 h. After removal the samples were immersed in cold water, frozen and treated with 5 ml of buffer (4g of NH_3HCO_3 and 35g of $NaHCO_3$ per L of distilled water) per nylon bag, in a Seward Lab Blender. The samples were washed in a washing machine (3×10 min rinse cycle) and oven dried at 40°C for 48 h. Analysis of N was carried out using a Leco FP-528 N analyser. The rumen degradability parameters a (rapidly degradable), b (potentially degradable) and c (rate of degradation of b) of the degradability curves and the effective degradability (ED) of DM (ED-DM) and N (ED-N) for each grass were determined assuming a rumen outflow rate of 6%/h (Ørskov and McDonald 1979). Data were analysed using analysis of variance and the PROC GLM statement of SAS. The model for the analysis included year, time, heading date, ploidy and the interaction time × heading date × ploidy as sources of variation.

Results Average *in vitro* organic matter digestibility was 882 (S), 880 (Nn), 892 (P) and 889 (M) g/kg DM for the cultivars. Average crude protein content was 204 (S), 207 (Nn), 195 (P) and 204 (M) g/kg DM and varied similarly for the cultivars throughout the season. The a value was higher (p<0.001) and the b value lower (p<0.001) for ED-N in I compared to L cultivars. The ED-DM was higher (p<0.05) for L than I cultivars. There was a significant effect of time (p<0.01) on ED-DM for all cultivars where ED-DM was lower in August than in May or July. Neither heading date nor ploidy had a significant effect on ED-N.

Table 1 Effect of heading and ploidy on effective degradability of DM

	Heading date		Ploidy		
	I	L	D	T	SEM
a	53.5	54.2	53.4	54.2	0.34
b	42.8	42.8	43.4	42.2	0.49
c	0.09	0.09	0.09	0.09	0.01
ED	77.7^a	79.2^b	78.1	78.8	0.51

Table 2 Effect of heading and ploidy on effective degradability of N

	Heading date		Ploidy		
	I	L	D	T	SEM
a	58.4^a	54.9^b	56.5	56.8	0.48
b	39.7^a	43.5^b	41.7	41.5	0.58
c	0.12	0.14	0.13	0.13	0.01
ED	84.4	84.4	84.0	84.9	0.47

$^{a, b}$ Means within rows, not sharing a superscript, differ significantly (p<0.05)

Conclusions Ploidy had no significant effect on ED–DM or ED-N in grass. Late heading cultivars had a significantly higher ED-DM but heading date did not influence ED-N. Given the similar ED-N for the four cultivars studied here it is unlikely that they would result in substantial differences in N excretion.

References

Ørskov and McDonald (1979). The estimation of protein degradability in the rumen from incubation measurements weighted according to rate of passage *Journal of Agricultural Science Cambridge,* 92, 499-503

Caucasian clover is more productive than white clover in temperate pastures

A.D. Black[1], R.J. Lucas[2] and D.J. Moot[2]
[1]*Teagasc, Grange Research Centre, Dunsany, Co. Meath, Ireland, Email: ablack@grange.teagasc.ie,*
[2]*Agriculture and Life Sciences Division, Lincoln University, Canterbury, New Zealand.*

Keywords: Caucasian clover, establishment, Kura clover, *Trifolium ambiguum*, white clover

Introduction White clover (wc) (*Trifolium repens)* is present and is often the dominant legume in the >11 m ha of grassland in New Zealand (NZ). However, wc has limitations and normally contributes less than 20% of total annual pasture dry matter (DM) production. The use of a wider range of legume species is one way to increase legume percentage in wc/grass pastures. Caucasian (Cc) or Kura clover (*Trifolium ambiguum)* is a persistent legume which is slower to establish than wc but can increase total legume production (Cc plus volunteer wc) and hence N_2 fixation and animal productivity. This paper compares the productivity of Cc and wc in irrigated and dryland environments, and relates their relative establishment success to differences in seedling development.

Materials and methods Three experiments were conducted at Lincoln University. 1) *Irrigated:* ewe hoggets were rotationally grazed on irrigated Cc/perennial ryegrass (rg) or wc/rg pastures; liveweight gain (LWG), total clover % of DM and N_2 fixation were measured in years 3 to 5 after seeding (Black *et al.*, 2000; Widdup *et al.*, 2001). 2) *Dryland:* Cc and wc monocultures were grown in dryland conditions (mean rainfall of 648mm/yr); DM production and net leaf photosynthesis rate (Pn), when light was non-limiting, were measured in years 2 to 4 (Black *et al.*, 2003). 3) *Establishment:* In growth chambers, Cc, wc and rg were grown at four mean temperatures (6.5, 10, 15 and 20°C); seedling development was recorded until 1400°C-days (Black *et al.*, 2002).

Results *Irrigated:* Cc/rg pastures produced more annual total clover (18 v. 11%, SED 0.6%) and greater hogget LWG (1140 v. 1040 kg/ha per year, SED 13kg) than wc/rg pastures over years 3 to 5. Both legume species had similar nutritive values (DOMD 78%, crude protein 29%). The proportion of N_2 fixed by the clovers was similar (0.56 of herbage N%) but N_2 fixation/ha was proportional to clover herbage production/ha.
Dryland: Cc produced less DM than wc in year 2 (3.6 v. 7.0 t/ha, SED 0.28 t) but more in years 3 (9.4 v. 7.0 t/ha, SED 0.84 t) and 4 (7.9 v. 4.6t/ha, SED 0.39 t). Photosynthesis rate for Cc was 6 μmol CO_2/m^2 per s higher than wc irrespective of temperature or soil moisture. Both clovers had similar temperature and soil moisture optima for Pn (21–25°C, 1.00–0.86 of soil water-holding capacity). Thus, for any given leaf area index, the canopy Pn rate of Cc can be expected to exceed that for wc and give more assimilate/unit leaf area. This may explain the productive advantage of Cc once established. Caucasian clover was less productive than wc in autumn which suggests that photosynthates were directed to roots and rhizomes in this season. Rotational grazing to encourage a build up in root reserves in late summer/autumn may therefore be beneficial for Cc.
Establishment: Reasons for poor competitive ability of Cc seedlings were quantified. Germination, emergence and initial rates of leaf production for Cc, wc and rg were similar but secondary development in rg (375°C-days to first tiller) and wc (532°C-days to first stolon) were much faster than Cc which did not produce a secondary crown shoot until >1000°C-days and the first rhizome was not initiated until >1400°C-days.

Conclusions Once established, Cc is a persistent productive clover which is complementary to wc in temperate perennial pastures. This results in greater N_2 fixation and animal productivity/ha. The best strategy for Cc establishment is to sow in spring with minimal competition. Fast establishing species such as rg should be avoided. Low seeding rates of slow establishing grasses (e.g. timothy, tall fescue, cocksfoot) or summer brassicas (e.g. rape) are recommendations acceptable to farmers.

References

Black, A.D., K.M. Pollock, R.J. Lucas, J.M. Amyes, D.B. Pownall & J.R. Sedcole (2000). Caucasian clover/ryegrass produced more legume than white clover/ryegrass pastures in a grazed comparison. Proceedings of the New Zealand Grassland Association 62: 69–74.
Black, A.D., D.J. Moot & R.J. Lucas (2002). Seedling development and growth of white clover, Caucasian clover and perennial ryegrass grown in field and controlled environments. Proceedings of the New Zealand Grassland Association 64: 191–196.
Black, A.D., D.J. Moot & R.J. Lucas (2003). Seasonal growth and development of Caucasian and white clovers under irrigated and dryland conditions. In: D.J. Moot (ed.) Legumes for dryland pastures. Proceedings of a New Zealand Grassland Association symposium, Lincoln University, 18–19 November 2003. Grassland Research and Practice Series No. 11: 81–89.
Widdup, K.H., R.J. Purves, A.D. Black, P. Jarvis & R.J. Lucas (2001). Nitrogen fixation by Caucasian clover and white clover in irrigated ryegrass pastures. Proceedings of the New Zealand Grassland Association 63: 171–175.

Grazing behaviour of beef steers grazing Kentucky 31 endophyte infected tall fescue, Q4508-AR542 novel endophyte tall fescue, and Lakota prairie grass

H.T. Boland[1], G. Scaglia[1], J.P. Fontenot[1], A.O. Abaye[2] and R. Smith[2]
[1]*Department of Animal and Poultry Sciences,and* [2]*Department of Crop and Soil Environmental Sciences, Virginia Polytechnic Institute and State University, Blacksburg, Virginia 24061, USA, Email: hterry@vt.edu*

Keywords: grazing behaviour, beef steers, tall fescue, prairie grass, fescue toxicosis

Introduction Tall fescue is the most dominant grass used for pasture in the U.S. covering over 14 million ha. As a result, fescue toxicosis is a major concern among producers, especially during the summer months when the symptoms, such as reduced weight gains, are most pronounced. Producers need alternative forages for grazing cattle that do not have the negative effects associated with endophyte infected tall fescue. The objective of this experiment was to determine the grazing behaviour of cattle on Kentucky 31 endophyte infected (E+) tall fescue (*Festuca arundinacea Schreb.*), Q4508-AR542 (Q) novel endophyte tall fescue, and Lakota (L) prairie grass (*Bromus catharticus*).

Materials and methods Eighteen Angus-crossbred steers (279±8 kg) were allotted to the three different pastures (in two replicates, three steers per pasture). Twelve steers (two/pasture) were halter broken and trained to wear electronic behaviour data recorders (Rutter *et al.*, 1997; Champion *et al.*, 1997). Each pasture was 1.11 ha, equally divided into six paddocks. Paddocks were rotationally grazed for 7 d. Steers wore the device for five consecutive days in two sampling periods in June and August 2004. Recorders were placed on the animals at 0800 on d 0 and removed at 1300 on d 5, ensuring uninterrupted 24 h blocks of data recordings, which were analyzed using GRAZE (Rutter, 2000). Data were statistically analysed using SAS (SAS Institute, Cary, NC). Forage samples were taken for quality analysis just prior to entrance of the steers into the experimental paddocks.

Results Forage nutritive value among treatments was similar for June (CP, 16% ± 0.9; NDF, 56% ± 1; ADF, 31% ± 0.4), and August (CP, 14% ± 1.4; NDF 63%, ± 1.1; ADF, 35% ± 0.9). As shown in Table 1, time spent grazing in June was lowest in Q. In August, grazing time in Q was greater than E+ but similar to L. In both periods, time spent ruminating was lower in E+ than in L or Q, with ruminating time in June being greatest in Q, but Q was similar to L in August. Time spent idling (defined as time with no jaw movements) during both periods was higher in E+ compared to L, with Q similar to E+ in June, but lower than E+ in August. During the months of June and August steers grazing E+ spent less ($p<0.05$) time lying (515 and 564 min) than Q (737 and 697 min) and L (713 and 714 min), with Q and L being similar ($p>0.05$). These results are probably due to the fact that in August steers were suffering more from the effect of fescue toxicosis than in June. Higher relative humidity and temperatures in August, as well as being further into the grazing season may have exacerbated the symptoms of toxicosis, thus affecting the grazing behaviour of the steers.

Table 1 Daily activities of steers grazing different forages (min)

		Treatment[1]			
Period	Activity	E+	L	Q	s.e.m.
June	Grazing	594[a]	634[a]	559[b]	8.05
	Ruminating	525[c]	568[b]	616[a]	7.85
	Idling	325[a]	226[b]	279[a]	9.10
August	Grazing	547[b]	673[a]	623[a]	10.14
	Ruminating	514[b]	606[a]	606[a]	4.13
	Idling	401[a]	182[c]	280[b]	11.75

[1]E+=endophyte infected tall fescue; L=Lakota prairie grass; Q=Q4508-AR542 tall fescue
[abc]Means within row with different superscripts differ ($p<0.05$)

Conclusions In this study, L, Q, and E+ had similar quality characteristics. However, L and Q offer benefits during summer months in terms of animal behaviour, with steers spending more time grazing and ruminating, and less time idling compared to steers grazing E+ pastures. The use of L and Q may benefit beef producers currently using E+.

References

Champion, R.A., S.M. Rutter & P.D. Penning (1997). An automatic system to monitor lying, standing, and walking behaviour of grazing animals. *Applied Animal Behaviour Science,* 54, 291-305.

Rutter, S.M., R.A. Champion & P.D. Penning (1997). An automatic system to record foraging behaviour in free-ranging ruminants. *Applied Animal Behaviour Science,* 54, 185-195.

Rutter, S.M. (2000). GRAZE: A program to analyze recordings of the jaw movements of ruminants. *Behaviour Research Methods, Instruments & Computers,* 32, 86-92.

Yield components in a Signal grass-Clitoria mixture grazed at different herbage allowance

R. Jiménez-Guillen[1], S. Rojas-Hernández, J. Olivares-Pérez, A. Martínez-Hernández and J. Pérez-Pérez
[1]*Campo Experimental Chilpancingo, Col. Burócratas s/n, Chilpancingo, Guerrero, México, rjguillen@hotmail.com*

Keywords: mixed pasture, grazing intensity, leaf and stem yield

Introduction A Signal grass-Clitoria mixture provides good quality forage in the dry tropic of southern Mexico. However, its response in leaf and stem yields to grazing at different daily herbage allowances is not well documented. The objective of this study was to determine available and residual leaf and stem yields in a Signal grass (*Brachiaria decumbens*)-Clitoria (*Clitoria ternatea*) mixture grazed at different daily herbage allowance.

Materials and methods Field work was undertaken at the Coastal Experimental Station of the University of Guerrero (17° 20'N, 100°02'W). The experimental design was a randomized complete-block with four replications in an experimental unit of 400 m^2 of pasture. Grazing was rotational consisting of 1 and 35 days of occupation and rest, respectively. Four grazing cycles were completed within the rainy season. Seeding was carried out the previous year consisting of three rows of Signal grass (SG) and two of Clitoria (C). At the onset of the rainy season a cut was taken and grazing commenced 35 days later. Heifers were used as grazers and daily herbage allowances were: 2.5, 4.0, 5.5 and 7.0 kg dry matter (DM) /100 kg living weight (LW). For available and residual herbage 8 sampling units (0.5X3 m) on a transect were located at fixed intervals and perpendicular to rows allowing for two rows of each species to be inside the sampling unit and cuts were taken to ground level. Weights of leaf and stem DM for both SG and C were measured.

Results Covariate, herbage allowance x grazing cycle interaction and grazing cycle were not significant ($P>0.05$) so main effects of herbage allowance were compared averaged over grazing cycles (Tables 1 and 2). Herbage allowance determined ($P<0.05$) on-offer leaf and stem yields in SG but not ($P>0.05$) in C, the 2.5 kg DM/100 kg LW allowance gave the lowest on-offer yields in SG. In both species, the 2.5 kg DM/100 kg LW allowance had the lowest residual yields; however, in C, residual leaf yield was different only between 2.5 and 7 kg DM/100 kg LW allowance, and in SG the 2.5 kg DM/100 kg LW allowance had lower yields than the 5.5 kg DM/100 kg LW allowance. Favorable moisture, temperature and photoperiod conditions prevented an effect of grazing cycle. The higher sensitivity of SG to herbage allowance in both on-offer and residual leaf and stem yields compared to C could be explained on the higher biomass of the former.

Table 1 Available leaf and stem yields in a Signal grass-Clitoria pasture grazed at four herbage allowances

Herbage allowance (%)	Signal grass leaf +	Signal grass stem	Clitoria leaf	Clitoria stem
	Kg ha^{-1}			
2.5	514 $^{b\,\delta}$	477 ^{b}	295	383
4.0	750 ^{ab}	940 ^{a}	401	546
5.5	860 ^{a}	1192 ^{a}	447	616
7.0	867 ^{a}	1230 ^{a}	455	685

\+ Mean of three grazing cycles
δ Means within columns with one letter in common different (α = 0.05; Tukey)

Table 2 Residual leaf and stem yields in a Signal grass-Clitoria pasture grazed at four herbage allowances

Herbage allowance (%)	Signal grass leaf +	Signal grass stem	Clitoria leaf	Clitoria stem
	Kg ha^{-1}			
2.5	111 $^{b\,\delta}$	260 ^{b}	43 ^{b}	314
4.0	198 ^{ab}	549 ^{ab}	87 ^{ab}	392
5.5	275 ^{a}	756 ^{a}	94 ^{ab}	446
7.0	294 ^{a}	766 ^{a}	144 ^{a}	523

\+ Mean of three grazing cycles
δ Means within columns with one letter in common are not are not different (α = 0.05; Tukey)

Conclusions Available leaf and stem yields decreased when grazing at 2.5 kg DM/100 kg LW allowance because of the lower amount of residual leaf as the grazing intensity increased. SG was more sensitive to herbage allowance than C.

References

Steel, R.G.D. & J. H. Torrie (1998). Principles and procedures of statistics. McGraw-Hill, New York, U.S.A.

Response of warm-season grass pasture to grazing period and recovery period lengths

B.E. Anderson and W.H. Schacht
Department of Agronomy and Horticulture, University of Nebraska, Lincoln, Nebraska 68583-0910 USA, Email: banderson1@unl.edu

Keywords: basal density, species composition

Introduction Grazing period and recovery period lengths are key variables influencing grassland production and composition. Systems with short grazing periods and lengthy recovery periods require numerous pastures. Relatively high facility and management costs associated with multiple-pasture systems can be justified only if plant response is favorable and/or if livestock production is improved. This study determined the effects of 4 different combinations of grazing period/recovery period lengths on percentage basal cover (PBC) and relative species composition (RSC) of seeded, warm-season grass pasture.

Materials and methods Four grazing units (2.24 ha) were seeded in 1990 to a mixture of switchgrass (*Panicum virgatum* L.), big bluestem (*Andropogon gerardii* Vitman), indiangrass [*Sorghastum nutans* (L.)], little bluestem [*Schizachyrium scoparium* (Michx.) Torr.], and sideoats grama [*Bouteloua curtipendula* (Michx.) Torr.] at the University of Nebraska Agricultural Research and Development Center near Mead, Nebraska. Grazing units were fertilized annually with 90 kg N ha^{-1} in late May and burned in early May of 1995 and 1999. Each grazing unit was fenced into 4 paddocks representing 1/2, 1/4, 1/6, and 1/12 of the grazing unit area to simulate 2, 4, 6, and 12-paddock grazing systems, respectively. From 1995 through 2002, 10 crossbred steers grazed the 4 paddocks in each grazing unit rotationally from mid-June to late August. Stocking rate on all paddocks was equal and was controlled by the number of days in a paddock. Each year, steers were moved through the 4 paddocks of a unit in 3 grazing cycles of 12, 36, and 24 days in length. In June 1995 and 2003, PBC by species and RSC were estimated in each paddock using a modified step-point method (Owensby 1973). Non-seeded species, primarily annual bromegrasses (*Bromus spp.)* and smooth bromegrass (*Bromus inermis* Leyss.), were placed in an 'other' category. Experimental design was a modified latin square. Major sources of variation were grazing unit (replication), grazing strategy (grazing period/recovery period length combinations), and species.

Results Averaged across all grazing strategies, total basal cover of the 5 seeded species declined from 17% prior to grazing in 1995 to 7.4% in 2003. The decline in PBC during the study was greatest for big bluestem, switchgrass, and indiangrass; PBC of sideoats grama and little bluestem did not change; and PBC of other species increased (Table 1.) The decline in cover of the 3 warm-season tall grasses occurred coincidentally with the increase in basal cover of the other species. The lack of spring prescribed burning in the last 4 years of the study probably favored the annual bromegrasses and smooth bromegrass. These bromegrasses rapidly invade pastures in eastern Nebraska when not controlled. Little bluestem and sideoats grama were minor components throughout the study period as their RSC and PBC were consistently low. The RSC of the other category increased while RSC of big bluestem, switchgrass, and indiangrass decreased over the 8 years of the study. Grazing strategy did not affect RSC nor the change in percentage total basal cover of the 5 seeded species and the other category from 1995 to 2003.

Table 1 Changes in PBC and percent RSC of 5 warm-season tall grasses and the other grass category following 8 seasons of grazing

Species	PBC	RSC
Big Bluestem	-3.9	-18.3
Switchgrass	-3.1	-21.2
Indiangrass	-2.5	-15.4
Sideoats grama	0.1	0.4
Little bluestem	-0.1	-0.9
Other	14.1	56.2

Conclusions Results of this study suggest that recovery periods associated with a 12-paddock system did not affect persistence of a warm-season tall-grass mixture any differently than shorter recovery periods associated with a 2-paddock system. Bromegrasses became the dominant species in all paddocks regardless of grazing strategy used. Bromegrasses are very competitive in pastures of eastern Nebraska and may have overwhelmed differences in species composition that would have developed in their absence. Results indicate that grazing strategy alone does not affect the extent of invasion of warm-season tall-grass mixtures by the bromegrasses.

References

Owensby, C.E. (1973). Modified step-point system for botanical composition and basal cover estimates. *Journal of Range Management*, 26, 302-303.

Renovation-year forage quality of grass pastures sod-drilled with Kura clover

P.R. Peterson[1], P. Seguin[2], G. Laberge[2] and C.C. Sheaffer[1]
[1]*University of Minnesota, Dept. of Agronomy & Plant Genetics, St. Paul, MN 55108, USA, Email: peter072@umn.edu,* [2]*McGill University – Macdonald Campus, Dept. of Plant Science, Sainte-Anne-de-Bellevue, QC H9X 3V9, Canada*

Keywords: Kura clover, forage quality, sod seeding, pasture renovation

Introduction Including legumes can enhance yield, quality, and animal performance potential of grass pastures. Kura clover is an exceptionally winter hardy forage legume with high forage quality (Taylor & Smith, 1998). However, its seedling vigor is poor. Herbicide sod suppression prior to sod drilling enabled kura clover to establish in the north central USA and eastern Canada (Cuomo *et al*., 2001; Laberge *et al*., 2005), but its percentage of renovation-year forage yield was less than for sod-seeded red or white clover. The objective of this study was to determine the influence of herbicide suppression and clover species on renovation-year forage quality of grass pastures sod-drilled with Kura clover versus red or white clover.

Materials and methods Three pasture renovation factors were imposed on grass pastures in a split-split plot restriction in five environments in 2001 and 2002; three in the north central USA (Minnesota) and two in eastern Canada (Quebec). Factors included clover species ('Cossack' Kura clover vs. 'Scarlett' red or 'Shasta' white clover), herbicide sod suppression intensity (paraquat at 0.9 kg a.i./ha or glyphosate at 0.8 or 3.3 kg a.i./ha), and renovation-year N fertilization (0 or 110 kg N/ha). Sward crude protein (CP), neutral detergent fiber (NDF), and *in vitro* true digestibility (IVTD) concentrations were determined via NIRS in autumn after spring sod drilling.

Results Forage quality was positively correlated with clover DM production. The high rate of glyphosate resulted in lower NDF concentrations than paraquat suppression in most environments. Kura clover had less impact on sward forage quality than white and red clover since its renovation-year DM production was less. In Minnesota, swards sod-seeded with Kura clover averaged 40 g/kg less IVTD than swards with red or white clover in 10 of 18 site-herbicide-N treatment combinations. Swards with kura or white clover had 35 g/kg less CP than red clover swards in 13 of 18 site-herbicide-N treatment combinations. Where renovation-year precipitation was limiting (Quebec), there was little effect of clover species on forage quality.

Table 1 Forage quality (g/kg) in autumn of the renovation year following spring sod drilling of clovers into grass pastures for selected treatments and environments (0 kg N/ha level)

Treatments		Minnesota 2001			Quebec 2002		
Clover	Herbicide	CP	NDF	IVTD	CP	NDF	IVTD
Kura	Paraquat	114	593	646	220	460	739
	Glyphosate Low	113	586	655	264	399	806
	Glyphosate High	115	518	654	265	386	794
Red	Paraquat	189	447	746	217	493	728
	Glyphosate Low	184	458	747	245	399	795
	Glyphosate High	196	429	763	282	360	839
White	Paraquat	174	483	742	209	481	722
	Glyphosate Low	172	453	760	260	412	789
	Glyphosate High	163	456	760	261	390	792
LSD(0.05)	Species	13	31	18	50	63	89
LSD(0.05)	Herbicide	12	29	17	51	64	89

Conclusions Kura clover had less influence on renovation-year forage quality than red or white clover because its DM production was less. Thus, little renovation-year forage quality improvement should be expected after spring sod drilling of Kura clover.

References

Cuomo, G.J., D.G. Johnson & W.A. Head, Jr. (2001). Interseeding Kura clover and birdsfoot trefoil into existing cool-season grass pastures. *Agronomy Journal* 93, 458-462.

Laberge, G., P. Seguin, P.R. Peterson, C.C. Sheaffer, N.J. Ehlke, G.J. Cuomo & R.D. Mathison (2005). Establishment of Kura clover no-tilled into grass pastures with herbicide sod suppression and nitrogen fertilization. *Agronomy Journal* 97, 250-256.

Taylor, R.W., & R.R. Smith (1998). Kura clover (*Trifolium ambiguum* M.B.) breeding, culture and utilization. *Advances in Agronomy* 63, 153-178.

Forage yield and quality of Signal grass-Clitoria mixture grazed at different frequencies

R. Jiménez-Guillen, J. Olivares-Pérez, S. Rojas-Hernández and A. Martínez-Hernández
Campo Experimental Chilpancingo, Col. Burócratas s/n, Chilpancingo, Guerrero, México, rjguillen@hotmail.com

Keywords: tropical pasture, grazing frequencies, botanical composition

Introduction A Signal grass-Clitoria mixture provides good quality forage in the dry tropic of southern Mexico. Grazing frequency is a management tool that determines yield, botanical components and quality of pastures. The objective of this study was to determine forage yield, quality and botanical components in a Signal grass (*Brachiaria decumbens*)-Clitoria (*Clitoria ternatea*) mixture when grazed at different frequencies.

Materials and methods Field work was undertaken at the Coastal Experimental Station of the University of Guerrero (17° 20'N, 100°02'W). Grazing frequencies were once every 5, 6, 7 and 8 weeks. Experimental design was a randomized complete-block with four replications in an experimental unit of 400 m^2 of pasture. The grazing period lasted 24 hours and in each the number of heifers used was such as to keep herbage allowance 4.0 kg of dry matter (DM)/100 kg live weight across all frequencies. Five and four grazing periods were completed in the 5 and 6 week frequencies, respectively, and three grazing periods for the 7 and 8 week frequencies. Seeding was carried out the previous year consisting of three rows of Signal grass and two of Clitoria. Available and residual forage were determined, for both, on a fixed transect and distance using 8 sampling units (0.5X3 m) located perpendicular to rows, allowing for two rows of each species to be inside the sampling unit. Cuts were taken to ground level. Total yield and components were estimated on a DM basis. Crude protein (CP) and dry matter digestibility (DMD) were determined in available forage. For statistical analyses (Steel and Torrie, 1998) available and residual forage yields were divided by the number of days of the rest period and the average for the season was calculated. The CP and DMD were also averaged for the season. Mean separation was by Tukey with α=0.05.

Results Available forage yield (total and components) was not affected ($P>0.05$) by grazing frequency. Forage quality decreased ($P<0.05$) after the 6-week grazing frequency. The 8-week grazing frequency showed the lowest residual forage and the lowest amount of Signal grass in the residual forage (Table 1, Figure 1). The heavy trampling of dead material and the difficulty of recovering this material at the time of sampling might explain this result. Active forage accumulation in Signal grass-Clitoria peaked between the fifth and sixth week of regrowth, which explains the decrease in forage quality and similar yield in the available forage with longer rest periods.

Table 1 Forage yield and quality in a Signal grass-Clitoria pasture grazed at four frequencies

	Forage yield		On-offer forage quality	
Frequency (weeks)	On-offer	Residual	Protein	DMD
	kg DM/ ha/day		%	
5	45.2	23.7 [ab 1]	6.5 [a]	42.3 [a]
6	44.7	23.0 [ab]	6.4 [a]	41.4 [a]
7	44.3	24.8 [a]	5.4 [b]	39.0 [b]
8	51.6	20.8 [b]	6.2 [ab]	41.2 [ab]

[1] Means within columns with one letter in common are not different (α = 0.05; Tukey).

Figure 1 Botanical components in a Signal grass-Clitoria pasture grazed at four frequencies

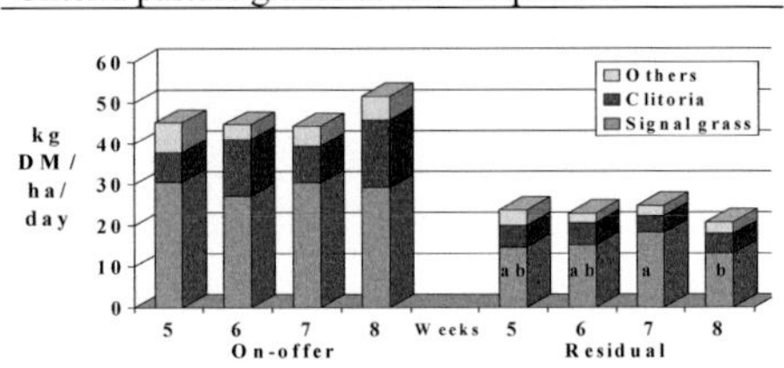

[1] Within bars with one letter in common are not different (α = 0.05; Tukey).

Conclusions In Signal grass-Clitoria, grazing frequency should be decided on the basis of forage quality rather than on forage yield.

References.

Steel, R.G.D., Torrie J.H. (1998). Principles and procedures of statistics. Mc Graw-Hill. New York, NY, U.S.A.

Section 2

Appropriate animals for grazing

Farm performance from Holstein-Friesian cows of three genetic strains on grazed pasture

K.A. Macdonald[1], B.S. Thorrold[1], C.B. Glassey[1], J.A.S. Lancaster[1], G.A. Verkerk[1], J.E. Pryce[2] and C.W. Holmes[3]
[1]*Dexcel, Private Bag 3221, Hamilton. 2001, New Zealand. Email: Kevin.macdonald@dexcel.co.nz*
[2]*Livestock Improvement Corporation, Private Bag 3016, Hamilton, New Zealand.* [3]*IVABS, Massey University, Private Bag 11222, Palmerston North, New Zealand*

Keywords: genetic strain, Holstein-Friesian, pasture, performance

Introduction Dairy selection objectives and farm production systems in USA and Europe are different from those in New Zealand (NZ). The use of overseas semen in NZ in the last 20 years has changed the genetics of the former NZ Holstein-Friesian (HF) strain. This trial was designed to demonstrate the genetic progress in the NZ HF dairy herd in the last 25 years and how high production potential North American HF cows perform under pasture-based feeding systems.

Materials and methods Three strains of Holstein-Friesians were farmed in a range of feeding systems for 3 years. These included two high breeding worth ($BW; year 1999) strains of either North American origin (OS90; $BW 84) or NZ origin (NZ90; $BW 86), and a low $BW strain of 1970 NZ Friesians (NZ70; $BW 10). Systems were designed to provide feed allowances of 4.5 to 7.0 t DM/cow per year, based on different stocking rates and supplement inputs. When feed allowance was higher than 5.5 t DM/cow, additional feed above pasture grown was brought in. The trial started in 2001 with all the cows being first lactation animals. An annual replacement rate of 25% with 1st calving heifers was used. This paper reports on the third year in which there were 15 cows in each NZ70 farmlet and 20 in the NZ90 and OS90 farmlets. In this year the age structure was 55% (4 year), 20% (3 year) and 25% (2 year old) cows. Data were analysed using ASReml. The model used included the effects of age at first calving, strain, parity and feed offered (in t as a linear and quadratic covariate within strain). Farmlet and cow were fitted as random effects. As the variance of farmlet was difficult to estimate it was fixed to 0.10 of the cow variance.

Results Per lactation, NZ90 produced the greatest milk volume, with the highest concentrations of fat and protein, while OS90 had the shortest lactation, were the heaviest, and had the lowest body condition score (BCS; using the NZ 10 point BCS scale). However, OS90 produced slightly higher daily fat and protein (milksolids; MS) yields than NZ90 (1.75 versus 1.71 kg MS per day of lactation), while NZ70 produced only 1.42 kg MS per day. Although all cows calved at close to the optimum BCS of 5.5 the OS90 cows were at less than 4.0 during most of lactation (Fig 1). This lower BCS was a major determinant for their shorter lactations. Figure 1 presents the average BCS for the season and indicates that feeding level had no effect on OS90 whereas for the other 2 strains, BCS increased with increased feeding levels. Reproduction as defined by incalf rate was similar for all strains in the first 2 lactations (data not shown). In the third lactation 7, 15 and 24% of NZ70, NZ90 and OS90 cows, respectively, failed to get in calf. Days from planned start of calving to mean calving date for years 2 and 3 of the trial were 20, 21 and 27 for NZ70, NZ90 and OS90, respectively, indicating later conception dates for OS90 cows.

Table 1 Production data, bodyweight (BW) and BCS of three strains of Holstein-Friesian (2003/04)

	NZ70	NZ90	OS90	SED
Milk (kg/cow)	4812	5593	5479	209.1
Milk fat (%)	4.65	4.86	4.26	0.124
Protein (%)	3.41	3.71	3.43	0.066
Milksolids (kg/cow)	380	468	415	15.7
BW (kg)	473	487	503	10.5
BCS	5.06	4.51	4.13	0.101
Lactation length (days)	286	286	252	5.6

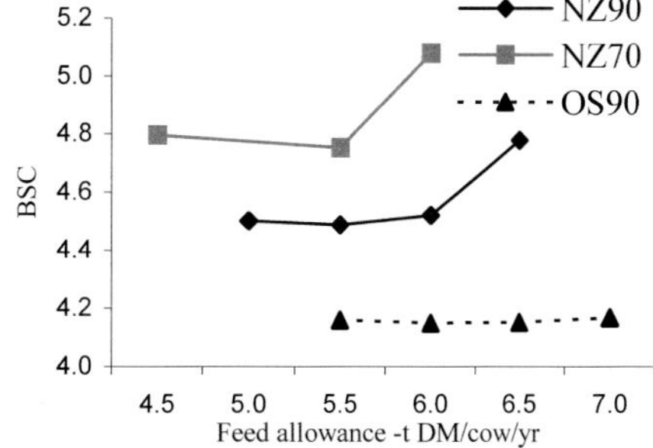

Figure 1 Average body condition score (BCS) of three strains of Holstein-Friesian (2003/04)

Discussion Compared with the 1970's, the 1990's NZ Holstein-Friesian cow produced 21 % more milk fat and 26 % more milk protein. An economic analysis of the data has shown that the productive gains made in the last 25 years of selective breeding are equivalent to an increase in Economic Farm Surplus of over $500/ha at the same feed allowance. The use of OS semen in NZ has increased the potential for milk production per day of lactation at a rate similar to selection within NZ, but because of the reduction in BCS and the associated shorter lactation length and lowered fertility, this is not expressed in production or profit.

Genetic alternatives for dairy producers who practise grazing

H.D. Norman, J.R. Wright and R.L. Powell

Animal Improvement Programs Laboratory, Agricultural Research Service, U.S. Department of Agriculture, Bldg. 005, Room 306, BARC-West, 10300 Baltimore Ave., Beltsville, MD 20705-2350, U.S.A,. Email: dnorman@ aipl.arsusda.gov

Keywords: fertility, grazing, mastitis, milk, protein

Introduction The decline in cow fertility has had a negative impact on all dairy producers, especially those that practise seasonal calving with pasture-based dairying. One alternative that is being tried in the United States (US) by a few graziers is to use bulls from New Zealand (NZ) because NZ producers have practised seasonal calving for some time. However, genotype-environment interaction is a concern; genetic correlations that were derived by the International Bull Evaluation Service (2004) between bull rankings from different countries were often lower for NZ than for other countries. The objective of this study was to compare the performance of daughters of NZ Friesian and Holstein artificial-insemination (AI) bulls with daughters of other Holstein AI bulls (predominantly from the US) that were in the same US herd and calved at the same time.

Material and methods Milk, fat, protein, somatic cell score (SCS, an indicator of mastitis) and days open were examined for the first three parities of Holstein cows. Traits were standardized for environmental effects in the same manner as in the current US Department of Agriculture genetic evaluation. Cows were required to have calved after December 1999 and before August 2004 and to have had the opportunity to express the performance trait; i.e. the herd remained on production testing. Data for first-parity yield traits and SCS were from 489 daughters of 14 NZ bulls and 5419 daughters of 1732 other bulls in 149 herds. Second- and third-parity yield traits represented 345 and 174 NZ daughters and 5057 and 2840 other daughters in 126 and 78 herds, respectively. Data for first-parity days open were from 450 daughters of 13 NZ bulls and 5036 daughters of other bulls in 138 herds. Number of NZ daughters per herd ranged from 1 to 36. The model included fixed effects for herd-year-season and strain. Strain difference for each parity-trait combination was tested for significance at $p \leq 0.05$, 0.01 or 0.001.

Results Strain differences in trait means are given in Table 1. Mean first-parity milk and protein yields were lower by 501 and 5 kg, respectively, for daughters of the NZ bulls than for daughters of other bulls. Mean second-parity milk and protein yields were lower by 467 and 5 kg, and third-parity means were lower by 448 and 4 kg. Fat yields were higher by 2 kg (nonsignificant). First-parity daughters of NZ bulls had higher mean SCS than did daughters of other bulls (3.2 versus 3.0). Daughters of NZ bulls had 7 fewer days open during first lactation ($p \leq 0.05$) than did daughters of other bulls but had 2 and 3 greater days open during second and third lactations (nonsignificant). Fewer traits showed significance for later parities because of fewer observations.

Table 1 Performance comparison of Holstein daughters of NZ AI bulls with daughters of other AI bulls by parity[1]

Trait	Parity 1	Parity 2	Parity 3
Milk (kg)	–501***	–467***	–448***
Fat (kg)	2	2	2
Protein (kg)	–5**	–5*	–4
SCS	0.2***	0.1	0.2
Days open	–7*	2	3

[1]Significance of strain difference (NZ minus other daughters) designated at $p \leq 0.05$ (*), 0.01 (**) and 0.001(***).

Conclusions Strain differences existed for several performance traits. Daughters of US bulls were more productive than daughters of NZ bulls for milk and protein. First-parity daughters of US bulls also had lower SCS, but daughters of NZ bulls had fewer days open. However, the individual bulls chosen to be sires from each country influenced all strain differences. Producers should consider the economic values of all the performance traits when making genetic choices between US and NZ bulls, and those values should be combined into an index appropriate for expected economic conditions. Producers who practise grazing and seasonal calving should place more weight on fertility traits than is recommended for the general dairy cattle industry because of their higher economic value in a seasonal grazing environment.

References

International Bull Evaluation Service (2004). Description of National Genetic Evaluation Systems for dairy cattle traits as applied in different Interbull member countries. http://www-interbull.slu.se/ national_ges_info2/ begin-ges.html, accessed November 30, 2004.

Suitability of small and large size dairy cows in a pasture-based production system

M. Steiger Burgos[1], R. Petermann[1], P. Hofstetter[4], P. Thomet[1], S. Kohler[1], A. Munger[2], J.W. Blum[3] and P. Kunz[1]
[1]*Swiss College of Agriculture, Laenggasse 85, CH-3052 Zollikofen, Switzerland, Email: peter.kunz@shl.bfh.ch*
[2]*Swiss Federal Research Station for Animal Production and Dairy Products, CH-1725 Posieux, Switzerland*
[3]*Inst. of Animal Genetics, Nutrition and Housing, Univ. of Bern, CH-3012 Bern, Switzerland*
[4]*Schupfheim Agricultural Education and Extension Centre, CH-6170 Schupfheim, Switzerland*

Keywords: pasture, cow size, body condition score, body weight, milk acetone

Introduction Pasture-based dairy production with greatly reduced supplemental feeding and block-calving in spring is increasingly applied in Switzerland. The prevalent cow type has been selected mainly for high individual production in a barn feeding system with balanced diet. This cow type has continuously increased in size over the last 30 years. The question arises whether this type is suitable for the new system, and particularly if cow size is a critical factor. Theoretically a large, heavy type of cow has a higher intake capacity, while the nutrient requirements for a small, light type are easier to satisfy.

Materials and methods In a 3-year trial, two herds of multiparous Red-Holstein-Simmental crossbred and Brown Swiss cows were formed (breed distribution: 50/50). Herd B consisted of 13 large and heavy cows [in 2002: 4.4 ± 0.9 years old (mean ± SD), 726 ± 62 kg body weight (BW) of the cows having calved at turnout to pasture, 147 ± 2 cm withers height (WH); in 2003: 4.6 ± 1 years old, 720 ± 63 kg, 146 ± 2 cm WH]. Herd S consisted of 16 smaller and lighter cows (3.8 ± 0.6 years old, 558 ± 34 kg BW, 138 ± 3 cm WH, resp. 4.4 ± 0.6 years old, 590 ± 38 kg BW, 137 ± 2 cm WH). Each herd had access to 5.8 ha pasture in a rotational system, so that the same overall stocking rate was obtained (1700 kg/ha). Each herd received 2115 kg of concentrate until the end of the 10-week breeding season. Individual milk production and contents of fat, protein and acetone as well as BW were recorded once a week, and body condition score (BCS) once a month. Concentration of acetone was determined by flow injection analysis as described by Reist *et al.* (2000). BW and BCS changes were analysed statistically with a t-test resp. Aspin-Welch test in case of unequal variances. Acetone values were analysed with a Mann-Whitney test.

Results Herd S produced more milk than herd B (82,527 *vs* 78,741 in 2002 and 91,173 *vs* 83,560 kg ECM in 2003). During the same time, the changes in BW and BCS were slightly higher in the B-cows than in the S-cows (Table 1). Acetone milk content was also higher in the B-cows, especially during the first year (Figure 1).

Table 1 Body weight and BCS changes (mean ± SD) of the large and heavy cows (B) and small and light cows (S) between calving and nadir

	Type B		Type S	
Body weight changes				
In 2002 (kg/cow)	-101	± 32	-53	± 35*
In 2002 (%)	-13.7	± 3.9	-10.4	± 4.9*
In 2003 (kg/cow)	-132	± 57	-91	± 24*
In 2003 (%)	-17.2	± 6.6	-14.8	± 3.2
BCS changes				
In 2002	-0.65	± 0.45	-0.32	± 0.34*
In 2003	-0.73	± 0.51	-0.55	± 0.32

*differences between types were significant (P<0.05)

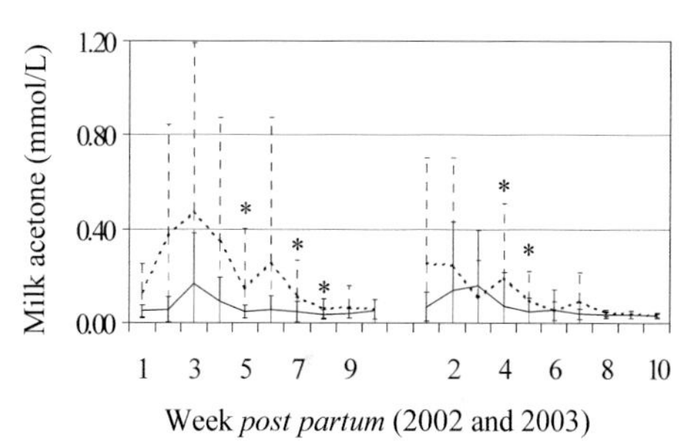

Figure 1 Acetone content in the milk (mean ± SD) of the large and heavy cows (---) and of the small and light cows (—)(* P<0.05 between types)

Conclusions The results of the first 2 years show that the small and light cows seem to be better adapted to the pasture-based milk production as practiced in this trial: as a herd they produced more milk but mobilized less body reserves than the large and heavy cows. Differences between types were more distinct in 2002 and became statistically less significant in 2003. If this trend is confirmed during the third year, this would suggest that the cows might have adapted to trial conditions over the years.

References

Reist, M., A. Koller, A. Busato, U. Kupfer & J.W. Blum (2000). First ovulation and ketone body status in the early postpartum period of dairy cows. *Theriogenology*, 54, 685-701.

The effect of stocking rate and lamb grazing system on sward performance of *Trifolium repens* and *Lotus corniculatus* in Uruguay

F. Montossi, R. San Julián, M. Nolla, M. Camesasca and F. Preve
Instituto Nacional de Investigación Agropecuaria (INIA), Ruta 5, km 386, PC: 45000, Tacuarembó, Uruguay, Email: fmontossi@tb.inia.org.uy

Keywords: stocking rate, grazing system, lambs, sward height

Introduction Lambs have a great potential to diversify and stimulate meat and wool production and economical returns within the industry. The main objective of this study was to evaluate different feeding and management alternatives for the production of high quality wool and meat as well as their effects on sward characteristics in the Basaltic region of Uruguay.

Materials and methods The experiment was carried out from May 22 till September 10, 2001, using a two year old mixed sward of *Trifolium repens* (cv. LE Zapicán) and *Lotus corniculatus* (cv. San Gabriel) grazed by 60 Corriedale lambs (8 months of age; 27 kg initial liveweight). The effects studied were stocking rate (SR; 12 and 24 lambs/ha) and grazing system (GS; continuous, CG; strip, SG; and 7 days rotational grazing, 7G). The experimental area was 3.68 ha, divided into 6 plots. The variables measured were (pre and post grazing): a) on sward (each 14 days): herbage mass (ton DM/ha -DM-, ton green DM/ha -GM- and ton green leaf DM/ha -GL-), sward height (cm, H), botanical composition (BC) and nutritive value (NV); and b) on animals (each 14 days): liveweight (LW) and LW gain (LWG); and c) on carcasses (at slaughter): cold weight (CCW) and fat cover (GR). The design was a complete randomised block arranged in a factorial structure.

Results Before and after grazing, the increased SR reduced DM and GM and H (Table 1; $P<0.01$) without affecting NV and BC. GS had a significant effect ($P<0.01$) on sward variables (DM and GM and GL and H) having in general, higher values for CG, and lower and similar for SG and 7G, particularly after grazing. BC and NV of post grazing forage were not affected by GS. Overall, SR had higher effect than GS on the sward characteristics. As the experiment progressed DM and H values remained very stable, evolving into a sward dominated by *Trifolium repens* with a substantial increase in NV. SR affected LWG (210 vs. 168 g/d, $P<0.01$), final LW (50.0 vs. 45.3 kg, $P<0.01$), CCW (23.4 vs. 20.9 kg, $P<0.01$) and GR (16.7 vs. 11.9 mm, $P<0.01$) for 12 and 24 lambs/ha, respectively. GS influenced LWG (197, 191 and 180 g/d, $P<0.01$) and final LW (48.5, 47.8 and 46.6 kg, P<0.05), for CG, 7G and SG, respectively. CG showed higher CCW (23.2 vs. 22.3 and 21.0 kg, $P<0.01$) and GR (17.1 vs. 14.2 and 11.6 mm, $P<0.01$) compared with 7G and SG. The increase in SR did not alter wool production and quality, but reduced CCW and GR and boneless leg. Implementation of more controlled grazing systems produced a progressive reduction in LWG and LW, without effecting wool production and quality. Rotational grazing systems produced lighter carcasses with low fat, without modifying the other evaluated carcass characteristics. Within the range of SR used and the sward maintained during this short fattening period, the implementation of a more controlled grazing system would not be justified biologically and economically.

Table 1 Effects of SR and GS on sward characteristics (pre and post grazing)

	Variables	SR			GS				SR*GS						
		12	24	P	C	S	7	P	12-CG	12-SG	12-7G	24-CG	24-SG	24-7G	P
Pre	DM (t DM/ha)	2.42a	2.13b	**	2.15b	2.40a	2.28ab	*	2.37ab	2.61a	2.28b	1.93c	2.20b	2.27b	*
Pre	GM (t DM/ha)	2.12a	1.85b	**	1.83b	2.16a	1.97b	**	2.06b	2.32a	1.99b	1.60c	2.01b	1.95b	*
Pre	GL (t DM/ha)	1.00a	0.92b	*	0.86c	1.07a	0.95b	**	0.96bc	1.09a	0.94c	0.75d	1.05ab	0.97bc	*
Pre	H (cm)	15.8a	13.2b	**	11.6c	17.0a	14.9b	**	14.9b	18.3a	14.9b	8.3c	15.6b	15.6b	**
Post	DM (t DM/ha)	2.17a	1.72b	**	2.07a	1.97ab	1.80b	**	2.34a	2.31a	1.86b	1.81b	1.62b	1.74b	**
Post	GM (t DM/ha)	1.83a	1.39b	**	1.78a	1.60b	1.45b	**	2.07a	1.84b	1.58c	1.49cd	1.36d	1.32d	ns
Post	GL (t DM/ha)	0.76a	0.57b	**	0.77a	0.64b	0.58b	**	0.88a	0.71b	0.68b	0.66bc	0.57c	0.47d	ns
Post	H (cm)	10.2a	6.2b	**	11.0a	6.1c	7.4b	**	14.0a	7.6c	9.0b	8.1bc	4.7e	5.8d	**

ns = $P>0.05$; *= $P<0.05$ and **= $P<0.01$. Means with different letters between columns differ significantly ($P<0.05$)

Conclusions This study shows the high productive potential of mixed swards of high nutritive value and stocking rate, where grazing system played a minor role in affecting animal productivity during this short autumn-spring lamb fattening system. All the lambs coming from the different treatments achieved the requirements of the Heavy Lamb Market of Uruguay, producing between 280 and 440 kg of animal liveweight/ha, in a 110 days period, demonstrating the great potential of white clover dominated swards for lamb production in the Basaltic production systems.

Effect of stocking rate and grazing system on fine and superfine Merino wool production and quality on native swards of Uruguay

I. De Barbieri, F. Montossi, E.J. Berretta, A. Dighiero and A. Mederos
Instituto Nacional de Investigación Agropecuaria (INIA), Ruta 5, km 386, PC: 45000, Tacuarembó, Uruguay, Email: idebarbieri@tb.inia.org.uy

Keywords: fine wool, native swards, stocking rate, grazing system

Introduction Modern textile tendencies show that consumers prefer light, soft, resistant, natural, and comfortable clothes, for which fine and superfine wools are in great demand, particularly at the high value markets (Whiteley, 2003). The main objective of the present study was to define sustainable stocking rates and grazing systems on native swards for fine and superfine wool production in the Basaltic region of Uruguay.

Materials and methods The trial was carried out for two years (October 2001 to October 2003), using the same 72 Australian Merino whethers in both years. The evaluated factors were stocking rate (SR, 5.3 and 8.0 animals/ha) and grazing system (GS, continuous -CG- and 21 days strip grazing -21G-), and their combination. The experimental area was 6.0 ha, (1.5 ha per treatment). The variables measured each 21 days were: a) on pre grazing swards: herbage mass (DM, kg DM/ha), sward height (H, cm), botanical composition (BC) and nutritive value; b) on all animals: liveweight (LW, kg) and condition score (CS, units); and c) on wool (annually): fleece weight (FW, kg), fibre diameter (FD, μ) and other quality characteristics. The design was a complete randomised block, arranged in a factorial structure, the main factors being: SR and GS at two levels each.

Results The SR of 5.3 animals/ha gave higher DM yield and H than the SR of 8.0 animals/ha (Table 1). The overall sward botanical composition and nutritive value were not affected by SR and GS. The BC (annual average) was: green grass leaf 48.0%, green grass stem 5.5%, dead material 40.0% and other species 6.5%; nutritive values were: CP 8.0%, NDF 68.5% and ADF .8%. BC and nutritive value were affected only by SR, essentially during spring and summer. A SR increase from 5.3 to 8.0 animals/ha resulted in higher sward quality. SR affected LW, CS and FW, associated with its effect on sward quantity and quality. Wool quality was not affected by SR (Table 2). GS did not significantly affect the variables related to animals and wool (Table 2).

Table 1 Effect of SR and GS on sward quantity characteristics

Variable	SR			GS			SR*GS				
	5.3	8.0	P	21G	CG	P	5.3*21G	5.3*CG	8.0*21G	8.0*CG	P
DM (kg MS/ha)	3043a	1745b	**	2734a	2054b	**	3593a	2493b	1875c	1615d	**
H (cm)	10.8a	6.1b	**	9.4a	7.5b	**	12.3a	9.3b	6.5c	5.7c	**

**= P<0.01. a, b, c, d = means with different letters between columns differ significantly (P<0.05).

Table 2 Effect of SR and GS on animal traits

Variable	SR			GS			SR*GS
	5.3	8.0	P	21G	CG	P	P
LW (kg)	51.2a	46.5b	**	48.5	49.2	ns	ns
CS (units)	3.5a	3.1b	**	3.3	3.3	ns	ns
FW (kg)	3.53a	3.29b	*	3.34	3.48	ns	ns
FD (μ)	18.5	18.4	ns	18.3	18.6	ns	ns

ns = P>0.05, *= P<0.05 and **= P<0.01

Conclusions Stocking rate had the major impact on forage production and quality, liveweight and fleece weight production and wool quality in fine and superfine Merinos. Grazing system had a minor effect on production and quality). These results suggest that the advantage of using controlled grazing systems is limited for quantity and quality of fine and superfine wool production on native swards. The information generated in this study in the Basaltic region, highlights the possible implementation of high quality wool production systems with an interesting economical return when controlled grazing system, suitable stocking rate and known animal genetics merits are used.

References

Whiteley, K. (2003). Características de importancia en lanas finas y superfinas. In: Seminario Internacional de Lanas Merino finas y superfinas. Roberto Cardellino(Ed.) SUL, INIA, CLU& SCMAU. p. 17-22.

Cattle and sheep mixed grazing 1: species equivalence

R.D. Améndola-Massiotti, S.J.C. González-Montagna and P.A. Martínez-Hernández
Universidad Autónoma Chapingo, Posgrado en Producción Animal, Programa de Investigación en Forrajes, Carretera México-Texcoco.Km 38.5, Chapingo, Edo. Méx., CP56230, México, Email: r_amendola@yahoo.com

Keywords: heifers, ewes, stocking rate, herbage intake

Introduction The effects of mixed grazing of cattle and sheep depend on stocking rate (SR) and species ratio (Nicol, 1997). Calculations of SR and species ratio require the use of species equivalence. Equivalents are often estimated in terms of intake requirements calculated on the basis of $LW^{0.75}$. Freer (1981) stated that $LW^{0.9}$ would be more appropriate for comparisons of intake requirements for maintenance of sheep and cattle. Nonetheless, Nolan & Connolly (1977) stated that the equivalence is system-specific and depends on the species being considered. The objective of this experiment was to estimate species equivalence for the evaluation of the introduction of sheep into a cattle dairy system based on grazing in temperate Mexico.

Materials and methods The experiment was carried out at Chapingo, Mexico (19°29' N, 98°54' W, 2240 m a.s.l.), between 15 March and 25 May 2000. Holstein heifers (initial LW 336±7 kg) and pregnant Suffolk ewes (initial LW 75.8±0.8 kg) were used; most ewes lambed during the experiment. There were 8 treatments, 3 heifers were used per mixture treatment and numbers of ewes varied to achieve the proportions of species shown in Table 1. Nine paddocks of *Medicago sativa* and *Dactylis glomerata* of 0.46 ha were strip-grazed for 7 days each. Three blocks of 3 paddocks were formed and treatments were randomly allotted to paddocks; only one paddock was grazed per week, the paddocks of the first and second week of each block were grazed by three treatments (each on separate areas), while the third-week paddock was grazed by the remaining two treatments. Fresh areas of pasture were allotted when already grazed areas reached 10 cm height (falling disc); no back fencing was used. The areas effectively grazed after 7 days were measured. Herbage dry matter (DM) intake was measured using chromium oxide as an external marker and *in situ* digestibility of hand plucked herbage samples.

Results Excluding the treatment of grazing by ewes only, a linear equation was calculated of area allotted (m^2/d) on number of ewes: $y = 93.4 + 4.43\ x$, $R^2 = 0.94$, ($P < 0.05$) (Table 1); the intercept and the regression coefficient represent the area allotted to 3 heifers and the additional area allotted per ewe, respectively. The equivalence based on area allotted resulted in 7.0 ewes per heifer [(93.4/3)/4.43]. The equivalence based on average herbage DM intake in Table 1 (8.76 and 2.31 kg DM/animal/d) for heifers and ewes, respectively) resulted in 3.8 ewes per heifer (8.76/2.31), corresponding with the estimate of intake requirements using $LW^{0.9}$ ($336^{0.9}/75.8^{0.9}$), as recommended by Freer (1981). This difference between estimates of equivalence, expresses the higher efficiency of herbage utilisation under mixed grazing than under single species grazing by heifers, which was due to grazing by sheep below 10 cm height (data not shown here).

Table 1 Area allotted and herbage intake of heifers and ewes under single species grazing and mixed grazing with different ewes to heifer's ratios

Nr. of heifers	Nr. of ewes	Area allotted (m^2/d)	Intake per species (kg DM/heifer/d)	Intake per species (kg DM/ewe/d)	Total intake (kg DM/m^2/d)
3	0	93.4	8.17 (0.49)		0.262
3	6	131.6	10.50 (0.33)	2.08 (0.15)	0.334
3	9	148.2	7.68 (0.49)	2.07 (0.15)	0.281
3	12	116.0	9.54 (0.33)	1.98 (0.15)	0.452
3	24	218.6	8.98 (0.33)	2.52 (0.15)	0.400
3	36	269.5	6.49 (0.49)	2.49 (0.15)	0.405
3	48	287.6	9.93 (0.33)	2.49 (0.15)	0.519
0	15	91.5		2.54 (0.15)	0.416

Conclusions The use of an equivalence based on DM intake (3.8 ewes/heifer), which concurs with intake requirements based on $LW^{0.9}$, would neglect benefits of more efficient pasture utilisation under mixed grazing; the use of the higher equivalent based on area allotted (7.0 ewes/heifer) should lead to better performance of the system.

References

Freer, M. (1981). The control of food intake by grazing animals. In: F. H. W. Morley (ed.). Grazing Animals. Elseviers Scientific Publishing Company, Amsterdam, 105-124.

Nicol A. M. (1997). The application of mixed grazing. In: J. G. Buchanan-Smith, L. D. Bailey and P. Mc Caughey (eds.) *Proceedings of the XVIII International Grassland Congress,* 525-534.

Nolan, T. & J. Connolly. (1977). Mixed stocking by sheep and steers. A review. *Herbage Abstracts 47,367-364.*

Cattle and sheep mixed grazing: 2: competition

S.J.C. González-Montagna, P.A. Martínez-Hernández and R.D. Améndola-Massiotti
Universidad Autónoma Chapingo, Posgrado en Producción Animal, Programa de Investigación en Forrajes, Carretera México-Texcoco.Km 38.5, Chapingo, Edo. Méx., CP56230, México, Email: r_amendola@yahoo.com

Keywords: heifers, ewes, pasture utilisation, replacement series

Introduction The outcome of mixed grazing depends on the degrees of complementarity and competition between animal species. Complementarity increases the utilisation of herbage resource but competition may be desirable when one grazing species has a higher priority ranking in the farming system. A species wins in the competition by harvesting a higher proportion of the available herbage than the other (Nicol, 1997). De Wit (1960) used the replacement series based on degrees of substitution of species, for the quantification of the outcome of mixtures experiments. The use of species equivalence is required in order to apply this approach to the analysis of the outcome of a mixed grazing experiment. This paper aims to discuss the competition between heifers and ewes in a dairy system, where heifers have a higher priority ranking. Data on species equivalence, reported in Paper 1, were used for this purpose.

Materials and methods The species equivalence based on average herbage dry matter intake (DMI) estimated in Paper 1 corresponded with the estimate of DMI requirements based on live weight (LW), using $LW^{0.9}$. Analysis using the replacement series approach was carried out *a posteriori*. Species proportions (SP) were calculated as SP= $\sum LW^{0.9}$ of the species/$\sum LW^{0.9}$ of both species (average LW 361 and 75.7 kg for heifers and ewes, respectively). DMI/ha of each species (DMI_a) was calculated as $II_a \times n_a / AA \times 10000$, where II_a is the DMI/animal of species a, n_a is the number of animals of species a and AA is the daily area allotted (m^2). The proportions DMI/ha of each species in total DMI/ha were calculated thereafter and linear regression equations of those proportions on the proportions in total $LW^{0.9}$ were developed.

Results Total DMI increased with the proportion of ewes, i.e., the inclusion of ewes increased the efficiency of herbage utilisation (Figure 1a). There was a narrow relationship (R^2=0.97, p<0.0003) between proportions of DMI and proportions of $LW^{0.9}$, with small deviations from the 1:1 relationship (Figure 1b). Both species appeared to be more competitive to some extent when in high proportions of total $LW^{0.9}$. Therefore, heifers were slightly more competitive than ewes if their proportion in total LW^{09} was >0.47 (Figure 1b).

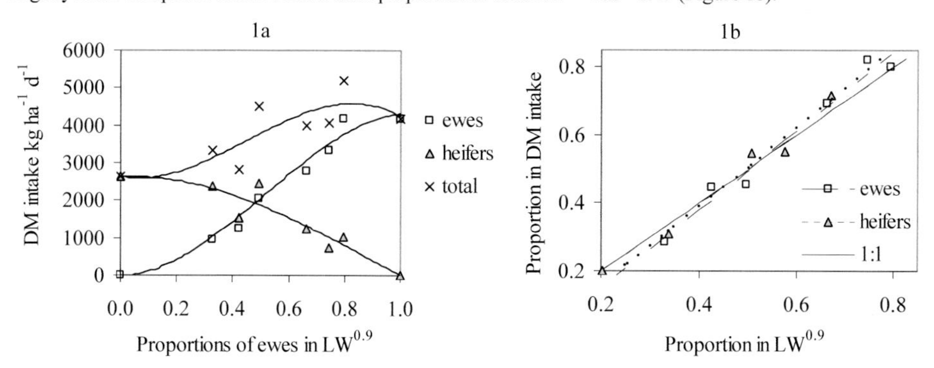

Figure 1 Herbage DMI of heifers and ewes related to proportions of ewes in total $LW^{0.9}$ (1a) and DMI of each species as proportion of total DMI related to proportions of the species in total $LW^{0.9}$ (1b)

Conclusions Ewe inclusion increased the efficiency of utilisation of herbage. One might recommend ewe inclusion at <0.53 of total $LW^{0.9}$ because that gave a slight competitive advantages to heifers (the species with higher priority ranking in the farming system).

References

de Wit, C. T. W. (1960). On competition. Verslag Landbouwkundig Onderzoek No. 66, Pudoc, Wageningen, The Netherlands, 1-82.

Nicol, A. M. (1997). The application of mixed grazing. In: Buchanan-Smith, J. G., L. D. Bailey & P. Mc Caughey (Eds.) *Proceedings of the XVIII International Grassland Congress*, Winnipeg and Saskatoon, Canada, 525-534.

Effect of strain of Holstein-Friesian cow and feed system on reproductive performance in seasonal-calving milk production systems over four years

B. Horan[1,2], J.F. Mee[1], M. Rath[2], P. O'Connor[1] and P. Dillon[1]
[1]*Dairy Production Department, Teagasc, Moorepark Production Research Centre, Fermoy, Co. Cork, Ireland, Email: bhoran@moorepark.teagasc.ie,* [2]*Department of Animal Science, Faculty of Agriculture, University College Dublin, Belfield, Dublin 4, Ireland*

Keywords: Holstein-Friesian, cow strain, feed system, reproductive performance

Introduction In Ireland most dairy farms operate seasonal calving grass-based milk production systems. Feed demand and supply are matched by having calving highly concentrated in spring. This requires high pregnancy rates within a short time following the start of mating in late April or early May, but has become increasingly difficult to achieve due to declining fertility in Irish dairy herds (Mee, 2004). In New Zealand, cows of North American Holstein-Friesian origin have poorer fertility than New Zealand Holstein-Friesians on pasture-based seasonal calving systems (Harris & Kolver, 2001). The present study sought to determine the effect of strain of Holstein-Friesian (HF) cow and feed system on reproductive performance within Irish milk production systems.

Materials and methods Three strains of Holstein-Friesian (HF) cows: high production North American (HP), high durability North American (HD) and New Zealand (NZ) were assigned, within strain, to one of three pasture-based feed systems: Moorepark (MP; 350 kg concentrate/cow, stocking rate of 2.47 cows/ha), high concentrate (HC; 1,500 kg concentrate/cow, stocking rate of 2.47 cows/ha), and high stocking rate (HS; 350 kg concentrate/cow, stocking rate of 2.74 cows/ha). The total number of animals in each of the four years ranged from 99 to 126. Cows were bred by artificial insemination (AI) over a 13-week period, starting in late April. Pregnancy detection was performed by ultrasound imaging at 30 to 37 d and again 60 to 67 d after AI. A final manual pregnancy examination was carried out 150 d after mating start date. Data for number of services were analysed using a non-parametric model (PROC NPAR1WAY) (SAS, 2002). Differences in 24-d submission rate and pregnancy rates were investigated taking account of year, parity, strain of HF and feed system using the PROC GENMOD (binomial distribution and logit link function) procedure of SAS (SAS 2002).

Results There was no significant interaction between strain of HF and feed system for any of the reproductive variables measured and therefore only the main effects are shown (Table 1). The HP strain received a greater number of services per cow and had a lower submission rate than the other strains. The NZ strain had a higher pregnancy rate to first service and six-week pregnancy rate compared to the HP strain, with the HD strain intermediate. The HP strain had the lowest overall pregnancy rate and the NZ strain had the highest. Feed system had no significant effect on reproductive performance.

Table 1 Effect of strain of Holstein-Friesian and feed system on reproductive performance (2001-2004)

	Strain of HF			Sig.†	Feed system			Sig.†
	HP	HD	NZ		MP	HS	HC	
24-day submission rate (%)	78[a]	90[b]	88[b]	**	90	79	87	NS
Conception rate to first service (%)	45[a]	54[ab]	62[b]	**	56	51	54	NS
Services per cow (no.)	2.07[a]	1.79[b]	1.61[b]	**	1.86	1.77	1.83	NS
6-week pregnancy rate (%)	54[a]	65[b]	74[b]	***	66	65	61	NS
Overall pregnancy rate (%)	74[a]	86[b]	93[c]	***	84	82	86	NS

HP = High production, HD = High durability, NZ = New Zealand, MP = Moorepark feed system, HS = High stocking rate feed system, HC = High concentrate feed system, [a b] Means with different superscripts within the same row are significantly different ($P<0.05$). Significance: ***= $P<0.001$, **= $P<0.01$, NS = Not significant

Conclusions Both the NZ and HD strains, selected for lower milk production and better reproductive traits, had better reproductive performance than a North American HF strain selected for high milk production. The results indicate that offering higher levels of concentrate supplementation will not alleviate the reduced reproductive performance of the HP strain in a pasture-based feed system.

References

Harris, B.L. & E.S. Kolver (2001). Review of Holsteinization of intensive pastoral dairy farming in New Zealand. *Journal of Dairy Science*, 84 (E. Supplement), E56-E61.

Mee, J.F. (2004). Temporal trends in reproductive performance in Irish dairy herds and associated risk factors. *Irish Veterinary Journal*, 57, 158-166.

Pasture intake and milksolids production of different strains of Holstein-Friesian dairy cows

J.L. Rossi[1], K.A. Macdonald[2], B.S. Thorrold[2] and C.W. Holmes[1]
[1]*Institute of Veterinary, Animal and Biomedical Sciences, Massey University, Private Bag 11-222, Palmerston North, New Zealad, Email:Jose.Rossi@Dexcel.co.nz,* [2]*Dexcel, Private Bag 3221, Hamilton, New Zealand*

Keywords: Holstein-Friesian, dairy systems, pasture intake

Introduction Cows of high yield potential require high daily dry matter intakes (DMI) to meet their increased energy demand. For this reason, DMI may be constrained in a pasture-based system. Daily milksolids yield and DMI of three strains of Holstein-Friesian dairy cows farmed at low and high feeding level during season 2002-2003 are reported.

Materials and methods Three strains (S) of Holstein-Friesian dairy cows [High breeding worth (merit) cows of overseas (OS90) or New Zealand (NZ90) origin and a 1970 NZ Friesian strain (NZ70)] were farmed in a range of feeding systems (self contained farmlets, 15-20 cows each). Feeding level (FL) in the systems ranged from 4.5 to 7.0 t DM/cow per year based on different stocking rates, supplement inputs (maize grain and silage) and the different adult liveweight of the strains (Rossi *et al.*, 2004). Daily milksolids production, body condition score (BCS) and DMI were recorded. Intake was estimated using the *n*-alkane and the $\delta^{13}C$ techniques (Dove & Mayes, 1991; Garcia *et al.*, 2000). Data collected in spring and autumn from the lowest (pasture only) and highest (pasture only in spring but supplemented in late lactation) FL is presented. Data were analysed as a mixed model (SAS) with S, FL and their interactions as fixed effects and cow as a random effect.

Results The NZ90 and OS90 strains had greater milksolids yield ($P<0.001$) and intake ($P<0.05$) than the NZ70 in spring (Table 1). In autumn, both high merit strains received more supplement at the high FL. Milksolids yields were higher ($P<0.001$) for them and an S*FL interaction for total DMI ($P<0.05$) was measured. There was a trend for a larger DMI for the NZ90 than for the OS90 in spring ($P=0.07$) but similar in autumn. In addition, the OS90 lost more BCS in early lactation (during September) ($P<0.001$). Milksolids yield and DMI were similar between FL in spring, however in autumn, milksolids yields were greater at the high FL ($P<0.05$). Pasture DMI across all strains was reduced at high FL in autumn ($P<0.001$) due to supplementation, however total DMI increased for the NZ90 and OS90.

Table 1 Daily milksolids yield and DMI (both in kg/cow) during early and late lactation

S	NZ70		NZ90		OS90					
FL	Low	High	Low	High	Low	High	sed	S	FL	S*FL
FL per cow (t DM /year)	4.5	6.0	5.0	6.5	5.5	7.0				
Early Lactation (spring)										
Milksolids yield	1.41	1.53	1.92	2.01	1.88	1.94	0.13	***	NS	NS
Pasture DMI	13.04	13.77	15.89	14.88	14.47	14.57	0.79	**	NS	NS
BCS change	-0.14	-0.15	-0.28	-0.12	-0.39	-0.38	0.10	***	NS	NS
Late Lactation (autumn)										
Milksolids yield	0.94	0.93	1.19	1.42	1.03	1.21	0.14	***	*	NS
Pasture DMI	12.64	9.78	14.05	11.14	14.42	10.04	0.83	*	***	NS
Supplement DMI	-----	3.00	-----	6.74	-----	6.64	0.80	***	-----	-----
Total DMI	12.64	12.78	14.05	17.88	14.42	16.68	1.06	***	***	*

sed: maximum; S: strain; FL: feed level. * $P<0.05$; ** $P<0.01$; *** $P<0.001$.

Conclusions Although both NZ90 and OS90 produced similar milksolids yield in early lactation, the greater pasture intake of the NZ90 provided a higher proportion of their daily requirements, which was associated with a lower loss in BCS. In late lactation, all the strains ate less pasture when supplemented, however, a lower reduction in pasture DMI was observed in the NZ90 strain. These results indicate a greater constraint for the OS90 strain under a New Zealand pasture-based system.

References

Dove H. & R. W. Mayes (1991). The use of the plant wax alkanes as marker substances in studies of the nutrition of herbivores: a review. *AustralianJournal of Agricultural Research*, 42, 913-952.

Garcia S.C., C. W. Holmes, J. Hodgson, & A. Macdonald (2000). The combination of the n-alkanes and ^{13}C techniques to estimate individual dry matter intakes of herbage and maize silage by grazing dairy cows. *Journal of Agricultural Science*, 135, 47-55.

Rossi J.L., K. A. Macdonald, B. S. Thorrold, C. W. Holmes, C. Matthew, & J. Hodgson (2004). Milk production and grazing behaviour during early lactation of three strains of Holstein-Friesian dairy cows managed in different feeding systems. *Proceedings of the New Zealand Society of Animal Production*, 64, 232-236.

Does the feeding behaviour of dairy cows differ when fed ryegrass indoors vs. grazing?

A.V. Chaves, A. Boudon, J.L. Peyraud and R. Delagarde
UMRPL – INRA, 35590, St-Gilles, France. Email: Alexandre.Chaves@rennes.inra.fr

Keywords: dairy cows, eating time, grass, ruminating time

Introduction Dairy cows eating ryegrass ingest smaller boli when grazing than when fed indoors (93 vs. 142 g; Boudon *et al*., 2004). To investigate whether this difference in bolus affects feeding behaviour of the cows, an automated system (chewing halters) was used to monitor feeding behaviour of cows given ad libitum access to perennial ryegrass in individual feed troughs (indoor feeding, IF) or at pasture (grazing, GR).

Material and methods In a study comprising four 14-d periods (d 1-7 for adaptation to halters; d 8-14 for measurements), eight lactating, ruminally fistulated Holstein cows were provided ryegrass by access to pasture (GR) or as forage cut and fed indoors (IF) in a 4 × 2 crossover. For GR, pastures with 21 d of re-growth were allotted to provide each cow with 45 kg/d of herbage DM when grazed to ground level. Dry matter intake (DMI) was estimated from total faecal output and organic matter digestibility (OMD) determined using intraruminally dosed dye (for faecal pat identification) and ytterbium oxide. For IF, herbage cut twice per day with a forage harvester, to the same height as GR pastures after grazing, was offered in seven meals (Figure 1). Refusals were collected at 1800 and 0830 h. Dry matter intake was calculated as DM offered minus DM refused, and by YbO as for GR. Chewing and ruminating activities of each cow were monitored for 24-h periods using halter-mounted devices linked to a data logger (Bechet *et al.*, 1989). Intake rates were calculated daily as DMI/eating time.

Results Ryegrass nutritive value (17% DM, 20% CP, 45% NDF, 0.81 OMD) was similar between IF and GR. Milk yield was higher ($p<0.02$) with IF than with GR (25.6 vs. 23.6 kg/d, respectively). Average DMI was 16.7 kg/d with IF, but estimates from the YbO marker were 18.6 and 16.5 kg/d for IF and GR, respectively ($p<0.01$). Times spent eating (604±94.5 min/d) or ruminating (456±27.7 min/d) were similar ($p\geq0.14$) between IF and GR, but feeding behaviours (min/h eating or ruminating) differed with treatment (Figure 1). The IF cows spent more time at night (2200 to 0600) eating than did GR cows ($p<0.02$), who spent more time ruminating. Meal duration was longer with GR than with IF (average 79 min vs. 44 min; $p<0.0001$), and rate of DM intake by GF cows was lower (27.1 g/min vs. 31.0 g/min with IF).

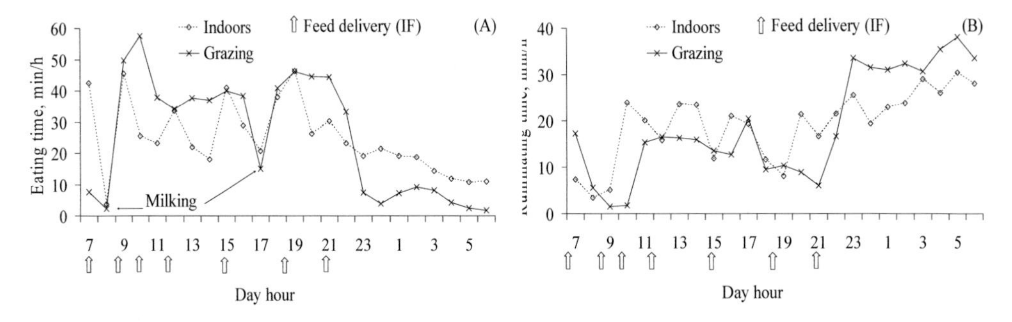

Figure 1 Proportions of time (min/h) cows spent eating (A) and ruminating (B) when grazing or fed indoors.

Conclusions Differences in feeding behaviour of cows with ad libitum access to good quality ryegrass either as cut forage (provided in 7 meals/d) or as pasture (grazing) were reflected in different DMI and rates of intake. Grazing cattle consumed 13% less DM, which appeared to be related to relatively less time spent eating at night. For grazing cows, intake per bite seemed to decline as ryegrass pasture availability decreased during the day, but increased time spent eating (during daylight) was insufficient to maintain intake compared to indoors feeding cows.

References

Boudon, A., A. Acosta, R. Delagarde & J.L. Peyraud (2004). Effect of grazing versus indoors feeding on the damage done to the grass during ingestive mastication. Animal Feed Science and Technology, 13 (Suppl. 1), 35-38.

Bechet, G., M. Theriez, & S. Prache (1989). Feeding behaviour of milk-fed lambs at pasture. Small Ruminant Research, 2, 119-132.

Variation between individuals in voluntary intake and herbage intake of grazing dairy cows

H.M.N. Ribeiro Filho[1], R. Delagarde[2], L. Delaby[2] and J.L. Peyraud[2]
[1]*Universidade do Estado de Santa Catarina, Lages-SC, Brazil, Email: a2hrf@cav.udesc.br,*
[2]*Institut National de la Recherche Agronomique, UMR Production du Lait, Saint-Gilles, France*

Keywords: dairy cows, grazing, herbage allowance, milk yield, voluntary intake

Introduction Herbage intake and milk yield of unsupplemented grazing dairy cows are highly variable between animals within a herd (Delaby et al., 2001). The objective of this experiment was to describe the relationship between the individual voluntary intake (VI) of dairy cows measured before turnout and their herbage intake at grazing, at two herbage allowances.

Materials and methods The individual voluntary intake of a total mixed ration (VI_{TMR}, maize silage/concentrate ratio: 70/30) was measured indoors for 3 weeks in March with 16 Holstein dairy cows that were 90 ± 11 days in lactation. After a transition period in April, cows strip-grazed exclusively pure ryegrass or ryegrass-white clover mixtures, at two herbage allowances (HA: Low 20 and Medium 35 kg DM/cow per day to ground level). Individual herbage intake at grazing was estimated in May and June in four 10-day periods, twice per cow for each HA, as described by Ribeiro Filho *et al.* (2005). The expected VI (eVI) during the grazing period was estimated assuming a theoretical decrease from the VI_{TMR} of 0.57 kg DM/day per month (Faverdin *et al.,* 1987). The expected milk yield (eMY) was estimated from the milk yield of the reference period and assuming a theoretical weekly persistency of 0.98 (Delaby *et al.,* 2001).

Results The herbage intake at grazing averaged 64% of the eVI (14.6 vs. 22.9 kg DM/day), and increased between cows by 0.34 kg DM/day for each kg of eVI. However, in the range observed, herbage intake at grazing decreased from 75 to 60% of the eVI with increasing eVI. The between-cow range of intake at grazing was independent of the herbage allowance. Even under very severe grazing conditions (Low HA), cows with highest VI_{TMR} ingested 3 kg DM/day more than cows with lowest VI_{TMR}. The herbage intake at grazing increased between cows by 0.27 kg DM/day for each kg of eMY.

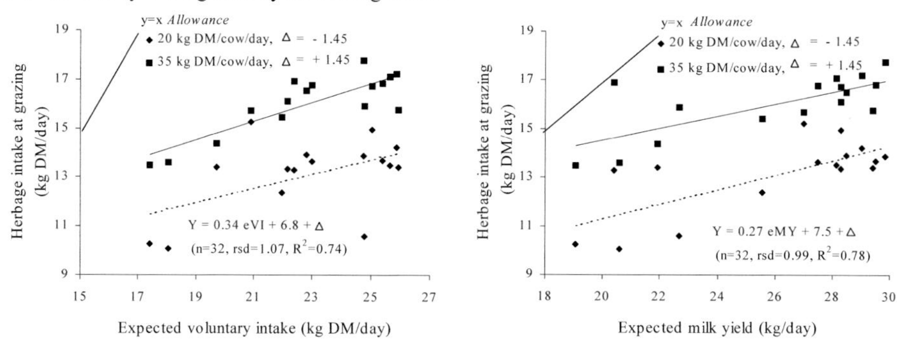

Figure 1 Relationship for individual cows between the expected voluntary intake and milk yield during the grazing period and the actual herbage intake at grazing.

Conclusions The herbage intake of grazing dairy cows largely depends on their individual voluntary intake capacity independently of the grazing conditions. The range in herbage intake due to the variability of voluntary intake between individuals is larger than that associated with differences in herbage allowance.

References

Delaby, L., J. L. Peyraud & R. Delagarde (2001). Effect of the level of concentrate supplementation, herbage allowance and milk yield at turnout on the performance of dairy cows in mid lactation at grazing. *Animal Science*, 73, 71-181.

Faverdin, P., A. Hoden & Coulon, J.B. (1987). Recommandations alimentaires pour les vaches laitières. INRA Bulletin Technique CRZV Theix, 70, 133-152.

Ribeiro Filho, H.M.N., R. Delagarde & J. L. Peyraud (2005). Herbage intake and milk yield of dairy cows grazing perennial ryegrass swards or white clover/perennial ryegrass swards at low and medium herbage allowances. *Animal Feed Science and Technology*, 119, 13-27.

Relationships between traits other than production and longevity in New Zealand dairy cows

D.P. Berry[1], B.L. Harris[2], A.M. Winkelman[2] and W. Montgomerie[3]
[1]*Teagasc, Moorepark Research Centre, Fermoy, Co. Cork, Ireland, Email: dberry@moorepark.teagasc.ie*
[2]*Livestock Improvement, Private Bag 3016, Hamilton, New Zealand* [3]*Animal Evaluation Unit, Private Bag 3016, Hamilton, New Zealand.*

Keywords: survival analysis, longevity, New Zealand, dairy

Introduction Reduced longevity in dairy cattle is recognised world-wide to be a considerable cost to the dairy industry, especially in seasonal calving grazing production environments. The objective of the present study was to investigate the relationships among traits other than production (TOP) and true and functional longevity in purebred and crossbred New Zealand cows from commercial herds operating seasonal calving grass-based systems of milk production. This study made use of survival analysis, a technique used to allow the inclusion of incomplete (i.e., censored) longevity data in the analysis while simultaneously accounting for the skewed distribution of longevity data and the changing environmental and genetic effects over time.

Materials and methods Data on 16 TOP (4 management traits and 12 conformation traits) on primiparous cows were extracted from the New Zealand national database. Snell's transformations were used to minimise skewness and kurtosis in TOP scores. Each TOP was pre-adjusted for age at calving (nested within breed) and stage of lactation. Residuals were normalised within contemporary group of herd-year-season of calving. Normalised residuals were coded as a qualitative variable with 20 levels. Longevity data, including calving dates, date of culling and reasons for culling were extracted from the national database on the 15 March 2004. Records were censored if the cow was spring calving and had an official record after 1 June 2003 but no subsequent culling record; records of autumn calving cows were censored if the animal had an official record after 1 January 2003 but no culling record. Animals were also treated as censored if they entered a contemporary group with less than four non-censored records. Only animals residing in sire proving herds (indicative of commercial herds) were retained in the analysis. In total, 259,280 animals were included in the analysis. The survival analysis was undertaken using a proportional hazards Cox model and was stratified by breed; breed was defined as Holstein-Friesian, Jersey or crossbred. The hazard function was described by the baseline hazard function, herd-year contemporary group (class variable [CL]; time dependent [TD]), pedigree registration status (CL; time independent [TI]), age at first calving (CL; TI), heterosis (CL; TI), proportion of overseas Holstein-Friesian, New Zealand Holstein-Friesian and Jersey genes (CL; TI), period of last calving (CL; TI), type score relative to the contemporary group (CL; TI). Production values (CL; TI) and milk production variables (CL; TD) were only included in the analysis of functional longevity. Tests of significance for explanatory variables were based on the Akaike's information criterion.

Results Risk of being culled was higher in later calving cows, non-registered cows with lower milk production. The risk of culling also increased as the proportion of overseas Holstein-Friesian genes increased, except at high levels >80% when it decreased again. All TOP affected ($P<0.001$) true and functional longevity. Of the TOP, farmer opinion of the cow had the greatest influence on true and functional longevity. Two composite traits, overall udder and dairy conformation, also exhibited some of the strongest influences on cow longevity. Of the individual TOP describing the conformation of the cow, the udder-related traits had the largest influence on longevity. Rump angle, rump width and legs had least influence on longevity. Across all management traits, cows with strongly undesirable scores were at a higher risk of being culled than cows of intermediate scores, but cows with high scores conferred no additional advantage in longevity. The risk of culling in cows of poor farmer opinion was 1.2 to 2.5 times that of a cow of intermediate farmer opinion. A similar trend of lower scores being associated with a higher risk of culling was observed for stature, udder support, fore udder attachment, rear udder height and dairy conformation. Legs showed an opposite trend; a higher risk of culling existed in cows with higher legs scores (i.e., more sickled legs). Capacity, rump angle, rump width and teat placement exhibited intermediate optima with an increased risk of culling observed in extreme scores.

Conclusions Results from this study indicate a strong association between farmer opinion, scored in first lactation, and longevity. Other farmer scored traits and udder-related traits were also strongly related to true and functional longevity in commercial herds. The results clearly demonstrate the suitability of TOP, especially farmer scored traits, as phenotypic indicators of the ability of the cow to delay voluntary and involuntary culling in grass based system of production. Several traits, most notably legs and teat placement, exhibited an intermediate optimum to avoid voluntary/involuntary culling.

Acknowledgements The Irish Holstein-Friesian Association is acknowledged for their support of this study.

Section 3

Developments in modelling of herbage production and intake by grazing ruminants

Modelling the effect of breakeven date in spring rotation planner on production and profit of a pasture-based dairy system

P.C. Beukes, B.S. Thorrold, M.E. Wastney, C.C. Palliser, G. Levy and X. Chardon
Dexcel Ltd, Private Bag 3221, Hamilton 2001, New Zealand, Email: pierre.beukes@dexcel.co.nz

Keywords: grass silage, farm cover, management

Introduction The breakeven date is the expected date when pasture supply exceeds cow demand. This date is used to plan the rotation rates, slow during the winter, when pasture growth is low and cows are dry, to a fast rotation in spring, when growth is accelerating and most cows lactating. This date is influenced by regional climate, mainly rainfall and soil temperature, which affects timing and rate of growth acceleration. The objective of this modeling exercise was to explore the effect of the breakeven date on milksolids (MS), grass silage, farm cover and economic farm surplus (EFS) over different climate years for the Canterbury region of New Zealand.

Materials and methods Observed starting farm covers, herd data, silage stacks, irrigation (585 mm) and fertilizer (200 kg N/ha) schedules from Lincoln University Dairy Farm (LUDF) near Christchurch, Canterbury for the 2002/03 season were used to initialize Dexcel's Whole Farm Model (WFM, Wastney *et al.*, 2002), first with the observed breakeven date of 25 September (Strategy 1) and then with a breakeven date of 20 October (Strategy 2). Both strategies were simulated for five seasons (1994/95 to 1998/99) using observed climate data from LUDF and a stocking rate of 3.65 cows/ha. Seasons were simulated individually (no carry-over) starting 1 June for 365 days. Economic results are presented assuming a payout of NZ$ 3.90/kg MS.

Results Averages of model predictions over five climate years are given in Table 1, and model predictions of seasonal changes in farm covers for the 1996 season are shown in Figure 1. An early breakeven date (Strategy 1) resulted in faster rotations earlier in the season with the consequence that lactating cows obtained a larger proportion of their demand from pasture and could be fed less grass silage compared to Strategy 2 (Table 1). The faster rotations earlier resulted in slower recovery of farm covers for Strategy 1, however, covers had recovered to Strategy 2 level before Christmas (Figure 1). With higher covers early in the season more silage could be made with Strategy 2 (Table 1). Generally grass silage has lower nutritional value than spring pasture (Holmes *et al.*, 2002), which explains why MS production per cow and per hectare was lower for Strategy 2 (Table 1). The effect of silage feeding on production was most pronounced in the early part of the season when the breakeven date determines rotation length and the proportion of the daily intake from pasture and from silage. The significantly higher EFS for Strategy 1 (Table 1) confirmed the benefit of feeding quality pasture during peak lactation, although it might be at the expense of farm covers and silage making.

Table 1 Predicted averages (±SD) with results of paired t-tests

Parameter	Strategy 1	Strategy 2	Significance (P)
Silage fed (kg DM/ha)	1882±702	2050±726	0.19
Silage made (kg DM/ha)	2557±535	2816±540	0.12
MS (kg/cow)	399±5.5	394±4.9	0.02
MS (kg/ha)	1455±20	1439±19	0.01
EFS (NZ$/ha)	2261±265	2180±256	0.02

Figure 1 Effect of breakeven date on average farm cover for the 1996 season

Conclusions The results of this modeling exercise show that for the Lincoln farm a breakeven date of 25 September for the spring rotation planner is more profitable than a date of 20 October. Optimum breakeven dates for different farms will depend on stocking rate and local climate.

References

Wastney, M.E., C.C. Palliser, J.A. Lile, K.A. Macdonald, J.W. Penno & K.P. Bright (2002). A whole-farm model applied to a dairy system. *Proceedings of the New Zealand Society of Animal Production,* 62, 120-123.
Holmes, C.W., I.M. Brookes, D.J. Garrick, D.D.S. Mackenzie, T.J. Parkinson & G.F. Wilson (2002). Milk production from pasture. Principles and practices. Massey University, Palmerston North, New Zealand, 601.

Development of a model simulating the impact of management strategies on production from beef cattle farming systems based on permanent pasture

M. Jouven and R. Baumont
INRA, Unité de Recherches sur les Herbivores, Centre de Clermont-Theix-Lyon, F-63122 Saint Genès Champanelle, France, Email: mjouven@clermont.inra.fr

Keywords: simulation, grazing systems, animal intake, animal performance, vegetation dynamics

Introduction Grazing systems in Europe increasingly have to meet environmental objectives, which influence management strategies. A deterministic model describing farming system dynamics is being developed in order to elucidate interactions between nature-friendly management practices, as for example late (after flowering) hay harvest or moderate stocking rate, and agricultural output.

Model presentation The model predicts, with a daily time step, farm system operation and agricultural output from a given farm structure and management strategy. It is built and calibrated for non-intensive French beef cattle farming systems based on permanent pasture. The model is made up of four interacting sub-models. A grassland resource sub-model adapted from Carrère *et al.* (2002) predicts grass growth and quality at the paddock level, from soil quality, vegetation functional traits and climatic data. A herd performance sub-model based on INRA (1989) calculates weight gain and milk production from energy intake, for an average cow and calf. A feeding sub-model predicts selective intake at pasture for an average cow and calf, and herd feeding indoors. A management sub-model decides on herd movements, concentrate supplementation to achieve intermediate objectives (calf weight at sale, cow body condition score at calving) and day of hay harvest. The whole model and individual sub-models are undergoing sensitivity analysis and validation.

Simulation example The effects of stocking rate and proportion of late hay harvest were examined in a 2x2 factorial design. The model was run for 6 climatic years, with stocking rate (SR) 1.4 or 1.2 LSU/ha, half the surface planned for hay harvest (half for 1 cut, half for 2 cuts), with 50% or 100% late 1st cut, and calves sold at 9 months, 320 kg. SR change was performed by altering the number of animals. Simulation results are given in Table 1; the model being deterministic, standard deviations refer to discrepancies between climatic years.

Table 1 Simulated annual agricultural output (mean ± SD); DOM=digestible organic matter

Management strategy	Days at pasture	Grazed DOM		Harvested DOM(t/cow)	Concentrate consumption	
		(t/calf)	(t/cow)		at pasture (kg/calf)	housed (kg/cow)
SR 1.4 – 50% late	172 ± 7	0.35 ± 0.08	1.69 ± 0.07	1.27 ± 0.12	180 ± 13	36 ± 33
SR 1.2 – 50% late	178 ± 10	0.34 ± 0.03	1.75 ± 0.10	1.47 ± 0.13	184 ± 13	20 ± 23
SR 1.4 – 100% late	170 ± 7	0.32 ± 0.02	1.68 ± 0.07	1.15 ± 0.16	204 ± 57	64 ± 40
SR 1.2 – 100% late	177 ± 10	0.34 ± 0.03	1.74 ± 0.09	1.30 ± 0.19	204 ± 50	53 ± 49

Lowering SR increases grass availability at pasture, and thus lengthens grazing season (+7 days); consequently, grazed DOM per cow is higher (1.75 vs 1.69 t/cow), even though grazed DOM per day is unchanged (9.86 kg/cow). Lowering SR also increases harvested DOM per cow (1.39 vs 1.21 t/cow); therefore, winter concentrate consumption per cow tends to be reduced (37 vs 50 kg/cow). 100% late 1st cut has little effect on grazed DOM at pasture, but reduces harvested DOM per cow (1.23 vs 1.37 t/cow) via a decrease in mean hay digestibility (0.59 vs 0.63) that is not compensated by an increase in dry matter yields (2.01 vs 2.02 t/cow). Thus, concentrate consumption for cows is higher for 100% late 1st cut (59 vs 28 kg/cow). Overall, whatever the management strategy, calves performance remains unchanged. Simulated agricultural output varies widely with climatic years, as in a real farm.

Conclusions This model may offer a useful tool to support discussion between research, advisory and environmental services. It can be enriched by new management rules and a biodiversity score sub-model.

References

Carrère P., C. Force, J.F. Soussana, F. Louault, B. Dumont & R. Baumont (2002). Design of a spatial model of perennial grassland grazed by a herd of ruminants: the vegetation sub-model. *Grassland Science in Europe*, 7, 282-283.

INRA (1989). Ruminant Nutrition: recommended allowances and feed tables. Jarrige, R. (ed.), John Libbey Eurotext, Paris, 389 pp.

Intake by lactating goats browsing on Mediterranean shrubland

M. Decandia, G. Pinna, A. Cabiddu and G. Molle
Istituto Zootecnico e Caseario per la Sardegna 07040 Olmedo, Italy, Email: mdecandia@tiscali.it

Keywords: intake, dairy goats, shrubland

Introduction In Mediterranean regions goat feeding systems are mainly based on shrubland that contain a wide variety of species. There are only a few equations for predicting feed intake of stall-fed goats (Luo *et al.*, 2004). The objective of this study was to develop a model for predicting the intake of lactating goats browsing on Mediterranean shrubland.

Materials and methods A database of mean treatment observations (N=44) from goat feeding studies was analysed. The studies were conducted with dairy goats that browsed for 5-7 hrs/day on 5 ha shrubland (Decandia *et al.*, 2000a,b). The goats received 200 g of concentrate and 200 g of ryegrass hay/day. The shrubland contained many tanniferous species and a low proportion of herbaceous species. The variables analysed were: body weight and body condition score (BCS); supplement intake; pasture intake, and botanical composition of the diet (Kababya *et al.*,1998); chemical composition of the diet (Meuret *et al.,* 1995); botanical composition of pasture (Daget & Poissonet, 1969); fat corrected milk (FCM) and milk urea.

Results The variables most strongly related to the intake were FCM, CP and polyphenolic tannin level in the diet. The effect of supplementation was not significant. Two prediction equations for pasture intake were (1) DMI = -18.63 + 6.75 CP + 0.02 FCM; N=38; R^2=0.77; P<0.001; (2) DMI = 52.54 +0.037 FCM – 16.44 PT/CP; N=40; R^2=0.59; P<0.001, where: DMI=DM intake (g /Kg $BW^{0.75}$); FCM=fat (4 %) corrected milk yield (kg); (% DM); PT/CP= ratio between polyphenolic tannins and CP in the diet. Equation 1 has a higher R^2 and is easier to implementation at farm scale than Equation 2. Comparing predicted with observed values, a strong relationship was found (R^2=0.77; P<0.001; a, b not statistically different from 0 and 1, Figure 1). CP in the diet was related to the percentage of grass (GRAP) in the pasture and milk urea level (mg 100 ml^{-1}; MU): (3)CP = 5.10 + 0.20 GRAP + 0.119 MU; N=31; R^2=0.82; P<0.001.Using this relationship, a two-step prediction model of DM intake of browsing goats is proposed. Step 1: on the basis of GRAP and MU, CP content in the diet can be estimated (Equation 3). Step 2: knowing FCM and dietary CP level, the DMI at pasture can be predicted (Equation 1).

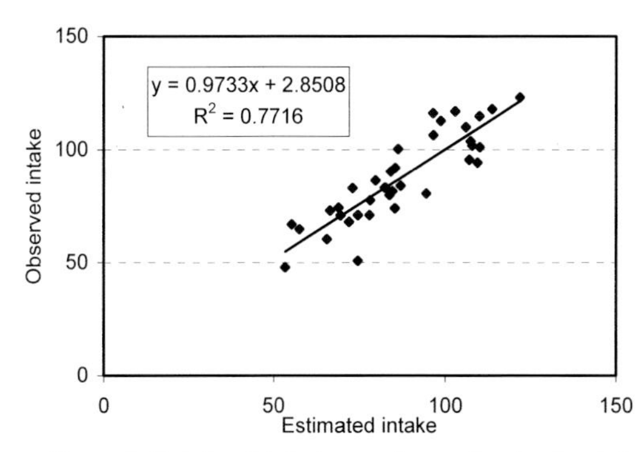

Figure 1 Relationship between observed and estimated pasture DM intake (g/kg $BW^{0.75}$)

Conclusions This model provides a useful tool for estimating the intake of goats browsing Mediterranean shrubland rich in tanniferous species, with a low percentage of herbaceous species.

Acknowledgement This study has been funded by Italian Ministry of Agriculture and Forestry.

References

Decandia M., Sitzia M., Cabiddu A., Kababya D., Molle G., (2000a). The use of polyethylene glycol to reduce the anti-nutritional effects of tannins in goats fed woody species. *Small Ruminant Research* 38, 157-164.

Decandia M., Molle G., Sitzia M., Cabiddu A., Ruiu P.A., Pampiro F., Pintus A., (2000b). Responses to an antitannic supplementation by browsing goats. *Proc of VII° International Conference on goats* 15-18 May 2000 Tours, France. pp 71-73.

Luo J., Goetsch A. L., Nsahlai, I. V., Moore, J. E., Galyean, M. L., Johnson, Z. B., Sahlu, T., Ferrell C. L., Owens F. N. (2004) Voluntary feed intake by lactating, Angora, growing and mature goats. *Small Ruminant Research* 53, 357-378.

Kababya D., Perevolotsky A., Bruckental I., Landau S., (1998). Selection of diets by dual-purpose mamber goats in Mediterranean woodland. *Journal of Agricultural Science Cambridge* 131, 221-228.

Meuret M., Bartiaux-Hill N. and Bourbouze A., (1985). Evaluation de la consommation d'un troupeau de chèvres laitières sur parcours forestier: - Méthode d'observation directe des coups de dents, - Méthode du marquer oxyde de chrome. *Annales de Zootechnie*, 34: 159-180.

The impact of concentrate price on the utilization of grazed and conserved grass

P. Crosson[1,2], P. O'Kiely[1], F.P. O'Mara[2], M.J. Drennan[1] and M. Wallace[2]
[1]*Teagasc, Grange Research Centre, Dunsany, Co. Meath, Ireland, Email: pcrosson@grange.teagasc.ie,* [2]*Faculty of Agri-Food and the Environment, University College Dublin, Belfield, Dublin 4, Ireland*

Keywords: beef production, systems, mathematical model, linear programming

Introduction A linear programming model was designed and constructed to facilitate the identification of optimal beef production systems under varying technical and policy scenarios. The model operates at a systems level and most activities that could occur in Irish spring-calving, suckler beef production systems are included. In this paper, the components of the model are described together with a simple application of the model involving changing concentrate prices.

Model description The model was developed using a mathematical programming methodology which encompasses the following characteristics: 1) a range of possible activities, 2) various constraints to prevent free selection from the range of activities, and 3) an objective which can be quantified (Dent *et al.*, 1986). It is a single year steady-state design. The fundamental unit on which the model is based is the cow unit. Due to the predominance of pasture-based systems in Ireland a detailed set of grazing options that are typical of those available to Irish cattle farmers is specified. Model details are specified on a monthly basis. This enables it to respond to monthly fluctuations in feed supply and animal requirements. Financial budgets (Teagasc, 2003) assign a cost or revenue to each activity and thus the program identifies the optimal net farm margin. Nutritional specifications are described in terms of net energy (NE) requirements subject to a maximum intake capacity.

Model application A scenario investigating the impact of a change in concentrate price on optimal systems is presented. Concentrates generally are the most costly feedstuffs and the cost of concentrates influences farm margin and system operated to a large extent. The influence of concentrate price on the optimal system is presented below (Table 1) together with the resulting impact on net margin (Figure 1). Above €140/tDM it was found that there was no response to further increases in concentrate price. Therefore, results are presented for the range €100/tDM to €140/tDM.

Table 1 Production results for concentrate price change scenario

Concentrate price (€/tDM)	100	105	110	115	120	125	130	135	140
Area for grazing (ha)	60.0	60.0	60.0	50.0	50.0	49.0	45.3	35.7	35.7
Area for grass silage (ha)	0.0	0.0	0.0	0.0	0.0	1.0	4.7	14.3	14.3
Land rented (ha)	10.0	10.0	10.0	0.0	0.0	0.0	0.0	0.0	0.0
Total N applied (kg/ha)	199.4	196.3	196.3	196.3	196.3	202.4	217.7	243.4	243.4
Concentrates fed (t)	94.5	84.5	84.5	70.4	70.4	58.8	39.1	6.0	6.0
Suckler cow numbers	61.3	61.2	61.2	51.0	51.0	53.7	56.4	56.2	56.2

Up to around €120/tDM all animals are finished on concentrate based diets but above this steer progeny are finished off grass and from about €130/tDM all progeny are finished off grass. Stock numbers initially decrease with an increase in concentrate price but recover somewhat after the change to grass-based finishing.

Conclusions Concentrate price impacts crucially on optimal systems driving both the operated finishing system and grass silage requirements. The model can be used also to analyse current or prospective scenarios. Future changes in agricultural policy can be investigated routinely. Whilst the production data are based mainly on performances obtained at Grange, the parameters can be modified to reflect other situations.

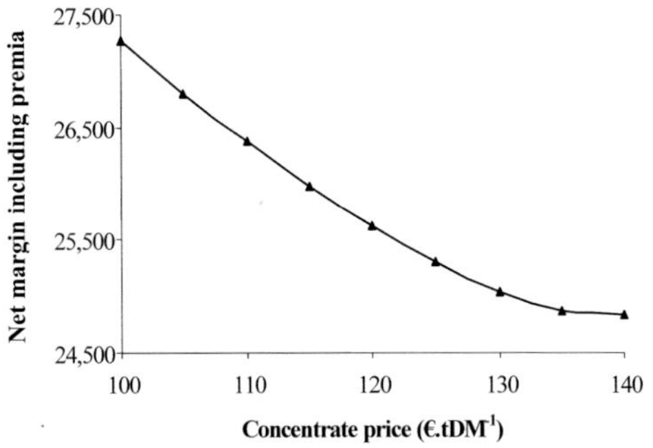

Figure 1 Change in net margin with increasing concentrate price

References

Dent J.B., S.R. Harrison & K.B. Woodford (1986). Farm Planning with Linear Programming: Concept and Practice. London: Butterworths, pp. 32-52.

Teagasc (2003). Management Data for Farm Planning, Teagasc, 19 Sandymount Avenue, Dublin 4, Ireland.

Adapting the CROPGRO model to predict growth and perennial nature of bahiagrass

S.J. Rymph[1], K.J. Boote[1] and J.W. Jones[2]
[1]*University of Florida Agronomy Department, P.O. Box 110500 Gainesville, Florida 32611 USA, Email: kjb@ifas.ufl.edu,* [2]*University of Florida Department of Agricultural and Biological Engineering, P.O. Box 110570, Gainesville, Florida 32611 USA*

Keywords: crop model, forage, tissue N concentration, tropical perennial grass, *Paspalum notatum*

Introduction The objective of this research was to modify an existing crop growth model for ability to predict growth and composition of bahiagrass (*Paspalm notatum* Flügge) in response to daily weather and management inputs. The CROPGRO–CSM cropping systems model has a generic, process-oriented structure that allows inclusion of new species and simulating cropping sequences and crop rotations. An early adaptation of CROPGRO-CSM "species files" for bahiagrass over-predicted growth during late fall through early spring, and totally failed in re-growth if all foliage was lost from freeze damage. Revised species parameters and use of "pest damage" offered only a partial solution. Three processes, absent from the annual CROPGRO-CSM model, contributed to prediction of excessive cool-season growth: (1) no provision for storage (reserve) structures, (2) lack of winter dormancy, and 3) freeze damage killed all leaves at once and resulted in crop death. In addition, the model lacked the CO_2-concentrating effect of C_4 photosynthesis in the leaf photosynthesis routines. Therefore, we modified the source code of CROPGRO to include these processes to improve biological accuracy of re-growth patterns and prediction of seasonal patterns of growth (Rymph *et al.*, 2004).

Materials and methods A new plant organ (STOR) was added to simulate stolons, thereby serving as a perenniating sink and source for storing carbohydrate and N. Partitioning of new growth among leaves, stems, roots, and STOR is assumed to shift toward STOR with increasing vegetative maturity. New functions were added to promote re-growth after harvest and in the spring. These include increasing the mobilisation of N and carbohydrate (CH_2O) from STOR progressively more rapidly as LAI falls below 3.0 and more rapidly as whole-plant N status increases from 30 to 70% of potential N-status. Functions for dormancy were added to regulate the degree of partitioning to STOR versus leaf and stem, and to regulate rate of mobilisation of CH_2O and N from STOR. Dormancy is initiated when day-length is <12.5 h, becoming progressively stronger (maximum) as day-length reaches 10.5 h. The strength of this dormancy signal acts to increase partitioning to STOR, decrease mobilisation of N and CH_2O from STOR and roots, and reduces herbage and root growth (because of partitioning shift). As day-length increases above 10.5 h in spring, the process is reversed.
The freeze damage process was modified to use a "death constant" that reduces leaf and stem mass 5% for each degree of minimum daily temperature <-5°C. A lethal freezing temperature threshold was defined as the low temperature required to kill the STOR organ.
A "CO_2-concentrating factor" was added to the leaf-level photosynthesis code to simulate C_4 photosynthesis, accounting for effect of high CO_2 concentrations in bundle sheath chloroplasts of C_4 plants. This factor reduces sensitivities of quantum efficiency and leaf photosynthetic rate to atmospheric CO_2 concentration.

Results Five experiments from Texas and Florida were simulated with the modified version of CROPGRO. Predicted herbage mass, herbage N concentration, and herbage N mass were compared to measured values (303 observations). Seasonal patterns of growth were more accurately predicted to mimic low leaf and stem growth during winter, and increase in STOR mass during fall through early spring. STOR mass is used (depleted) to support re-growth in spring and after each harvest. The crop now survives winter freeze events that kill all green foliage. Performance of the new model version was improved with better index of agreement (d-index) values for predicted herbage mass of 0.81 and 0.82 (out of a possible 1.0) using leaf-level and daily canopy photosynthesis options, respectively, compared to 0.71 and 0.73 for the older CSM version (slope, intercept, and r^2 were also improved). Prediction of herbage N concentration was similarly improved. Predicted leaf-level photosynthetic response to elevated CO_2 was of the same magnitude as observed in phytotron measurements. Predicted N-stress in the model is excessive, but the cause has not been identified. Additional work is needed on N-related parameters of the crop and soil to reduce the predicted N stress and refine model response to N.

Conclusions The modified forage version of CROPGRO marks a significant step in adapting this model to more accurately reflect the perennial and seasonal patterns of organ growth of bahiagrass. Summer vs. winter patterns of herbage production, herbage N concentration, leaf, stem, root, and stolon growth are more accurately predicted. Sensitivity analyses show reasonable leaf growth and stolon storage dynamics after cutting harvest.

References

Rymph S.J. (2004). Modeling growth and composition of tropical perennial forage grasses. Ph.D. diss. University of Florida, Gainesville.

Modelling urine nitrogen production and leaching losses for pasture-based dairying systems

I.M. Brookes and D.J. Horne

College of Sciences, Massey University, Private Bag 11 222, Palmerston North, New Zealand, Email: I.Brookes@massey.ac.nz

Keywords: dairy production, nitrogen losses, pasture feeding, supplementary feeds

Introduction Urine from dairy cattle grazing pastures with high crude protein (CP) concentrations is a major source of N lost in drainage water from New Zealand farms. This paper provides predictions of urinary N leaching losses for a range of stocking rates and levels of supplementation.

Method Urinary outputs (kg N/ha) were estimated for dairy herds on notional 100 ha farms stocked at 2.35 (L), 3.00 (M) or 5.00 (H) cows/ha. Farm L was assumed to grow 13,000 kg pasture dry matter (DM)/ha annually. For farms M and H, 100 and 200 kg fertiliser N/ha were applied, to give annual pasture yields of 14,080 and 15,160 kg DM/ha, respectively. A feed budget identified periods when pasture (P; 20% CP) required supplements of either maize silage (MS; 7% CP) or pasture silage (PS; 16% CP) to maintain at least 1,700 kg grass DM/ha. Supplements were fed from Feb. to Sept. on farm M and in all months on farm H, accounting for 16% and 47% of the total DM intakes. Urinary N outputs were estimated by the Cornell Net Carbohydrate and Protein System model (Fox *et al.*, 2004), and leaching losses by the Nitrogen Leaching Estimation (NLE) model of Di & Cameron (2000).

Table 1 Annual N intake, urine N output and N leached

Stocking rate/Supplement		L	MMS	MPS	HMS	HPS
Dietary CP	(% DM)	20.0	17.9	19.1	13.9	17.4
Dietary N intake	(kg/cow)	150	136	145	107	135
Urine N output	(kg/cow)	72	58	67	39	59
	(kg/ha)	169	174	200	196	296
N in urine patches*	(kg/ha)	889	725	833	490	740
N leached	(kg/ha)	7	16	21	23	44

* Assumes urine from a cow covers ~ 0.08 of the area

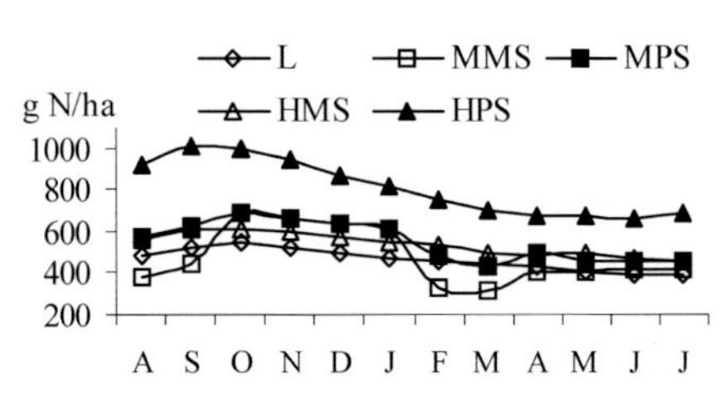

Figure 1 Monthly urine N outputs per ha

Results Dietary CP %, urinary N outputs and N leaching losses are given in Table 1 and Figure 1. Although the CNCPS model is reported to under-predict the ratio of urinary to faecal N, the results described here are supported by Mulligan *et al.* (2004), who showed a positive linear relationship between urinary N output and N intake, in cows fed supplements with high or low CP%. They also observed a decrease in the proportion of urinary N in the total excreta N when the supplement CP% concentration was reduced. Replacing pasture with a supplement of lower CP% can reduce the concentration of N under urine patches. On farm M, this occurs mainly in early spring and autumn, as no supplements were required in late spring and summer. The extent of N leaching was related to N fertiliser use and N concentration in the urine patches. Despite a two-fold increase in stocking rate and the higher N fertiliser usage, N concentration under urine patches was 50% lower on farm H with MS than on farm L. The reduction in urinary N production per cow and the increased area of urine deposition on the HMS farm results in N leaching losses that are not much greater than those for the M systems. However, when pasture silage was fed on farm H, N leaching losses were much higher.

Conclusions This analysis suggests that where increases in stocking rate are accompanied by the use of maize silage, intensification may not pose an increased risk of nitrogen leaching from dairy farms. There is a need for quantification of urine N output on farms as affected by stocking rate and diet, and for measurement of the exact area occupied by urine patches.

References

Di, H.J. & K.C. Cameron (2000). Calculating nitrogen leaching losses and critical nitrogen application rates in dairy pasture systems using a semi-empirical model. *N Z Journa lof Agricultural Research*, 43, 139-147.

Fox, D.G, L.O.Tedeschi, T.P.Tylutki, J.B.Russell, M.E.Van Amburgh, L.E.Chase, A.N.Pell & T.R.Overton (2004). The Cornell Net Carbohydrate and Protein System model for evaluating herd nutrition and nutrient excretion. *Animal Feed Science and Technology*, 112, 29-78.

Mulligan, F. J., P. Dillon, J.J. Callan, M. Rath & F.P. O'Mara (2004). Supplementary concentrate type affects nitrogen excretion of grazing dairy cows. *Journal of Dairy Science*, 87. 3451-3460.

A model to evaluate buying and selling policies for growing lambs on pasture

P.C.H. Morel, B. Wildbore, I.M. Brookes, P.R. Kenyon, R.W. Purchas and S. Ramaswami
College of Sciences, Massey University, Private Bag 11-222, Palmerston North, New Zealand, Email: P.C.Morel@massey.ac.nz

Keywords: lamb finishing, simulation modelling, pastoral systems

Introduction In pastoral sheep finishing systems, farmers aim to maximize profitability by deciding on when and how many animals to buy and/or sell, while taking into account feed availability and current prices. This paper describes a stochastic lamb growth simulation model with a set of heuristic rules, which has been developed to financially evaluate different management strategies for growing lambs on pasture.

Model inputs Feed supply as pasture dry matter (DM; kg) is described in terms of daily DM growth rates, metabolisable energy concentration in pasture (ME; MJ/kg DM), minimum and maximum pasture DM covers and pasture utilisation (%). The financial parameters used are lamb buying price ($/kg live weight), a selling price ($/kg CW) based on carcass weight (CW) and fatness (GR; mm) and pasture cost (c/kg DM). The stocking rate at the start of the period is also an input parameter. Simulations have been run for a 100 ha farm in New Zealand, starting with 25 kg live weight weaned lambs in January.

Lamb growth The stochasticity in the program is in the form of normally-distributed multiplication factors (1±SD) for: live weight (LW; kg; 1±0.12) at the start, dry matter intake (DMI; kg DM/d; 1±0.01), metabolisable energy for maintenance (MEm; 1±0.0033) and net energy per kg gain (NEg; 1±0.0066). Each of these characteristics is unique for each lamb. The daily growth rate (ADG; g/d) for a lamb is calculated as follows:

$$DMI = C \times 0.04 \times 70 \times (LW/70) \times (1.7 - (LW/70))$$

where 70 kg represents the mature live weight and DMI is reduced by factor C to ensure that pasture cover does not fall below the desired minimal value of 1200 kg DM/ha.

$$\text{ME intake (MEI)} = DMI \times ME \qquad MEm = LW^{0.75} \times [0.39 / (((0.35 \times ME) / 18.4) + 0.503)]$$

$$\text{ME for growth (MEg)} = MEI - MEm$$

$$NEg = 2.5 + 0.35 \times LW \qquad ADG = MEg \times ((0.0424 \times ME) + 0.006) / NEg$$

$$CW\ (kg) = -2.04 + 0.473 \times LW; \qquad GR\ (mm) = -10.8 + 1.2 \times CW$$

Random normally-distributed deviations with means of zero are added to the CW (SD=±0.67kg) and GR (SD= ±2.6mm) values (Garrick *et al.*, 1986).

Heuristic rules The lambs are sold at weekly intervals, but only when 50 or more have reached 45 kg LW. At the same time as lambs are sold, new lambs may be purchased, with the number bought being calculated as a function of pasture cover and the predicted length of time to grow lambs to 45 kg.

Model outputs The model calculates the predicted farm gross margin for a one-year period (returns from lamb sales minus the costs of lamb purchases and of pasture consumed) and can be used to investigate the effects of different management options on profitability. Results of a sensitivity analysis conducted with this model are presented elsewhere (Morel *et al.*, 2005).

References

Garrick, D.J., R.W. Purchas & S.T. Morris (1986). Consideration of alternative lamb drafting strategies. *Proceedings of the New Zealand Society of Animal Production, 46,* 49-54.

Morel, P.C.H., B. Wildbore, I.M. Brookes, P.R. Kenyon, R.W. Purchas & S. Ramaswami (2005). Sensitivity analysis of a growth simulation for finishing lambs. *Proceedings of the Twentieth International Grassland Congress.*

Sensitivity analysis of a growth simulation for finishing lambs

P.C.H. Morel, B. Wildbore, I.M. Brookes, P.R. Kenyon, R.W. Purchas and S. Ramaswami
College of Sciences, Massey University, Private Bag 11-222, Palmerston North, New Zealand, Email: P.C.Morel@massey.ac.nz

Keywords: lamb finishing, simulation modelling, pastoral systems

Introduction A stochastic lamb growth simulation model with a set of heuristic rules has been developed to evaluate management strategies for a solely pastoral grazing system in New Zealand (Morel *et al.*, 2005). In the present paper the results of a sensitivity analysis for this model are presented.

Method In the sensitivity analysis, only one parameter was changed at a time and the others were kept at their default values. For each parameter combination, the farm gross margin ($/yr per ha) for a one-year period (FGM= returns from lamb sales minus the costs of lamb purchases and of pasture consumed) was calculated 1000 times for a 100 ha farm. The parameters investigated (default value) were: lamb buying price (220c/kg live weight); selling price (450 c/kg carcass weight); pasture cost (11c/kg dry matter (DM)); annual pasture production (10,956 kg DM/ha), initial pasture cover (1,500 kg DM/ha), minimal pasture cover (1,200 kg DM/ha) and initial stoking rate (15 lambs/ha).

Results The gross margin per ha (FGM) with default values was $856.7 ±$13.36 (mean ± SD). The relationship between initial stocking rate and FGM was curvilinear, with FGM increasing from $826.6 to $856.7 as stocking rate increased from 12 to 15 lambs/ha and then decreasing to $825.3 for 18 lambs/ha. The changes in FGM (±SD) with changes in each of six parameters from the defaults values are presented in Figure 1. Changes in financial parameters had a greater impact on FGM than changes in pasture parameters. A 1% change in lamb buying price, selling price, or pasture cost were equivalent to ±$51.4, ±$71.5 and ±$11.4 changes in FGM, respectively.

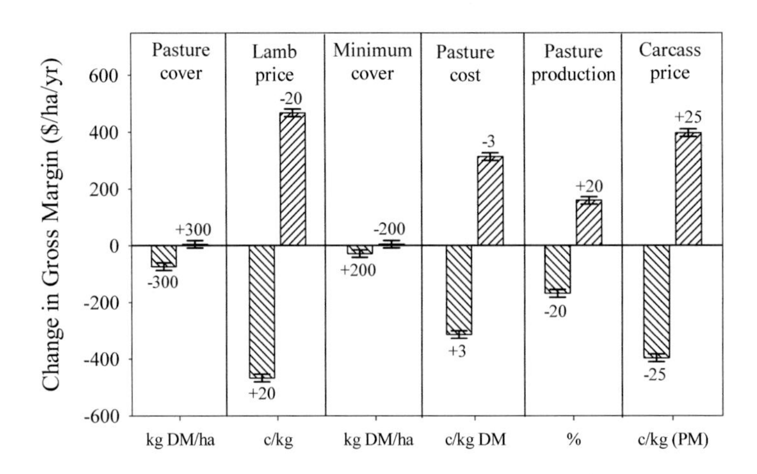

Figure 1 Changes in the gross margin per hectare with changes in each of six parameters with standard deviation bars for the default situation based on 1000 runs

A 1 % increase (decrease) in initial pasture cover, mimimum cover or total pasture production were equivalent to $0.22 (-$3.7), -$1.72 ($0.26) and +$8.4 (-$8.0), changes in FGM, respectively. The FGM decreased by $30.76 for each percentage point decrease in feed allowance from a default value of *ad lib* feeding.

Conclusions It is concluded that this model provides an efficient means of evaluating the relative importance of a number of changes to a system of lamb meat production on pasture.

Reference

Morel, P.C.H., B. Wildbore, I.M. Brookes, P.R. Kenyon, R.W. Purchas & S. Ramaswami (2005). A growth simulation model for finishing lambs in a pastoral system. *Proceeding of the XX Grassland Congress.*

Modelling winter grass growth and senescence

D. Hennessy[1,2,3], S. Laidlaw[3], M. O'Donovan[2] and P. French[2]
[1]*Teagasc Beef Research Centre, Grange, Dunsany, Co. Meath, Ireland, dhennessy@grange.teagasc.ie* [2]*Teagasc Dairy Research Centre, Moorepark, Fermoy, Co. Cork, Ireland* [3]*Queens University Belfast, Crossnacreevy, Belfast, BT6 9SH, Northern Ireland, UK*

Keywords: tissue turnover, winter, model, modelling

Introduction In temperate climates, because net grass growth in winter is low, most grass growth models deal with the main growing season (Mar-Oct in the N Hemisphere), with little emphasis on grass growth in winter (Nov-Feb). However, grass tissue turns over continuously (Hennessy *et al.*, 2004) and the fate of herbage entering the winter is important in extended grazing season systems. This study aimed to model winter grass growth for the period 15 Oct 2001 to 28 Jan 2002 for a range of autumn closing dates (1 Sep, 20 Sep and 10 Oct) by modifying an existing model, so that the amount of green leaf could be predicted at intervals over the winter.

Materials and methods The model of Johnson and Thornley (1983) was selected for modification. This is a vegetative grass growth model incorporating leaf area expansion and leaf senescence. It was run in Excel. The model suited the requirements of this study as it characterises leaves according to age (in line with tissue turnover concepts) and it was designed for an established vegetative grass crop supplied with unlimited nutrients and water. Tillers are mainly vegetative in Ireland during the winter, and water and nutrients are seldom limiting. As tissue turnover data were available from two sites, data from site 1 (Grange) were used to develop coefficients to modify the model, while data from site 2 (Moorepark) were used to validate the model. Mean daily air temperature and radiation, initial leaf area index (LAI) and amount of leaf in each age category were inputs to the model. The output predicted LAI at intervals over the winter. Daily meteorological data for the experimental period and latitude for the site, and initial lamina and sheath weight/unit area and LAI on 15 Oct were the inputs to the modified model. Leaf appearance rate (LAR) was modified for the winter period based on a simple regression equation between measured LAR and temperature. Coefficients for leaf senescence rate (LSR) of the 2nd and 3rd youngest leaves were derived from Grange data, based on the flux of material between leaf age categories. These coefficients were varied and the model run until the output (predicted LAI) was similar to the measured LAI for each of the closing date treatments at Grange. The output of the modified model was tested against actual LAI from Moorepark. Measured and predicted LAI were compared using mean squared prediction error (MSPE). Mean prediction error (MPE) was calculated also.

Results The model predicted the rapid decline in LAI on the 1 Sept and 20 Sept closing date treatments (Table 1.) However, the model predicted an increase in LAI on the 10 Oct treatment, which did not occur. Overall MSPE was 0.47. Most of the variation in the model was random (0.728) and LAI was not consistently over or under predicted (low mean bias value of 0.085). Where differences between measured and predicted LAI occurred they were short lived e.g. on 5 Nov on the two earlier closing treatments. This may be explained by an over-prediction of LAI as autumn moves into winter at the end of Oct/start of Nov.

Table 1 Measured and predicted LAI over the winter for 3 closing dates for Moorepark (the validation site)

Closing Date	1 September		20 September		10 October	
	Measured LAI	Predicted LAI	Measured LAI	Predicted LAI	Measured LAI	Predicted LAI
5 Nov	5.47	4.14	4.02	2.76	2.82	2.11
26 Nov	2.98	3.37	3.61	2.72	2.53	2.61
17 Dec	2.35	2.67	2.61	2.46	2.34	2.84
7 Jan	2.23	2.22	2.32	2.20	2.28	2.85
28 Jan	1.36	1.11	1.59	1.26	2.17	2.42

MSPE = 0.47; MPE = 0.25; R^2 = 0.65; Mean bias = 0.085; Line bias = 0.187; Random variation = 0.728

Conclusions This study suggests that it is possible to model winter leaf growth and senescence, and hence LAI, when swards are closed in autumn at a range of dates and at more than one sward state or site. This has potential for use as the basis of a winter grass growth model.

References

Hennessy, D., P. French, M. O'Donovan & A. S. Laidlaw (2004). Tissue turnover during the winter in a perennial ryegrass sward. *Grassland Science in Europe*, 9, 766-768.

Johnson, I. R. & J. H. M. Thornley (1983). Vegetative crop growth model incorporating leaf area expansion and senescence, and applied to grass. *Plant, Cell and Environment*, 6, 721-729.

The meal criterion estimated in grazing dairy cattle: evaluation of different methods

P.A. Abrahamse[1], D. Reynaud[2], J. Dijkstra[1] and S. Tamminga[1]
[1]*Animal Nutrition Group, Wageningen Institute for Animal Sciences (WIAS), Wageningen University, P.O. Box 338, 6700 AH Wageningen, The Netherlands, Email: Sander.Abrahamse@wur.nl,*
[2]*Laboratoire INRA-INAPG de Nutrition et Alimentation, 16 rue C.Bernard, 75231, Paris Cedex 05, France*

Keywords: grazing behaviour, meal criterion, modelling intake, dairy cows

Introduction The meal criterion (MC) has been found a useful tool to pre-treat intake behaviour data in dairy cows. It was defined as the longest interval between bouts that belong to the same meal (Tolkamp & Kyriazakis, 1999), necessary to cluster bouts to meals. The method of Yeates *et al.* (2001) calculating the $\log_e$-transformed intervals between bouts and using the Gaussian-Gaussian-Weibull (GGW) model to calculate the MC was found to provide the best estimation of the MC in biological as well as statistical terms. However, in grazing dairy cattle the MC-estimation has only been carried out by Rook & Huckle (1997) using a broken stick method. The aim of this study was to estimate the MC in grazing dairy cattle with the recently developed estimation methods.

Materials and methods Eight Holstein-Friesian dairy cows were allocated to two treatments: 1 day grazing in a plot of 0.125 ha (A) or 4 days grazing (B) in a plot of 0.5 ha. Four cows of group A were observed continuously, four cows in group B were observed during one day in each subperiod of 4 days. Four repetitions of a 4-day period were carried out in each of two periods after adapting the cows to grazing during 2.5 weeks.
A total number of 146 grazing behaviour recordings over 24h were collected using IGER Behaviour Recorders. Recordings were analysed using Graze-software version 8.0 (Rutter, 2000). Intervals between grazing bouts were calculated and Gaussian-Gaussian- (GG), Gaussian-Weibull- (GW), and Weibull-Weibull-populations (WW) were fitted to the frequency distribution of $\log_e$-transformed intervals between bouts using SAS. The quality of the fit of these models was calculated using the Minimal Function Value (MFV). The MC was calculated at the point where the two populations crossed.

Table 1 MFV values and MC-estimation of different subsets

	MFV			Meal Criterion WW
	GG	GW	WW	Minutes
Total dataset	34237	34188	33860	2.1
Group A[1]	26426	26388	26158	2.1
Group B[2]	7738	7714	7650	1.8
Period 1	16388	16364	16171	2.3
Period 2	17814	17788	17655	1.9
Period 1 A[1]	12021	12004	11872	2.3
Period 2 A[1]	14383	14362	14262	2.0
Period 1 B[2]	4323	4300	4267	2.5
Period 2 B[2]	3405	3399	3373	1.5

[1]Stripgrazing group; [2] 4-day grazing group

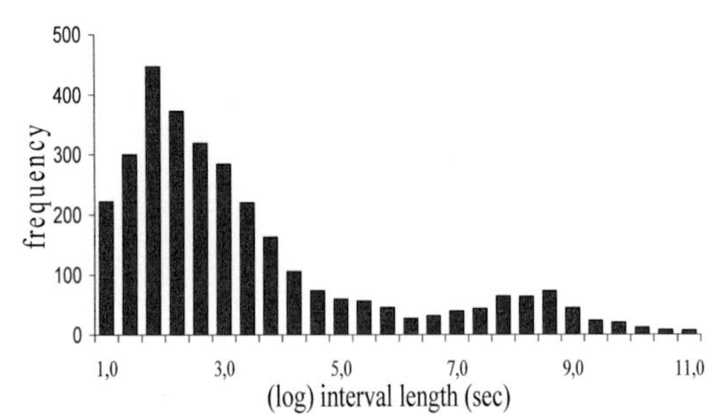

Figure 1 Frequency distribution of the $\log_e$-transformed interval between bouts

Results The MFV-values for the different models in Table 1 indicate that the WW-model gives the best fit to all different datasets.

Conclusions WW-model of the $\log_e$-transformed inter-bout intervals fits the frequency distribution better than the other models tested. The WW-model is superior to the broken stick model used by Rook and Huckle (1997) in biological as well as statistical perspective. The MC is estimated between 1.5 and 2.5 minutes and seems to differ between different subsets. Further analysis to test whether these are significant differences is necessary.

References

Rook A.J. & C.A. Huckle (1997). Activity bout criteria for grazing dairy cows. *Applied Animal Behaviour Science, 54, 89 – 96.*

Rutter, S.M. (2000). Graze: A program to analyze recordings of the jaw movements of ruminants. *Behaviour Research Methods, Instruments and Computers, 32 (1), 86 – 92.*

Tolkamp, B.J. & I. Kyriazakis (1999). To split behaviour into bouts, log-transform the intervals. *Animal Behaviour, 57, 807 – 817.*

Yeates, M.P., B.J. Tolkamp, D.J. Allcroft & I. Kyriazakis (2001). The use of mixed distribution models to determine bout criteria for analysis of animal behaviour. *Journal of Theoretical Biology, 213, 3165 – 3178.*

Effect of nitrogen on the radiation use efficiency for modelling grass growth

R. Lambert and A. Peeters
Laboratory of Grassland Ecology, Catholic University of Louvain, Place Croix du Sud 5 bte 1, B-1348 Louvain-la-Neuve, Belgium, Email: lambert@ecop.ucl.ac.be

Keywords: radiation use efficiency, photosynthetically active radiation, nitrogen, ryegrass

Introduction When nitrogen (N) is not at a sufficient level to permit maximum growth rate, dry matter production is reduced. Models of plant growth in relation to solar radiation intercepted by the crop have been largely used. According to these models, N deficiency can act on the leaf extension and thus on the quantity of radiation intercepted by the crop, but also by reducing the radiation use efficiency of the crop (RUE) (Bélanger, 1990). The effect of N on the RUE of ryegrass swards is determined and discussed.

Material and methods Radiation use efficiency is the ratio between the yield and cumulated photosynthetically active radiation absorbed by the crop (PARa) during a period of time. It is expressed in g DM/MJ of PAR. The PARa is determined using a model of leaf area index (LAI) change with temperature (Lambert *et al.*, 2004). Incident incoming radiation was measured during all the growing period using a photovoltaic sensor (Solar Haeni 130). The PAR fraction determined using a quantum sensor (Delta-T device LTD) and the McCree (1972) conversion coefficient to translate photon flux into energy units was found to be 0.47. Dry matter (DM) yields were measured during spring growth (generative growth) in 4 week old regrowths of *Lolium perenne* swards (cv Meltra) receiving different N fertilisation levels. Each yield value is the mean of four replicates. The N nutrition index (NNI) was calculated using the dry matter yield and the N content value (Lemaire *et al.*, 1989). Data from six different site-year combinations were used to establish the relation between RUE and NNI.

Results As it was assessed by Belanger (1990), N deficiency affected negatively both the RUE of the grass crop and the leaf extension. When N was not limiting for growth (NNI = 100), the RUE was 2 g DM/MJ during the spring (Figure 1). For modelling grass growth, yield can be predicted by multiplying the quantity of radiation intercepted by the crop and the RUE of the crop. RUE can be determined from the NNI of the grass using a linear relation (Figure 1). This model coupled with a model of LAI change with temperature (Lambert *et al.*, 2004) and PAR measurements allows yield accumulation during spring to be predicted. The effect of N on grass growth can be analysed in terms of radiation interception by the crop and in terms of transformation of intercepted radiation in biomass.

Conclusions Nitrogen supply acts on the RUE of the crop. A linear equation describes the relationship between RUE and NNI. This relationship can be used for grass growth modelling according to N status and radiation intercepted by the crop.

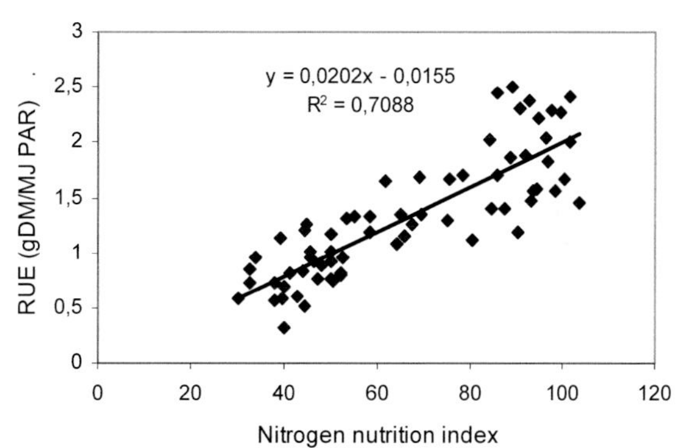

Figure 1 Relationship between radiation use efficiency and nitrogen nutrition index

References.

Bélanger G. (1990). Influence de la fertilisation azotée et de la saison sur la croissance, l'assimilation et la répartition du carbone dans un couvert de fétuque élevée en conditions naturelles. *Thèse Université de Paris-sud, Centre d'Orsay.* 170p.

Lambert, R., A. Peeters & B. Toussaint (2004). Modelling leaf area index of perennial ryegrass according to nitrogen status and temperature. In *Land Use Systems in Grassland Dominated Regions. Grassland Science in Europe Vol.9. EGF.* 793-795

Lemaire, G., F. Gastal & J. Salette (1989). Analysis of the effect of N nutrition on dry matter yield of a sward by reference to potential yield and optimum N content. *XVI International Grassland Congress, Nice, France*, 179-180.

McCree, K.J. (1973). The measurement of photosynthetically active radiation. *Solar energy* 15, 83-87.

Radiation use efficiency of ryegrass: determination with non cumulative data

R. Lambert and A. Peeters

Laboratory of Grassland Ecology, Catholic University of Louvain, Place Croix du Sud 5 bte 1, B-1348 Louvain-la-Neuve, Belgium, Email: lambert@ecop.ucl.ac.be

Keywords: radiation use efficiency, photosynthetically active radiation, nitrogen, ryegrass

Introduction The growth of a crop is generally described as biomass accumulation per unit time. Monteith (1977) developed a model of growth where biomass accumulation is related to solar radiation intercepted by the crop. This model has been largely used for different crops. The conversion factor between radiation absorbed or intercepted by the crop and the biomass production is called "radiation use efficiency" or "dry matter radiation quotient". Radiation use efficiency (RUE) is usually calculated as the regression coefficient of the linear relationship between crop biomass measured repeatedly during growth and cumulated intercepted or absorbed solar radiation. Demetriades-Shah *et al.* (1992) criticised this method because the strong correlation between two sets of cumulative data has no statistical value. Other authors did not agree with these critics. In this paper, we tried to measure the significance of the relation between biomass and photosynthetically active radiation absorbed (PARa) by a grass crop using a non-cumulative and independent data set.

Material and methods Photosynthetically active radiation absorbed by the crop (PARa) is determined using a model of LAI evolution with temperature and considering that interception efficiency obeys the Lambert-Beer law. Parameters of these relations were calculated previously (Lambert *et al.*, 1999). Incident incoming radiation was measured during the whole growing period using a photovoltaic sensor (Solar Haeni 130). The PAR fraction was 0.47. It was measured using a quantum sensor (Delta-T device LTD) and the McCree (1972) conversion coefficient to change photon flux in energy units. Dry matter (DM) yield was measured during spring growth on 4-week-old regrowths of *Lolium perenne* swards (cv Meltra). Each yield value is the mean of four replicates. A nitrogen (N) fertilisation level of 60 or 80 kg/ha was applied. The N nutrition index was calculated using the dry matter yield and the N content value (Lemaire *et al.*, 1989). Only regrowths with a high N nutrition level (NNI >80) were taken into account to assess the RUE of the crop. The RUE is the ratio between yield and PARa. It is expressed in g DM/MJ of PAR. Data from six different site-year combinations were used to establish the regression.

Results Yields varied between 1864 and 4475 kg DM/ha depending on site and year. RUE calculated for each regrowth was between 1.4 g DM/MJ and 2.5 g DM/MJ. The mean RUE was 1.9 g DM/MJ. These results with ryegrass are quite similar to those found by other authors with other sward grasses receiving a high N fertilisation. The absence of correlation between NNI and yield shows that N is not a limiting factor for growth (data not shown). The relationship between the yield and PARa is highly significant (r=0.549, n=24), but only explains 30% of the total variation (Figure 1).

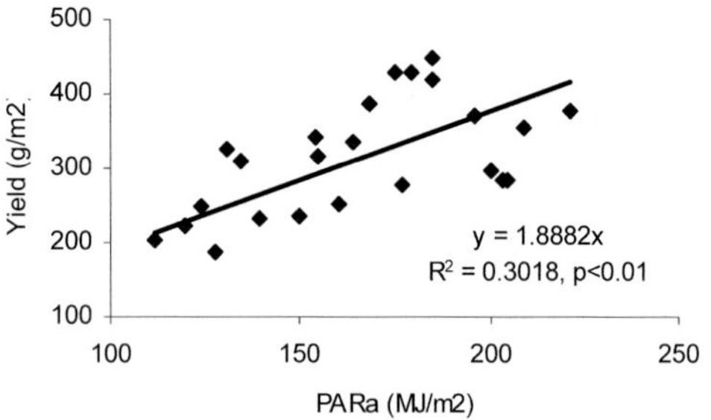

Figure 1 Relation between yield and PAR absorbed by the sward

Conclusions When N is not a limiting factor for growth, yield can be correlated to PAR absorbed by the crop. The mean RUE of *Lolium perenne* sward is 1.9 g DM/MJ. Radiation use efficiency can be used for grass growth modelling, but the precision is not very high.

References.

Demetriades-shah, T., M. Fuchs, E. Kanemasu & I. Flitcroft (1992). A note of caution concerning the relationship between cumulated intercepted solar radiation and crop growth. *Agricultural and Forest Meteorology,* 58, 193-207.

Lambert, R., A. Peeters & B. Toussaint (1998). LAI evolution of a perennial ryegrass crop estimated from the sum of temperature in spring time. *Agricultural and Forest Meteorology,* 97, 1-8.

Lemaire, G., F. Gastal & J. Salette (1989). Analysis of the effect of N nutrition on dry matter yield of a sward by reference to potential yield and optimum N content. *XVI International Grassland Congress, Nice*, 179-180

McCree, K.J. (1973). The measurement of photosynthetically active radiation. *Solar energy,* 15, 83-87.

Monteith, J.L. (1977). Climate and the efficiency of crop production in Britain. *Philosophical Transaction of the Royal Society, Serie B,* 281, 277-294.

Modelling the digestibility decrease of three grass species during spring growth according to the age of the grass, the thermal age and the yield

M.E. Salamanca, R. Lambert, M. Gomez and A. Peeters
Laboratory of Grassland Ecology, Catholic University of Louvain, Place Croix du Sud 5 bte 1, B-1348 Louvain-la-Neuve, Belgium, Email: lambert@ecop.ucl.ac.be

Keywords: digestibility, nutritive value, *Dactylis glomerata, Phleum pratense, Lolium perenne*

Introduction The nutritive value of forage changes during growth. For the protein content, a general evolution curve was found with the yield increase (Salette & Lemaire, 1984). The digestibility of the organic matter decreases during growth as cellulose and lignin content increase. Regrowth age is the main factor, which explains the digestibility decrease (Demarquilly & Jarrige, 1981). The crop age can be expressed in number of growth days but also in thermal age (cumulated temperature). We compared the digestibility change of three grass species during spring growth for two years as a function of yield increase, thermal age or number of days. Relationships were computed and compared to find the best one for predictive use.

Material and methods Digestibility and yields were measured (4 replicates) at different times during an uninterrupted spring growth on three common grass forage species (*Dactylis glomerata* cv Baraula; *Lolium perenne* cv Bastion; *Phleum pratense* cv Erecta) for two years. Nitrogen fertiliser was applied in one dressing of 100 kg N/ha in March. Digestibility was then plotted as a function of yield, temperature sum and days since the beginning of yield accumulation after winter. The beginning of yield accumulation was determined by the intercept with the X axis of the linear regression between yield and temperature sum (Lemaire & Salette, 1982). Digestibility of organic matter was determined *in vitro* by pepsin-cellulase digestion (Hayward & Jones, 1975; Rihs, 1985). Digestibility changes were compared between species and years with tests of equality of the slope (parallelism) and tests of equality of the constant (Dagnelie, 1986). A model of digestibility change is proposed.

Results Digestibility decreases linearly with yield, temperature sum or days whatever the species or the year, but for *Dactylis* and *Phleum*, linear relations are not parallel between years when digestibility is expressed as a function of yield or temperature sum. When the digestibility is expressed as a function of the number of growth days, linear relations established for each species individually are parallel between years (Table 1). The mean daily rate of digestibility decrease (slope) is -0.34 %/day for *Dactylis*, -0.35 %/d for *Lolium* and -0, 36%/d for *Phleum*. These daily rates are not significantly different. We also compared the constants of linear regression of digestibility as a function of days between species. Maximum digestibility value of *Dactylis* is 76.0 %. It is significantly lower than the digestibility of *Phleum* (80.3 %) and *Lolium* (81.8%).

Table 1 Linear relationships between digestibility (Y) and days since the beginning of the growth (X)

Species	Digestibility = fct (days since beginning of growth)	R^2	n
Dactylis glomerata	Y = 76.012 - 0.3415 X	0.7568	10
Phleum pratense	Y = 80.263 - 0.3587 X	0.7742	10
Lolium perenne	Y = 81.787 - 0.3514 X	0.7806	10

Conclusions Digestibility decrease of grass species during an uninterrupted growth period in spring is better modelled as a function of the number of days since the beginning of biomass accumulation (linear relation) than as a function of thermal age or yield. Maximum digestibility value is lower for *Dactylis glomerata* than for *Phleum pratense* and *Lolium perenne*. Daily rates of digestibility decrease are not significantly different between years or between the three species and are around –0.35%/day.

References

Dagnelie, P. (1986). Théorie et méthodes statistiques. Vol 2. Les Presses Agronomiques de Gembloux.

Demarquilly, C. & R. Jarrige (1981). Panorama des méthodes de prévision de la digestibilité et de la valeur énergétique des fourrages. *Prévision de la valeur nutritive des aliments des ruminants*. INRA publications, Versailles, 41-49.

Hayward, M.W.& D.H.I.Jones (1975). The effect of pepsine pre-treatment of herbage on the prediction of dry matter digestibility from solubility in fungal cellulase solution. *Journal Science Food Agrciulture* 26, 711-718.

Lemaire, G. & J. Salette (1982). Analyse de l'influence de la temperature sur la croissance de printemps de graminées fourragères. *C.R. Acad. SC. Paris, t.292 (30 mars 1981) Série III*, 843-846, 1981

Rihs, T. (1985). Determining the in vitro digestibility of organic matter of maize by a cellulase method. *In Vitro Newsletter, 1*, 4-5.

Visual Modelling of Alfalfa Growth and Persistence under Grazing

S.R. Smith, Jr.[1], L. Muendermann[2] and A. Singh[3]
[1]N222-E Ag. Science North Dept. Plant and Soil Sciences University of Kentucky Lexington, KY 40390-0091, USA, Email: raysmith1@uky.edu, [2]Biomedical Eng. Division, Stanford Univ., Stanford, CA 94305-4038, USA, [3]Agrapoint, Truro, Nova Scotia B2N 6Z4, Canada.

Keywords: alfalfa, lucerne, virtual plants, modelling, visual models, grazing, persistence

Introduction A 'virtual' alfalfa plant model was developed at the University of Manitoba in Canada as part of a comprehensive grazing research project. This model shows an alfalfa plant 'growing' on a computer screen and the plant's response to grazing (similar to time-lapse photography). The original model was constructed by Singh (2005) to show the research potential of visually modelling alfalfa plant growth. The ability to visually 'grow' a plant on a computer screen also offers tremendous opportunities for teaching and extension. Detailed morphological measurements were used in the construction of Singh's model, based on single plants subjected to the following management strategies: 1) no grazing; 2) rotational grazing; and 3) continuous grazing. The modelled growth of these three plants is accurate and can be modified, but has not been rigorously verified in comparison to other alfalfa plants. Singh's model can be downloaded as a video clip at http://www.cpsc.ucalgary.ca/~lars/models/. The objective of this project was to modify the current single plant alfalfa model to simulate an alfalfa sward under various grazing management scenarios.

Materials and methods The morphogenetic rules describing alfalfa development were written using the plant modelling language of L-systems and were interpreted by the program *cpfg* (continuous plant and fractal generator). L-studio™ (University of Calgary, Calgary, Alberta, Canada) is a software package that creates a user-interface between L-system based modelling and the simulation program, *cpfg* under Windows (Microsoft Corporation, Redmond, WA). The current project represents ongoing work to modify the existing 'virtual' alfalfa plant model from a research model into a user-friendly extension tool. Software development and programming changes include the following: 1) easier modification of individual alfalfa plants and alfalfa swards in response to different management scenarios, and 2) simpler format for data entry and manipulation.

Results The revised visual model incorporates 20 alfalfa plants growing simultaneously in a stimulated alfalfa sward. One scenario shows stand persistence over time between grazing tolerant and intolerant cultivars. Another scenario shows stand persistence of 'Florida 77' from the seedling stage (Figure 1), after two years of continuous grazing (Figure 2), and after two years of rotational grazing (Figure 3) using data from research conducted at the University of Georgia (Bouton and Gates, 2003). The actual model shows plant growth and mortality on a daily basis over the two-year stand life. Visual quality is crisp in the original working model or with large file size video clips. The visual model is not limited to these scenarios. Simple modifications in plant growth parameters allow simulation of numerous grazing or clipping scenarios.

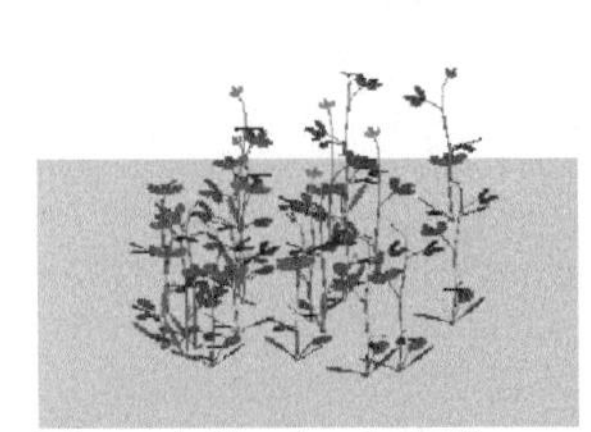

Figure 1 Simulated alfalfa sward at the seedling stage

Figure 2 Florida 77 following two years continuous grazing

Figure 3 Florida 77 following two years rotational grazing

Conclusions These results show the value of visual modelling as a research tool, but its greatest value may be in showing livestock producers the effect of mismanagement and/or poor cultivar choice. A visual model allows producers to observe the implications of different management decisions without having to wait for several years and suffer economic losses due to poor decisions.

References

Bouton, J.H. & R.N. Gates (2003). *Agronomy Journal* 95, 1491-1494.
Singh, A. (2005) PhD Dissertation, University of Manitoba, Winnipeg, MB, Canada

A new agro-meteorological simulation model for predicting daily grass growth rates across Ireland

R.P.O. Schulte
Teagasc, Johnstown Castle, Wexford, Ireland, Email: rschulte@johnstown.teagasc.ie

Keywords: grass, model, simulation, generative, vegetative

Introduction Grass growth rates and herbage yields depend on weather conditions, soil characteristics and grassland management and differ from year to year and from site to site. In the past, grass growth has been predicted using both mechanistic and statistical models. The accuracy of mechanistic models is commonly insufficient for practical application, while statistical models generally apply to one test site only (e.g. Han *et al.*, 2003). In this paper a semi-empirical grass growth model is presented which is numerically accurate, but which can be applied to contrasting sites across Ireland at the same time.

Materials and methods The new model computes changes in herbage biomass as the difference between above-ground primary production and respiration. The potential primary production P_{pot} (kg DM/day) is simulated in a new light utilisation function, with Leaf Area Index (LAI) and instant radiation flux as input variables. The actual primary production P_{act} (kg DM/day) is subsequently computed as:

$$P_{act} = P_{pot} \cdot F_T \cdot F_W \cdot F_N \cdot F_P$$

in which F_T is a sigmoid temperature function. F_W is a water function, which is driven by the Hybrid model for predicting moisture conditions in Irish grasslands (Schulte *et al.*, accepted). F_N accounts for the effects of nitrogen (N) fertiliser inputs, while F_P is a new submodel describing tiller dynamics during the generative phase as temporal functions of vernalisation, heading, apex removal during harvest and subsequent regrowth from dormant buds. The generative growth of vernalised tillers and new tillering from dormant buds are described by interrelated normal distributions over time, while apex removals are implemented as discrete events. Finally, above-ground respiration is modelled as a function of above-ground biomass, temperature and soil moisture.

Results The model was calibrated and evaluated, using data from the Teagasc "FAO grass growth network". This network measures the 4-weekly grass yield on a weekly basis on replicated (n=5) plots, subjected to three rates of N application, on contrasting sites across Ireland. As a pilot study, the model was calibrated for the Solohead site using the yield data from 1998 to 2000, by minimising the residual sum of squares between the observed and predicted yields, using the OpStat module of acslXtreme. Model performance was evaluated independently using the Solohead yields during the year 2001. Overall, the model explained 74% of the variation in yields, with neither consistent nor relative bias ($p>0.05$). Subsequently, the final values of the model parameters were calibrated using data of all years. Figure 1 shows that, with the exception of early spring 2001, the model accurately predicts the temporal patterns of herbage yields. In particular, the main novel tiller dynamics function satisfactorily accounts for the growth patterns during the generative phase.

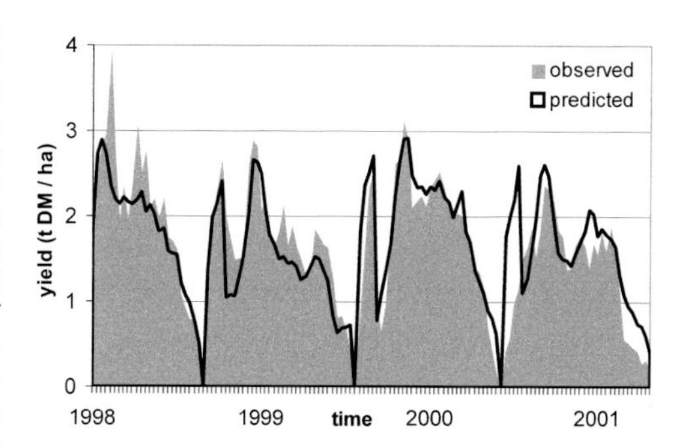

Figure 1 Observed and predicted yields for Solohead

Conclusions The new grass growth model accurately predicts grass yields throughout the year. It accounts for differences between sites that arise from meteorological conditions, soil drainage capacity, grass variety and grassland management. As a result, it allows yield predictions to be customised for individual farms.

Acknowledgements Grass yields and weather data provided by Mr. Frank Kelly (Teagasc, Kilmaley) and Met Éireann, respectively. Members of Agmet are gratefully acknowledged for their feedback.

References

Han, D., P. O'Kiely, & D. W. Sun (2003). Linear models for the dry matter yield of the primary growth of a permanent grassland pasture. Irish Journal of Agricultural and Food Research, 42, 17-38.

Schulte RPO, Diamond J, Finkele K, Holden NM, Brereton AJ, accepted for publication. Predicting the soil moisture conditions of Irish grasslands. Irish Journal of Agricultural and Food Research (accepted)

Section 4

Decision support systems for grazing

Pâtur'IN: a user-friendly software tool to assist dairy cow grazing management

L. Delaby, J.L. Peyraud and P. Faverdin
INRA – UMR Production du Lait, 35590 Saint Gilles, France, Email: luc.delaby@rennes.inra.fr

Keywords: grazing management, decision support system, dairy cows

Introduction The feeding of dairy cows at pasture presents many technical, economic and environmental advantages, while benefiting from a very favourable image. However, the management of grazed land is a complex game of strategy in which the farmer applies decisions in order to manage two unstable and uncertain fluxes of change: growth of grass and intake of the herd. Many tools (platemeter, etc.) and overall methods (local stocking rate references, farm cover, etc.) have been developed as aids to grazing management. Nevertheless, few decision-support systems are currently available that make it possible to anticipate and assess the consequences of a given decision in a dynamic way (Peyraud *et al.*, 2004). The objective of this article is to present the structure and functions of Pâtur' IN, a software tool designed to help the management of dairy cow grazing.

Structure and functions of Pâtur'IN After defining the structure of the farm, which comprises the plots, herds and available feed supplements, the user can, on the one hand, record all the events taking place on the farm during the grazing season and then, on the other hand, simulate various scenarios of sward use. A scenario consists of a succession of events (grazing, cutting, fertilisation, etc.) that act on the "Growth" and "Intake" functions according to rules of decision defined by the user (Figure 1). The "Growth" function makes it possible to calculate the biomass present on each day starting from a standard growth grid that can be modified by the user according to the local context. This grid is then adapted to each plot for each day using various models integrating the effects of climatic conditions, N fertilisation, amount of biomass present, etc. The "Intake" function calculates the quantities of grass ingested by the herd based on the intake capacity of the cows (INRA, 1988) and adapted to the characteristics of the pasture (sward height, pasture allowance, etc.). During a simulation, the user defines a set of rules of decision as well as the order of use of the plots. From this, the software program can determine, for each grazed plot, either the residence time in a rotational grazing system, or the land area to be offered each day in a strip grazing system. The user interface (Figure 1) allows a visualisation of the condition of each plot for each day, by means of various colours and the display of simulated events in the form of a grazing schedule. The whole set of results for a given simulation is available in the form of text file. Pâtur' IN can be run on a PC and is now available in an English version.

Conclusion The advantage of Pâtur'IN is that it makes use of data available at farm level, thus enabling farmers to carry out regular updating of the data according to the actual progression of the grazing season. In this way, users are able to develop simulations adapted to the particular context of each farm.

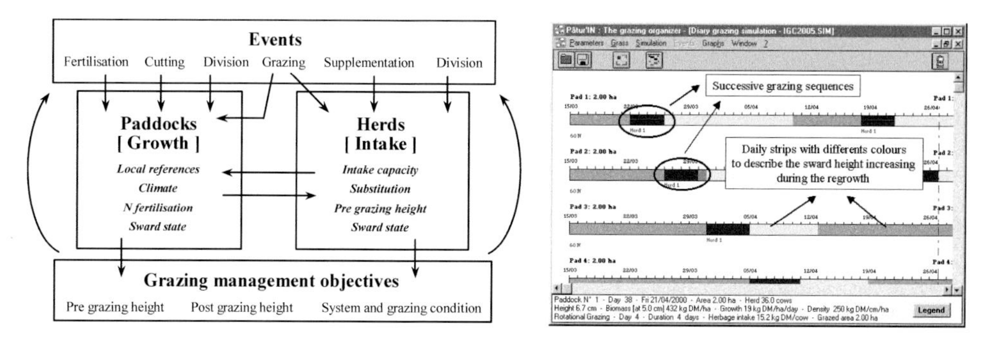

Figure 1 A graphical representation of the events and functions used in Pâtur'IN software tool and a screen copy of the user interface

References

INRA (1988). Ruminant Nutrition, R. Jarrige (ed.). INRA & John Libbey Ltd, Paris and London, 389 pp.

Peyraud, J.L., R. Mosquera-Losada & L. Delaby (2004). Challenges and tools to develop efficient dairy systems based on grazing: how to meet animal performance and grazing management. *Grassland Science in Europe*, 9, 373-384.

A farmer-based decision support system for managing pasture quality on hill country

I.M. Brookes and D.I. Gray
College of Sciences, Massey University, Private Bag 11 222, Palmerston North, New Zealand, Email: I.Brookes@massey.ac.nz

Keywords: feed budgeting, micro-budgeting, tactical management, sheep production, pasture quality

Introduction Despite considerable effort to promote formal feed budgeting in New Zealand, survey data suggests it is only adopted by 20% of farmers (Nuthall & Bishop-Hurley, 1999). Recent work (Gray *et al.*, 2003) has identified that farmers may use a different approach - micro-budgeting - to manage feed. Rather than operate at a whole farm level, micro-budgeting focuses at the paddock level. This paper describes micro-budgeting as used by a high performing hill country sheep and cattle farmer to manage pasture quality over spring and a decision support model developed to help other farmers undertake this process.

Farmer practice During spring, the farmer identifies separate sheep and cattle blocks. Different areas of the sheep block are allocated to mixed-age and two-tooth ewes, separated into triplet-, twin- and single-bearing ewe blocks, and a hogget block containing dry and lambing hogget areas. The cattle block is separated into rising one-year (R1) and rising two-year (R2) areas. The sheep are set-stocked from Sept. to early Jan. and the cattle are grazed on a 12-15 d rotation within each block. The farmer monitors pasture cover in each paddock fortnightly. Feed demand for the next two weeks is compared with forecasts of pasture growth for each paddock, and stock numbers adjusted to match feed demand with supply. The aim is for sward height on the sheep and cattle blocks to not exceed 1200 kg DM/ha and 1500 kg DM/ha respectively, thereby maintaining pasture quality and minimising within- and between-block variation. Options include shifting stock from areas with a feed deficit to those with a feed surplus; buying additional cattle for paddocks with a surplus; increasing stocking rate; and freeing up areas for additional stock, a forage crop or production of grass silage.

Decision support system A decision support system has been developed on an Excel spreadsheet (Table 1). Paddock number and area are entered in columns 1-2. The stock class is identified in columns 3-5, by age, birth rank and lambing date, and stock numbers in column 6. Feed demand per head is calculated from notional live weight, docking percentage and mean lambing date for each stock class (column 7). Per hectare demand in each paddock (column 8) is calculated from the number of grazing stock multiplied by feed demand (kg DM/head) divided by the paddock area (ha). Estimated pasture growth rates (PGR) are entered in column 9, and pasture cover at the start of the period in column 10. Final pasture cover is calculated (column 11) and compared with the target covers entered in column 12. The paddock area or stock numbers required to meet target cover are calculated in columns 13-14. This information is used to decide if an area can be freed for other stock, or if additional stock can be shifted from paddocks that are short of feed into paddocks with a surplus. Columns 15-16 provide an estimate of the stock numbers to be added to or removed from each paddock to ensure target pasture cover levels are reached. In this example, additional single-bearing ewes or R1 cattle may be placed in paddocks in this block. The cattle may come from the cattle block, if it is short of feed, or they may be purchased.

Table 1 A decision support system for managing pasture quality on hill country

1	2	3	4	5	6	7	8	9	10	11	12	13	14	15	16
						Intake		PGR	Pasture cover					Extra stock	
Paddock		Stock				kg DM		kg DM	kg DM/ha			Required		Ewes	Cattle
No.	ha	Class			No.	/head	/ha	/ha per d	Start	Final	Target	ha	No.	S	R1
1	5.0	MA	S	E	53	2.57	27.2	31.0	1170	1223	1200	4.72	56	3	1
2	4.5	MA	S	E	60	2.57	34.3	31.0	1230	1184	1200	4.65	58	-2	-1
3	5.2	MA	S	E	70	2.57	34.6	31.0	1250	1200	1200	5.20	70	0	0

Conclusions This decision support system for feed management is modelled on the practice of an expert farmer. The approach is quite different from the methods normally advocated by extension agents, but may prove more attractive for use on farm. This work suggests that the development of effective decision support systems for farmers requires an in-depth understanding of how they currently manage their feed.

References

Gray, D.I., W.J. Parker, E.A. Kemp, P.D. Kemp, I.M. Brookes, D. Horne, P.R. Kenyon, C. Matthew, S.T. Morris, J.I. Reid, & I. Valentine (2003). Feed planning – alternative approaches. *Proceedings of the New Zealand Grassland Association,* 65, 211-218.

Nuthall, P.L & G.J. Bishop-Hurley 1999. Feed planning on New Zealand farms. *Journal of International Farm Management. 2*, 100-112.

Understanding Livestock Grazing Impacts: a decision support tool to develop goal-oriented grazing management strategies

S.J. Barry, K. Guenther, G. Hayes, R. Larson, G. Nader and M. Doran
University of California Cooperative Extension, 1553 Berger Drive, Bldg. 1, San Jose, CA 95112, Email: sbarry@ucdavis.edu

Keywords: grazing management, endangered species, water quality, rangeland health

Introduction Managing grasslands in the western United States has become much more complex over the last few decades. A century ago the goal was to survive as a livestock producer, and grassland management involved using forage effectively and overcoming obstacles such as predators and shortages of water and feed. Today the successful grassland manager also needs to consider the diversity and health of the ecosystem as a whole. Livestock grazing can negatively and/or positively affect riparian areas, sensitive plants, and endangered wildlife. Since the impact on a specific factor will vary depending on the timing, intensity and class of livestock grazed, land managers need a decision support system that will help them simultaneously evaluate the affect of different grazing management strategies on a variety of environmental and economic factors. *Understanding Livestock Grazing Impact,* an interactive website, assembles and presents information on the impacts of livestock grazing in a way that is both comprehensive and accessible. This makes it easier for ranchers and land managers to analyse, compare and choose the grazing strategies that best achieve the goals for a given grazing unit.

Materials and methods Thirty individual environmental, economic and social indicators that are affected by livestock grazing on California's annual grasslands were identified. Environmental indicators include habitat for endangered species such as the kit fox, burrowing owl, checkerspot butterfly, and tiger salamander. Environmental indicators also include vernal pool habitat, hardwood riparian habitat, oak regeneration, water quality and water infiltration. Economic indicators include forage production, livestock gain per acre, need for supplemental feed and calving rate. Social indicators include recreation use. For each indicator statements of description, significance, rationale regarding potential grazing impact, monitoring methods and references are provided.

The grazing impact for each indicator is depicted graphically with a bar chart to illustrate positive/ negative or no impact with regards to grazing intensity, season of use and species of grazing animal. The bar charts were developed based on existing research findings. The interactive website allows managers to select the appropriate indicators for a site and evaluate the impact of grazing strategies that vary intensity, timing and species of livestock. Two workshops were held for grassland managers to test the use of this decision support tool before its development was finalized. This decision support tool has just debuted on the website www.grazingimpacts.info, and is in the University of California peer review process.

Results Grassland managers can use this decision support tool to determine the best grazing management strategy to work toward established objectives for a particular grassland site. Although the grassland manager will have to assign relative value to a list of selected indicators, he can use this tool to effectively evaluate multiple and conflicting impacts related to livestock grazing.

Conclusions *Understanding Grazing Impacts* uses the indicator concept (BLM, 2000) to predict grazing impacts from different grazing strategies (grazing intensity, time of use, and species of livestock) in annual grassland and oak woodland. Environmental, economic and social indicators could be added to make this decision support tool relevant in other regions and/ or across other ecotypes.

References

U. S. Department of the Interior, Bureau of Land Management (2000). Interpreting indictors of rangeland health. Interagency Technical Reference 1734-6. U.S. Department of Interior, Bureau of Land Management, National Science and Technology Centre Information and Communications Group, Denver, Colorado.

Enhancing grasslands education with decision support tools

H.G. Daily, J.M Scott and J.M. Reid
University of New England, Armidale, New South Wales 2351, Australia, Email: Helen.Daily@une.edu.au

Keywords: Internet, simulation, wool industry, grazing systems, problem solving

Introduction We have successfully used Decision Support Tools (DST) relevant to the management of grazing enterprises to enhance problem solving skills of undergraduates in Australia. Tools such as GrassGro™ (Moore *et al.*, 1997) and GrazFeed™ (Freer *et al.*, 1997) are accessed from a central server by authorised users at many widely dispersed Universities across Australia using remote access to thin-client technology via an Internet portal. This has been supplemented with training for lecturers. Experience in developing appropriate teaching and learning materials and the reliable delivery of simulation software to many clients has enhanced learning outcomes at tertiary level. We are also trialling the use of DST to other learning sectors.

Materials and methods Support from the Australian government's Department of Education Training and Youth Affairs, and Australian woolgrowers through Australian Wool Innovation Ltd have been used to extend and consolidate the use of thin-client technology for distributing grazing models to undergraduate students at institutions around Australia on request from their lecturers on a fee-for-service basis. In collaboration with CSIRO, we have trained over 30 lecturers in the use of these DST and their application in tertiary education programs such as rural science, agricultural economics and natural resource management. A Decision Support Specialist provides assistance to lecturers in the development of training materials based on the grazing management DST, GrassGro™, and the grazing ruminant nutrition DST, GrazFeed™. Registered lecturers and students log on to the eD-Serve portal (http://ed-serve.une.edu.au) to access a range of DST and other customised relevant information (Daily *et al.*, 2003a). We are piloting approaches to create awareness of grazing management decision support tools amongst wool producers and secondary school agriculture students.

Results We have increased the awareness of key profit drivers, risk management and the role of DST in decision making in the grazing industries, particularly the wool industry in Australia over 5 years. Students from 11 campuses of 8 Australian universities have accessed grazing DST through eD-Serve, and a survey of lecturers using these DST in their teaching provided evidence of their support for the system, and the benefit it provided to their teaching. Student surveys have revealed that they recognise that the acquisition of skills with these commercially available DST is especially relevant training for their future employment. Climate datasets for other international localities are currently being assembled so that simulations related to a range of international sites can be conducted by students, thus broadening their global perspective. This thin-client delivery system is also being developed for accessing other legacy software related to applications across the agricultural industries.

Conclusions This e-learning project using DST has demonstrated that excellent grasslands science can be readily acquired by undergraduate students and others interested in learning about complex grassland ecosystems through the use of DST in education. The exercises are tailored to explore scientific principles, illustrate profit drivers and/or test risk management in grazing enterprises, and overall, to develop problem solving skills and "systems" thinking in students (Daily *et al.*, 2000). Internet access to DST widens the experience of students and will increasingly provide opportunities for collaboration across Australia and internationally.

References

Daily, H.G., G.N. Hinch, J.M. Scott & J.V. Nolan (2000). The use of a decision support program to facilitate the teaching of biological principles in the context of agricultural systems. http://www.tedi.uq.edu.au/conferences/teach_conference00/abstractsA-H.html#Daily

Daily, H.G., J.M. Scott, & J.M. Reid (2003a). Internet delivery of Decision Support Tools for teaching nationally. In G. Crisp, D. Thiele, I. Scholten, S. Barker and J. Baron (Eds), *Interact, Integrate, Impact: Proceedings of the Twentieth Annual Conference of the Australian Society for Computers in Learning in Tertiary Education.* Adelaide, 7-10 December, 2003.

Freer, M., A.D. Moore, & J.R. Donnelly (1997). GRAZPLAN: Decision Support Systems for Australian Grazing Enterprises. II. The Animal Biology Model for Feed Intake, Production and Reproduction and the GrazFeed DSS. *Agricultural Systems,* 54, 77-126.

Moore, A.D., J.R. Donnelly & M. Freer (1997). GRAZPLAN:Decision Support Systems for Australian Grazing Enterprises. III. Growth and Soil Moisture Submodels, and the GrassGro DSS. *Agricultural Systems,* 55, 535-582.

Simulation of pasture phase options for mixed livestock and cropping enterprises

L. Salmon, A.D. Moore and J.F. Angus
CSIRO Plant Industry, GPO Box 1600, Canberra, ACT, Australia, 2601; Email: libby.salmon@csiro.au

Keywords: decision support tool, lucerne, subterranean clover, wheat, mixed farming

Introduction In southern Australia, 50% of grain-producing farms also run beef and/or sheep enterprises. Legume pasture leys are used to replace soil nitrogen and manage crop disease risks. Deep-rooted perennials, predominantly lucerne (*Medicago sativa*), are replacing annual *Trifolium subterraneum*-based leys to increase pasture production. They also have the environmental benefits of limiting soil acidity, rising water tables and dryland salinity. After recent droughts depletion of soil water by lucerne has penalised wheat yields. Decision support tools can help farmers evaluate the long-term effects of grazed annual and perennial leys on animal and crop production at the whole farm level.

Materials and methods FarmWi$e, a modular decision support tool (Moore 2001) was used to simulate wheat yield, pasture production and animal production in rotations of 3 years of wheat and 3 years of either lucerne or self-regenerating sub clover at Lockhart, NSW (mean annual rainfall 500mm). The water balance model used in the simulations was an updated version of that described by Moore *et al.*, (1997). Continuous simulations from 1954-2003 used daily weather records and a description of a Red Dermosol soil from the region. The management module of FarmWi$e was used to simulate the recommended practices of undersowing lucerne in the 3rd year of wheat, lucerne removal in mid-October of the 3rd year of the pasture phase and residue cultivation prior to sowing wheat. Simulations assumed maximum root depths in this soil of 0.7m for sub clover, 1.3m for wheat, 2.4m for lucerne and that soil nitrogen and disease did not limit pasture or crop growth. The effects of each pasture type on water balance were modelled but any differences in the effect of lucerne and sub clover on soil structure or soil nutrient dynamics were not. In each system the pasture phase was grazed by weaned 30 kg 4 month-old Border Leicester × Merino sheep to achieve a live weight of 45 kg. Six simulations of each of the rotations were run, each commencing in a different year of the rotation. The difference in wheat yield for the two pasture rotations was calculated for the each of the 50 years simulated (1954-2003).

Results Average wheat yields over 50 years of 3.90 and 3.75 t/ha for the sub clover and lucerne wheat rotations, respectively, agree with district performance. However yield differences between rotation types depended on the year of the crop phase (Figure 1). In Year 1 of the wheat phase, yields from crops sown after lucerne were, on average, 0.14 t/ha greater than crops sown after sub clover but were similar in Year 2 and 0.56 t/ha less in Year 3. Lucerne removed more water from the soil profile than sub clover but summer rainfall during the long fallow period before the first wheat crop often replenished the soil profile. In Year 3, wheat yields in lucerne rotations were penalised by competition from the undersown lucerne. Distributions of yield differences show that in 2% of years the penalty in Year 1 of the wheat phase after lucerne exceeded 1.2 t/ha. These years were the severe droughts of 2002 and 2003; the recent bad experiences of local farmers are therefore exceptional and are not necessarily a sound basis on which to choose pasture species.

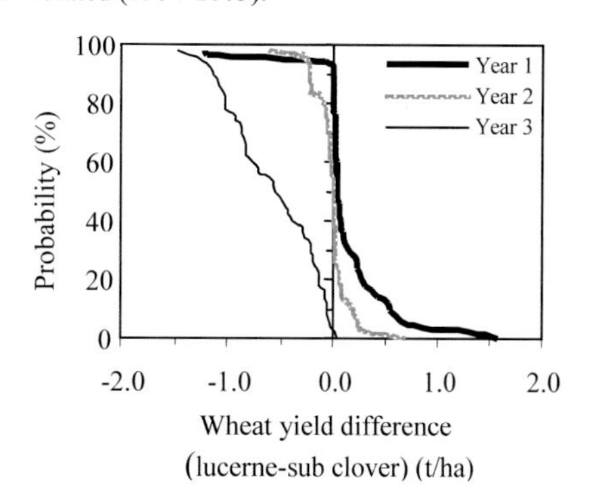

Figure 1 Distribution of differences between simulated lucerne and sub clover rotations in wheat yields for each year of the wheat phase

Conclusions Decision support tools enable a whole farm approach to assess the relative benefit of lucerne to both crop and livestock enterprises on farms and environmental sustainability.

References

Moore, A.D. (2001). FarmWi$e: a flexible decision support tool for grazing systems management. In: Gomide, J.A., W.R.S. Mattos & S.C. da Silva (eds.), *Proceedings of the Nineteenth International Grassland Congress*, Sao Pedro, Brazil. FEALQ, Sao Pedro, Brazil, pp. 1045-1046.

Moore A.D., J.R. Donnelly & M. Freer (1997). GRAZPLAN: decision support systems for Australian grazing enterprises. III. Pasture growth and soil moisture submodels, and the GrassGro DSS. *Agricultural Systems* 55, 535-582.

A farmer friendly feed budget calculator for grazing management decisions in winter and spring

M. Curnow and M. Hyder
Department of Agriculture Western Australia, 444 Albany Hwy, Albany 6330 Western Australia, Email: mcurnow@agric.wa.gov.au

Keywords: feed budget, electronic calculator, grazing, sheep

Introduction The Western Australian (WA) environment is Mediterranean with annual legume/grass pastures and a 6 month growing season. In autumn where over grazing can impact pasture establishment and in spring, prior to senescence, when under grazing can mean significant losses of efficiency are crucial times for grazing management. Pasture utilisation is typically low (25-35%) due to conservative stocking regimes; key to increasing productivity is increasing pasture utilisation (Grimm, 1998). Increased levels of productivity require farmer sophistication in the way they feed budget. To this end, satellite technology is being used to provide farmers in southern Australia with weekly estimates of pasture growth rate (PGR; kg DM/ha/d) and monthly estimates of Feed on Offer (FOO; kg DM/ha) (Kelly *et al.*, 2003). In addition the Green Feed Budget Paddock Calculator (GFBC) was developed to provide a simple and accessible electronic calculator, which utilises this new information to assist farmers to feed budget and to make more accurate and timely stocking rate decisions.

Materials and methods The GFBC is a computer-based tool that reflects the key tactical decisions that graziers have to make throughout the growing season. It uses feed intake data generated from Grazfeed® where the inputs have been modified to fit clover-dominant annual pastures. It has six scenarios, outlined in the main menu (Figure 1). The scenarios are divided into two phases - establishment (autumn- winter) and vegetative (late winter-spring) - to take account of the differences in pasture morphology (e.g. height, % dry matter). Each scenario provides a single point calculation and has an archive that allows paddock level recording. With the aim of developing a tool specifically for WA legume/grass pastures it was 'product tested' by farmer groups participating in the 'Pastures from Space' project (Kelly *et al.,* 2003), who are keenly involved in improving pasture assessment accuracy and lifting pasture utilisation. Feed back on the design was incorporated and the original calculation screens expanded to include feed mix calculators and other improvements.

Accessibility is a key design factor and due to slow rural internet line speeds the calculator is available as a runtime CD. The calculator is also free on the Department's website where additional information can also be sourced. Further research into feed intake, energy requirements and production targets of the sheep are being regularly incorporated into the calculator via new versions or updated screens on the web site.

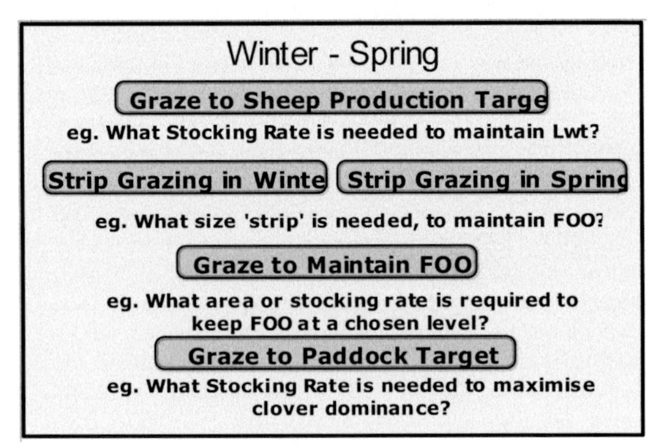

Figure 1 Choice of Winter-Spring scenarios in the calculator

Discussion The challenge was to build a simple, yet accurate product that allows the best tactical decisions to be made, according to the grazing system. Frequently simple decision tools suffer from not adequately reflecting the complexity of the biological system and therefore suffer inaccuracy. However, this potential source of error is small in comparison with the error inherent in pasture assessment. The benefits of this product are that it encourages frequent use, can be used in the paddock, and the skill and confidence gained by regular feed budgeting encourage farmers to explore whole farm budgeting (which is the ultimate end point). Other products available on the market tend to be complex and ask the grazier to enter a lot of data prior to calculation, which often contributes little to the output, leaving many farmers with a time consuming and intimidating product.

References

Grimm, M., (1998). Tactical grazing strategies for annual pastures, *Proceedings Grassland Society of Victoria*, Victoria, Australia, 67.

Kelly, R., A. Edirisinghe, G. Donald, C. Oldham & D. Henry, (2003). Satellite based spatial information on pastures improves Australian sheep production. *1st European Conference on Precision Livestock Farming*, June 15-18: 93-98.

GrassCheck: monitoring and predicting grass production in Northern Ireland

P.D. Barrett[1] and A.S. Laidlaw[2]
[1]*Agricultural Research Institute of Northern Ireland, Large Park, Hillsborough, Co. Down, BT26 6DR, UK,*
[2]*Department of Agriculture and Rural Development for Northern Ireland, Plant Testing Station, Crossnacreevy, Belfast, BT6 9SH, UK, Email: scott.laidlaw@dardni.gov.uk*

Keywords: herbage, growth, model, dairy, management

Introduction Grass budgeting is a key management practice on dairy farms to balance grass supply on paddocks with grass demand by the grazing herd. Grass budgets must be pre-emptive to be effective. The uncertainty of grass production and the difficulty in quantifying both current and forecasted rates of growth hamper effective budgeting and paddock management. Grass growth rates are highly variable both in time and space. Therefore, they vary greatly between locations at any given time and also across the season at any given location. Figure 1 shows the pattern of growth rates recorded at the Agricultural Research Institute of Northern Ireland (ARINI) in the two seasons before this project. The GrassCheck project was established in Northern Ireland to quantify current rates of grass growth and grass quality and to predict growth rates for up to 2 weeks in advance. The project will run from 2004 until 2006. This paper outlines the project and reports on its findings after one year.

Methodology A total of 6 sets of perennial ryegrass plots were established at 3 Department of Agriculture and Rural Development for Northern Ireland (DARDNI) sites in Northern Ireland: at ARINI, Hillsborough; Greenmount Campus, Antrim and The Plant Testing Station, Crossnacreevy. Plots, circa 1.5 x 5.0m, were cut 4cm above ground with a motor scythe. A total of 365kg N/ha was applied over the growing season. Each set of plots consisted of 9 plots comprising 3 series of 3 replicates. Only one series was cut/week under a sequential weekly cutting regime. Therefore, to simulate rotational grazing, all plots were given 21 days regrowth. Also, growth rates were predicted for the next 2 weeks using the ARINI GrazeGro growth model (Barrett *et al.*, 2004). Growth rate and grass quality were determined rapidly from the plots. They, plus the 2-week predicted growth, were reported to the farmers in weekly bulletins in the local farming press and on DARDNI websites.

Results Growth rates for 2004 varied considerably across the season and deviated consistently from the 5-year mean growth rate line determined over the 5 years preceding the project. Figure 2 shows the pattern of growth rates for 2004. Accumulated grass production from Mar to Sep 2004 was 15.2% above the mean for the previous 5 years. The GrazeGro growth model provided a reliable indication of future rates of grass growth. Figure 2 also shows predicted growth rates. R^2=0.78 for predicted output regressed against observed growth rates until the middle of Sep 2004.

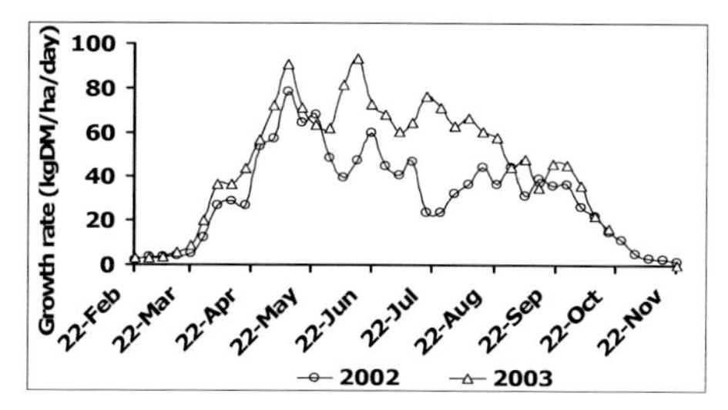

Figure 1 Variation in growth rate observed at ARINI 2002-03

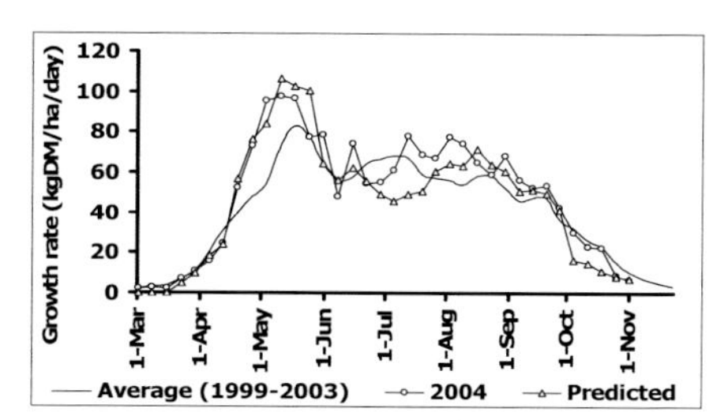

Figure 2 GrassCheck 5-year mean, actual and predicted growth rate for 2004

Conclusion Given the variability of growth, the need for a monitoring programme for grass growth and quality was demonstrated well. Weekly GrassCheck bulletins throughout the season provided accurate and timely indication of growing conditions to farmers and advisors in Northern Ireland. Given the good precision of the GrazeGro model, accurate estimates of predicted growth rates were given to aid in decision-making for grassland management and grass budgeting procedures.

References

Barrett P. D., A.S. Laidlaw & C. S. Mayne (2005). GrazeGro: A European herbage growth model to predict pasture production in perennial ryegrass swards for decision support. *European Journal of Agronomy,* In Press

Grass growth profiles in Brittany

P. Defrance[1], J.M. Seuret[2] and L. Delaby[3]
[1]*CRA de Bretagne, CS 74223, 35042 Rennes cedex, France, Email: defrance@st-gilles.rennes.inra.fr,*
[2]*Chambres d'Agriculture de Bretagne, avenue du chalutier "Sans Pitié", BP540, 22195 Plérin Cedex, France*
[3]*INRA, Unité Mixte de Recherche sur la Production du Lait, 35590 Saint-Gilles, France*

Keywords: grass growth, Brittany, grazing management

Introduction For farmers, knowing the local grass growth profile and the possible variations between years is very helpful in managing grazing. Indeed, the comparison with herd needs and anticipated farm cover change allows decisions to be made that will maintain the cover at the desired level. This paper proposes a ten-days grass growth profile corresponding to Brittany's different conditions of soil, climate and pasture management.

Materials and methods Grass growth data were used from seven years (1997-2003) of "Pâture Plus", a communication campaign on grazing management run in Brittany. On 20 farms per year (47 total), the height of all the paddocks (75% of perennial ryegrass and white clover mixtures and 25% of pure perennial ryegrass) was measured weekly from February to November with a rising plate meter. Grass growth in a paddock between two measurements was defined as $GG_w = (\Delta H * D_w) / \Delta t$ where GGw = grass growth of the paddock in week w (kg DM/ha per day), ΔH = Height difference between two measurements (cm), Dw = sward density in week w (kg DM/cm per ha) (Defrance et al., 2004) and Δt = time between two measurements.
The database was composed of 3397 pieces of growth data, defined as the average grass growth on all the paddocks of a farm neither grazed nor cut between two measurements, weighted by the paddock areas. Farms were classified into two zones in spring (February to May), favourable or unfavourable, and into three zones in summer (June to September), humid, intermediate or dry, according to the comparison between the farm growth profile and the average profile. No zones have been distinguished in autumn (October and November). Each period was analysed separately, taking in account the year, zone, farm within zone, ten-days period, the interaction between zone and ten-days period and the initial grass height centred within ten-days period. Except for the interaction in spring, all these factors were significant at 1% level.

Results and discussion The average growth is 39.1 kg DM/ha per day for an initial height of 7.2 cm. Models explain 60, 51 and 34 % of the total variability in spring, summer and autumn, respectively. In spring, growth increases from 11 in February to 71 kg DM/ha per day in mid-May with an average difference of 11 kg between the two zones. For both of them, mid-April is characterized by a growth deficit in accordance with a temperature deficit usually observed at this time. In summer, growth decreases in the three zones but much more strongly in the dry zone (reaching 0 at the beginning of August) than in the humid zone. The growth recovery in September is delayed in the dry zone. Average initial height at farm level is positively correlated with grass growth. This is 2.6, 3.2 and 1.2 kg DM/ha per day higher per additional centimetre in spring, summer and autumn, respectively.

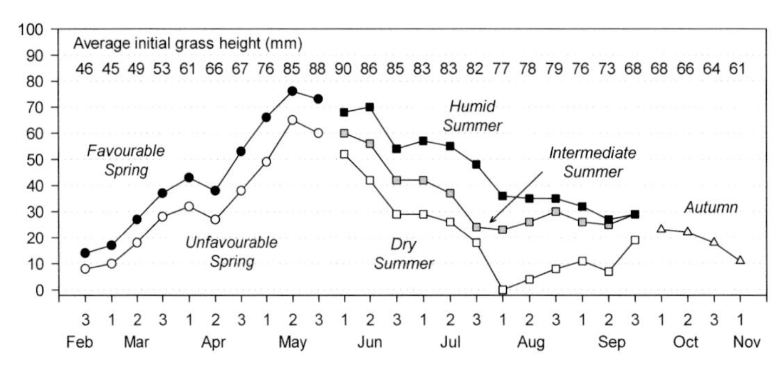

Figure 1 Ten-days grass growth (kg DM/ha per day) profiles in Brittany according to different soil and climate conditions in spring, summer and autumn and corresponding initial grass height (mm)

Conclusions This work has revealed the different grass growth potentials existing in Brittany, from 6.7 t/ha on average, in a farm with unfavourable conditions in spring and dry conditions in summer to 10.8 t/ha in a farm with favourable conditions in spring and humid conditions in summer. Furthermore, these profiles will be used within the "Agrotransfert Bretagne" project to develop user-friendly tools for grazing management.

References

Defrance P., L. Delaby, & J. M. Seuret (2004). Mieux connaître la densité de l'herbe pour calculer la croissance, la biomasse d'une parcelle et le stock d'herbe disponible d'une exploitation. *Rencontres Recherches Ruminants 11, 291-294.*

Effect of strategy of forage supplementation and of turnout date in a medium stocking rate system on the main characteristics of dairy cows grazing

P. Defrance[1], L. Delaby[2], J.M. Seuret[3] and M. O'Donovan[4]
[1]*CRA de Bretagne, CS 74223, 35042 Rennes cedex, France, Email: defrance@st-gilles.rennes.inra.fr,*
[2]*INRA, Unité Mixte de Recherche sur la Production du Lait, 35590 Saint-Gilles, France*
[3]*Chambres d'Agriculture de Bretagne, avenue du chalutier "Sans Pitié", BP540, 22195 Plérin Cedex, France*
[4]*Teagasc, Moorepark Production Research Centre, Fermoy, Co. Cork, Ireland*

Keywords: decision support system, grazing management, dairy cows

Introduction Having a stocking rate of 2.9 cows per hectare of grassland (35 ares/cow) in Brittany offers many options for turnout date and forage supplementation strategies. For a farmer, knowing the consequences of the different options during the course of the grazing season makes grazing management decisions easier. As experiments on grazing management require considerable resources and are hardly generalisable, various spring scenarios have been tested using a dynamic decision support system, Pâtur'IN (Delaby *et al.*, this volume).

Materials and methods The model herd was composed of 40 dairy cows (mean calving date, 01/10; peak milk yield, 35 kg). Fourteen ha of grassland were available for grazing, divided into ten paddocks, half sown in perennial ryegrass (PRG) and the other half in perennial ryegrass and white clover mixtures (PRG-WC). They were characterized by Brittany's average grass growth (Defrance et al., this volume). Average sward density is 235 kg DM/cm per ha for PRG and 205 for PRG-WC. Six scenarios were tested: three strategies of forage supplementation (maize silage fed until 1 April, until 20 April and throughout the entire season) combined with two turnout dates (15 February or 15 March). In each of the six scenarios, cows were supplemented with 2 kg concentrates per day. Until 14 March, animals were offered 16 kg DM maize silage per day in scenarios 2.1, 2.2 and 2.3 and 12kg DM/day in the other scenarios. From 15 March, the amount of maize silage decreased to 8 then 4 and possibly 0 kg DM/day according to the strategy constraints.

Results and discussion With turnout on 15 February (scenario 1.1), farm cover was too low on 1 April to terminate silage feeding (Table 1). This scenario was not sustainable. In the other five scenarios, cows consumed approximately the same amount of forage (2100 kg DM), but the proportion of grazed grass decreased when turnout date or silo closing date were delayed. Scenarios 1.2 and 2.1 had very similar results with a large proportion of grazed grass in the diet (73%) and little harvested grass (20 to 30% of the area). Delaying turnout date allowed forage supplementation to be stopped earlier. Continuous silage supplementation (scenarios 1.3 and 2.3) led to a small proportion of grazed grass in the diet and a high number of harvests.

Table 1 Main characteristics of the six scenarios examined (from 15/02 to 30/06)

Scenario		1.1	1.2	1.3	2.1	2.2	2.3
Turnout date		15th Feb.	15th Feb.	15th Feb.	15th March	15th March	15th March
Silo closing date		1st Apr.	20th Apr.	/	1st Apr.	20th Apr.	/
Silage ingested (kg DM/cow)			564	852	568	676	968
Grass ingested (kg DM/cow)			1521	1265	1515	1406	1145
Number of paddocks cut			2	5	3	4	7
Grass harvested (kg DM/cow)			213	488	260	353	624
Area grazed /rotation (ares/cow)			35/35/28/28	35/35/17.5/24.5	31.5/28/28	28/28/28	28/24.5/21
PreGH / PostGH (cm)			10.0 / 5.4	10.2 / 5.4	11.8 / 5.7	11.3 / 5.6	11.7 / 5.8
	15th Feb.	7 / 340	7 / 340	7 / 340	7 / 340	7 / 340	7 / 340
Days Ahead /	15th March	8 / 388	8 / 388	9 / 390	18 / 823	18 / 823	18 / 823
Farm Cover	1st April	10 / 456	12 / 568	12 / 571	21 / 967	23 / 1061	23 / 1061
(days / kg DM/ha)	20th April	5 / 239	13 / 606	13 / 610	19 / 848	25 / 1158	26 / 1168
	30th June		18 / 836	16 / 754	17 / 779	15 / 707	15 / 702

Conclusions The optimum grazing strategy is determined by the objectives of each individual farmer. To allow supplementation to be stopped quickly, a late turnout combined with a short transition appears best. However, many farmers prefer an early turnout to break away from winter drabness. In this case, full grass date is delayed.

References

Defrance P., L. Delaby, & J. M. Seuret (2005). Grass growth profile in Brittany. *This volume.*

Delaby L., J. L. Peyraud, & P. Faverdin (2005). Pâtur'IN : a user-friendly software tool to assist dairy cow grazing management. *This volume.*

Using the GrassGro decision support tool to evaluate the response in grazing systems to pasture legume or a grass cultivar with improved nutritive value

H. Dove and J.R. Donnelly
CSIRO Plant Industry, GPO Box 1600, Canberra, ACT 2601, Australia, Email: hugh.dove@csiro.au

Keywords: cultivar evaluation, nutritive value, simulation

Introduction Decision support tools (DST) based on models of grazing systems allow the evaluation of changes in enterprise management on productivity and profitability. The Grassgro DST (Moore *et al.*, 1997) uses historical weather data on a daily time step to simulate pasture growth and the resultant productivity of either grazing sheep or cattle. Different pasture species are represented within a parameter set that describes the response of pasture species to their environment. Manipulation of these parameters provides a means of evaluating, *a priori*, the likely responses of livestock production to 'improved cultivars'. We report the results of simulations conducted within grazing enterprises at three locations in southern Australia: a breeding ewe enterprise at Benalla; a wool-producing enterprise at Hamilton; and a beef breeding enterprise at Corryong.

Materials and methods All simulations were conducted using GrassGro and daily weather data for the above locations for the period 1957-1996. Simulated pastures contained either an unimproved grass cultivar, with or without legume (UC+, UC-) or an improved cultivar with or without legume (IC+, IC-), each grazed at five stocking rates (SR). The legume was subterranean clover. The legume content of the pasture was predicted by the simulations. The 'improved' cultivar was generated by altering parameters for the rate of pasture digestibility decline over summer so that IC herbage was 3-4% more digestible than UC herbage during summer. All results were evaluated by expressing enterprise gross margins/ha relative to that obtained at the lowest SR, with UC-.

Results Predicted enterprise gross margins responded in curvilinear fashion to increases in SR (Figure 1; Table 1). The higher summer digestibility of the IC- resulted in marked increases in gross margin. However, the advantage of having legume in the pasture was always much greater than the advantage of having the IC (e.g. Table 1, column 5 v. column 4). The results also indicated that the IC and especially the legume allowed increases in enterprise SR. Simulations based on IC+ suggested that there was little or no advantage relative to having legume alone, partly because the IC no longer made up all of the sward, but also because the legume helped to fill gaps in digestible pasture supply from UC grass.

Table 1 Relative gross margins/ha in response to SR, grass cultivar and legume for sheep enterprises in Australia

Location	SR/ha	Cultivar (UC, IC), legume status (±)			
		UC, -	IC, -	UC, +	IC, +
Benalla	5.0	1.00	1.12	1.39	1.37
	7.5	1.49	1.73	2.08	2.13
	10.0	1.81	2.19	2.7	2.78
	12.5	1.83	2.35	3.03	3.13
	15.0	1.48	2.15	2.96	3.18
Hamilton	5.0	1.00	1.5	1.92	1.68
	7.5	1.66	2.59	3.28	2.98
	10.0	1.58	3.04	4.36	4.24
	12.5	0.61	2.51	4.68	4.57
	15.0	-1.44	1.11	3.14	3.09

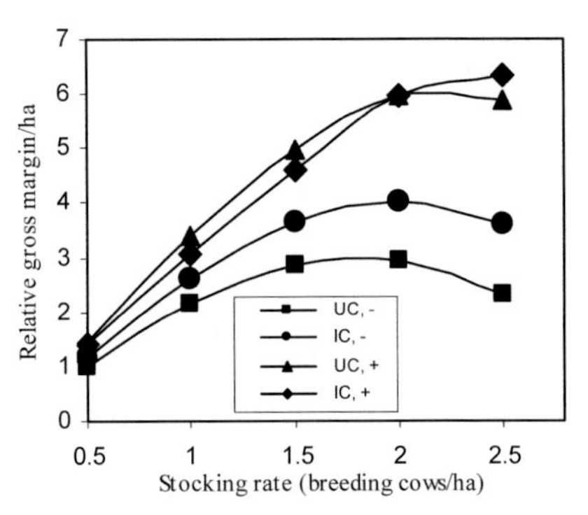

Figure 1 Relative gross margins/ha for a beef breeding enterprise in Corryong, Australia

Conclusions The use of the GrassGro DST allowed evaluation of the relative effects of either an 'improved' grass cultivar or legume inclusion on the long-term profitability of the chosen enterprises. In all systems, the presence of legume was more important than improved grass digestibility over summer and the combination of legume and IC grass was no more profitable in the long term than legume alone. The use of such DST is thus a powerful tool for making *a priori* comparisons of the profitability of different plant breeding and management objectives.

References

Moore, A.D., J.R. Donnelly & M. Freer (1997). GRAZPLAN: decision support systems for Australian grazing enterprises. III. Pasture growth and soil moisture submodels, and the GrassGro DSS. *Agricultural Systems*, 55, 535-582.

Pasture land management system decision support software

G.E. Groover[1], S.R. Smith[2], N.D. Stone[3], J.J. Venuto, Jr[3] and J.M. Galbraith[2]

[1]*AAEC Department, MC-0401, VA Tech, Blacksburg, VA 24061, USA,* Email: *xgrover@vt.edu*, [2]*CSES Department, MC-0404, VA Tech, Blacksburg, VA 24061, USA.* [3]*ISIS Labs, PO Box 1423, Norfolk, VA 23518-4234, USA*

Keywords: alternatives, DSS, grazing, simulation, cattle

Introduction Controlled or rotational grazing provides benefits to producers and society through profitable and sound management of grazing land and livestock. Pasture land management system (PLMS) is a decision support system developed to help university, government, and professionals provide technical pasture management assistance to beef and dairy producers. The PLMS focuses on the balance between seasonal forage supply and nutrient demand in a dairy or beef cattle operation. It allows users to explore and compare alternatives (dividing fields into multiple paddocks, changing stocking rates, and forage species) through a visual display and embedded simulation. Users enter a description of the farm by drawing a map. Maps can be drawn freehand, traced over a scanned image, or GIS data may be incorporated. Once map and field data are entered the grazing options are specified via input screens. Grazing systems can be easily compared without economic risk and with almost immediate feedback on how these alternative systems affect variables like milk production and pounds of beef sold. PLMS serves as both an educational tool and a strategic planning tool for evaluating alternative grazing operations and management related investments (website: http://clic.cses.vt.edu/PLMS/).

Materials and methods The PLMS has been developed using standard software engineering practices with active participation by users in the design and iterative development of the prototype system. The program incorporates GIS components using MapObjects™, which gives it the ability to display spatially referenced data stored as shape-files. The PLMS was written in Java using Microsoft's Visual J++™, both to enhance cross-platform development (a possible web version) and to simplify the integration of visual and model components. The design goals of PLMS are to: 1) increase adoption of management intensive grazing, 2) aid users in exploring alternative grazing management options, 3) allow graphical display of animal demands and farm forage supply to support an iterative process to evaluate alternatives, 4) simulate a representative 365-days with up to half-day rotations, 5) provide rule-based animal movements between pasture (see next section), 6) harvest surplus forage as hay, 7) feed hay during deficits (raised or purchased) and 8) provide supplements to dairy cattle.

The following rules are used for each simulation: 1) Animals enter pastures at 1,794 kg/ha and exit at 897 kg/ha of DM, 2) The entry threshold is reduced by up to 25% for a field growing maximally, 3) Exit thresholds are reduced when paddock growth rate declines as much 50%, 4) Hay is harvested when total DM in a field exceeds 3,587 kg/ha and 673 kg of DM per hectare is left as residue, 5) Hay is also harvested when forages in a field have stopped growing and more than 1,794 kg/ha of DM (half the hay harvest trigger of 3,587 kg/ha) remain.

Results Users receive immediate visual feedback using a pair wise comparison of any 2 alternatives (Figure 1). Results are depicted via line graphs (Figure 1) of animal demand, forage supply, hay fed, and via bar charts for major factors, e.g. total tons of hay. In addition to the visual output each simulation produces a detailed output file (all output variables over the 365-days simulated) that can be viewed in an Excel® spreadsheet. The output file is used to fine tune the final management plans and/or for checking errors.

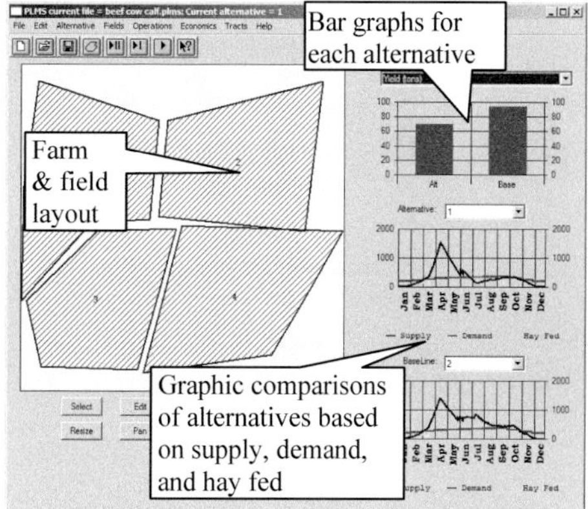

Figure 1 PLMS main screen and output

Conclusions The program was released in 2001 and has been used by extension, government, and other personnel to design grazing systems in the U.S. Mid-Atlantic region. End users have provided constructive criticism that is being incorporated into the model, resulting in 3 major revisions since 2001. It should not be evaluated as a planning tool for its ability to accurately reproduce field data collected in specific grazing trials. Rather, its value is in its ability to make reasonable and accurate comparisons between alternative grazing management plans.

Forecast of herbage production under continuous grazing

K. Søegaard[1], J. Berntsen[1], K.A. Nielsen[2] and I. Thysen[1]

[1]*Danish Institute of Agricultural Sciences, DK-8830 Tjele, Denmark, Email: Karen.Soegaard@agrsci.dk,*
[2]*Danish Agricultural Advisory Service, Udkærvej 15, DK-8200 Århus V*

Keywords: forecast, grazing, grass/clover, growth rate

Introduction The utilization of pasture is very sensitive to oscillations in herbage growth. The farmer's daily planning involves decisions on pasture use as well as on the amount and composition of supplement feeding. In this planning, expected daily growth rate is an important factor. Often the knowledge and experience about the growth rate is first available after changes in management should have been made. A different growth rate than expected should lead to changes in the grazing area or in the supplement feeding. Therefore, a simple model of grass/clover growth under grazing and irrigated conditions was developed. From spring 2004, the model was used to compute a forecast of grass/clover growth, which has been available to Danish farmers and advisors in the online crop information system PlanteInfo (for a version in English, see planteinfo.dk/english).

Materials and methods A model for grass/clover growth was developed based on existing data from grazing experiments with dairy cows on research stations and private farms in Denmark. The model uses solely climatic information as driving parameters. The experimental data represent a wide range of N-rates, clover contents, soil types etc. These conditions were quantified and the growth was described depending on general N-level, leaf area index (LAI), phenological development, radiation and temperature:
$\Delta = \varepsilon f(T)g(I)h(N)i(T_{sum})$ where f(T) is a function of temperature, g(I) is a standard function of radiation and LAI, h(N) a linear function of N-level and $i(T_{sum})$ describe the physiological development of the grass/clover.

Results Under Danish conditions, LAI in continuous grazed pastures on farms is between 0.6 and 2.3, and therefore the grazing pressure will affect the growth rate very much. Other parameters such as sward composition and age, soil type, fertilization rate and distribution throughout the season further affect the growth rate in the individual field. Therefore, it was decided not to estimate the growth for the individual field, but to estimate the growth rate under fixed conditions (LAI, N-rate, clover content and drought stress) and to show the growth in relative terms. The forecasted growth could then be used in relation to the knowledge about the normal growth in a certain pasture. An increase in the predicted growth rate could for example result in lesser amounts of supplement feed or changes in the pasture area. The web page is shown in figure 1. The line is the mean growth rate over 10 years based on the model. The dark area is the growth rate until the current day calculated by the growth model, and the light area is the herbage growth for the coming seven days based on the weather forecast. Further information about the increase/decrease in the forecast week in relation to the last week is shown. In this situation there is prognosticated 14% increase. The predicted growth rate in 2004 was validated against results from trial plots, which confirmed the growth curve for 2004.

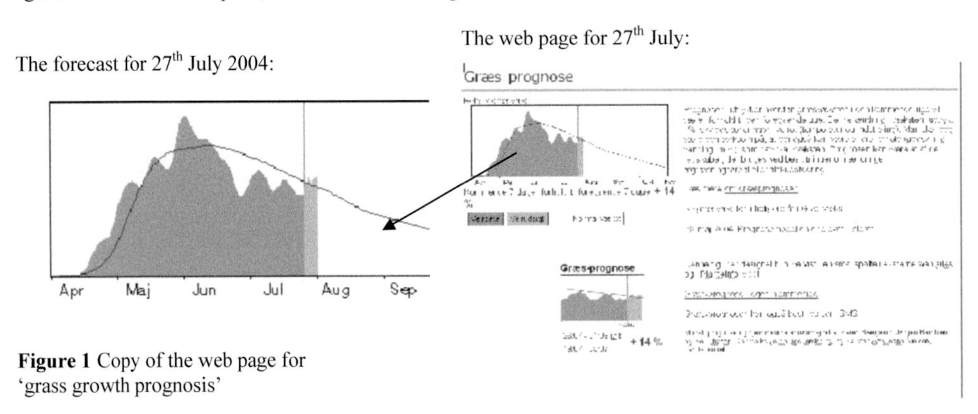

Figure 1 Copy of the web page for 'grass growth prognosis'

Conclusions It is planned to extend the forecast with herbage quality parameters. The work in 2005 will focus on water-soluble carbohydrates and crude protein especially in spring and early summer.

A decision support tool for seed mixture calculations

B.P. Berg and G. Hutton
Alberta Agriculture, Food and Rural Development, Agriculture Center, Lethbridge, Alberta, Canada, T1J 4V6
Email: bjorn.berg@gov.ab.ca

Keywords: seeding rate, seed mixture, calculator

Introduction Grassland species are normally seeded in mixtures rather than monocultures. In theory, seeding rates for mixtures are simply a sum of the amount of pure live seed (PLS) of each seed lot in the mix, an amount sufficient to ensure establishment and survival of each species. Mixtures can be complex because of the number of species used (especially in conservation and reclamation programs) and variations in seed purity and seed size. Soil limitations and seeding equipment settings need to be considered and in Canada, a metric conversion may be required. All these conditions make by-hand calculations of mixtures containing more than 3 species tedious and complicated. Thus, in practice, agronomists and growers use simple rules to set rates. The easiest rule is to estimate the mixture's components as a percentage by weight of a standardized total weight of the seed required (e.g. 10% of 10 kg/ha). The resulting errors can be observed in the predominance of thin stands, the unexpected dominance of small seeded species and the added costs of interseeding to compete with weeds and fertilizer to increase yield. The objective of this project was to develop a decision support tool, a seed mixture calculator to simplify conversion and improve the estimates of seed required for individual seeding projects.

Materials and methods Initial programming was done using Lotus 1-2-3®. Input data (such as seed density and purity) were entered step-wise in a standard array of interim calculations and results. In 2004 the program was converted to Microsoft Excel®. Toggle buttons were introduced to make unit conversions, and list boxes were used to limit the selection of species so that reference information could be placed in uniform tables. The old linear spreadsheet solution was rewritten and Microsoft VBA® subroutines were installed to handle input errors and manage tabular data as arrays. Thus, tables could be edited without changing the new programming solution and vice versa. The final version was exported to Java® as an interactive calculator on an Internet website.

Results A user of the original Lotus 1-2-3® program would follow customary practice and first choose a maximum seeding rate for bulk seed (MBR). This made economic sense because the user could compare project costs simply by multiplying the MBR by the bulk seed price. However, the linear programming solution that divided each species in the mix into its relative proportions was convoluted and overly complicated. More critically, if the user underestimated the MBR, (for example by using a MBR appropriate to a monoculture) the amount of seed of each species in the mixture (calculated as a proportion of the MBR) would be inadequate. Field experience suggests the MBR for mixtures of large seeded species should be much higher than for monocultures (S. Acharya, 1997, pers. comm.) and could be considerably lower for small seeded species or high quality seed. Low seeding rates impacted establishment and reduced yields for 2 years after seeding in the semiarid climate on the Canadian prairie (Leyshon *et. al.* 1981). Biodiverse mixtures should add stability to the stand but may be less stable than a comparable monoculture if important components in the mix fail to establish.

In the new Microsoft Excel® and Java® versions, the user manipulates the maximum PLS rates (MPR) for either broadcast or row seeding systems. Using MPR refocuses the solution onto issues about seed quality rather than the bulk seed cost. In fact, basing the new model on PLS makes it easier to compare and justify seed costs, refine mixtures and compare substitute cultivars. MBR is not ignored, just calculated in a different fashion. The model's output summarizes the amount of bulk seed required from each seed lot to achieve the MPR, calculates the MBR and gives the required seeding density on the ground (for seeder calibration) for each species. The model is also interactive because the user can readily change the mixture to observe the response of seeding density and seed cost. Selecting seed lots and mixtures with higher quality seed will invariably allow the user to reduce the total bulk seed purchased. Other refinements in the new model included converting database information to arrays to customize the worksheet for different institutional users.

Conclusions Seed mixtures are widely used in grassland seedings but are difficult to calculate. A customized, interactive seed mixture calculator was developed as a decision support aid for growers, seed sales agents, extension personnel and parks and reclamation projects. The calculator refocuses attention on issues of seed quality and seeding management for establishing grassland seed mixtures.

References

Leyshon, A.J., M.R. Kilcher, & J.D. McElgunn (1981). Seeding rates and row spacings for three forage crops grown alone or in alternate grass-alfalfa rows in southwestern Saskatchewan. *Canadian Journal of Plant Science* 61,711-717.

Section 5

Constraints and opportunities for animal production from temperate pastures

Environmental clustering of New Zealand dairy herds

J.R. Bryant[1], N. López-Villalobos[1], J.E. Pryce[2] and C.W. Holmes[1]
[1]*Institute of Veterinary, Animal and Biomedical Sciences, Massey University, Private Bag 11-222, Palmerston North, New Zealand, Email: J.R.Bryant@massey.ac.nz,* [2]*Livestock Improvement Corporation (LIC), Private Bag 3016, Hamilton, New Zealand*

Keywords: clustering, climate, environment, New Zealand

Introduction Previous studies have found that milk yield (a proxy for feeding level) and temperature-humidity index (THI) are important factors in explaining genotype x environment (G x E) interactions, indicating differences between the abilities of genotypes to forage or consume concentrates effectively or to cope with thermal stress (Ravagnolo and Misztal, 2000; Zwald *et al.*, 2003). The objective of this study was to quantify and cluster (CL) herd environments within New Zealand (NZ) based on production levels, a summer heat load index (HLI) and geographical location.

Materials and methods Production data consisted of 497,433 total lactation milk solids (fat + protein; MS) yields from LIC sire proving scheme animals from 1989-2003. Each data record had a map reference identifying the herd's location. Daily climatic data for the same period were obtained from 89 meteorological stations throughout NZ. Herd-year (HY) groups were clustered based on adjusted MS yield, summer HLI and latitude using the FASTCLUS procedure of SAS (SAS, 1999). Adjusted MS yield for each HY was obtained using the GLM procedure of SAS (SAS, 1999) fitting a model, which included the effects of HY, age, breed and days in milk. Summer HLI for each year was calculated for each weather station (where possible) using the HLI described by Castaneda *et al.* (2004). The HLI includes relative humidity, temperature (similar to THI), but also solar radiation and wind speed. The summer HLI average from the weather station (within a 50 km radius) nearest to each farm (found using ArcView GIS 3.2 (ESRI, 1999)) was used as the measure of climatic environment.

Results and discussion Five clusters were formed based on the Cubic Clustering Criterion. CL1 and CL2 consisted of low and medium MS yield herds, respectively in warm regions of the North Island (Table 1). CL3 was comprised of medium to low MS yield herds from Auckland (North Island) to the base of the South Island with a mild climate. CL4 herds had high MS yields and were located throughout NZ in mild climates. CL5 herds had the highest mean MS yield, and were primarily from the south of the South Island. There was significant variability in herd MS yield and in HLI between (and within) clusters. The top herds averaged in excess of 500 kg MS/cow in CL4 and CL5, whereas few herds averaged in excess of 350 kg MS/cow in CL1. Cluster means for HLI ranged from 60.9 to 67.9 in CL5 and CL1, respectively. A HLI of 67.9 is equivalent to 24 °C, a relative humidity of 80 % and a THI of 72.5, conditions that are considered thermo-neutral. However, individual farms throughout NZ experienced conditions (HLI>70 and THI>74), which correspond to some degree of heat stress.

Table 1 Summary of cluster means (s.d.)

	CL1	CL2	CL3	CL4	CL5
Number of HY	2611	3216	2554	1884	753
HLI	67.9 (1.7)	67.9 (1.4)	64.1 (1.5)	65.2 (1.6)	60.9 (1.8)
MS yield (kg cow/year)	240 (41.6)	293 (52.5)	269 (41.4)	337 (56.2)	343 (62.9)
Latitude	-37.1 (1.1)	-37.4 (1.0)	-39.8 (1.0)	-40.1 (1.5)	-45.4 (1.3)

Conclusions The results of the study demonstrate there is significant variability between clusters for MS yield and HLI within NZ. Subsequent analyses will identify sires, which are specifically suited to high or low MS yield or thermal stress environments.

References

Castaneda, C. A., J. B. Gaughan & Y. Sakaguchi. (2004). Relationships between climatic conditions and the behaviour of feedlot cattle. In: Proceedings of the 25th Biennial Conference of the Australian Society of Animal Production, vol. 25, pp. 33-36. Melbourne.

ESRI. (1999). ArcView GIS version 3.2: Environmental Research Institute, Inc., Redlands, CA, USA.

Ravagnolo, O. & I. Misztal. (2000). Genetic component of heat stress in dairy cattle, parameter estimation. Journal of Dairy Science. 83, 2126-2130.

SAS. (1999). SAS/STAT Users Guide. Cary, NC: SAS Institute.

Zwald, N. R., K. A. Weigel, W. F. Fikse & R. Rekaya. (2003). Identification of factors that cause genotype by environment interaction between herds of Holstein cattle in seventeen countries. Journal of Dairy Science. 86, 1009-1018.

Risk-efficiency assessment of haying

A.J. Romera[1,2], J. Hodgson[2], S.T. Morris[2], S.J.R. Woodward[3] and W.D. Stirling[2]
[1]*INTA, CC 276, (7620) Balcarce, Argentina, Email:ajromera@balcarce.inta.gov.ar,* [2]*College of Sciences, Massey University, Private Bag 11-222, Palmerston North, New Zealand.* [3]*Woodward Research Limited, P.O. Box 21160, Hamilton, New Zealand*

Keywords: cow-calf systems, computer model, long term simulation, haying policy

Introduction Pastoral livestock farms are complex, dynamic systems subject to many forms of external disturbance. Farm Management strategies are therefore typically designed to minimize system variation. The objective of this study was to explore the impacts of different haying policies, in terms of expected profit and risk, for cow-calf farms in the Salado Region of Argentina.

Materials and methods A computer model (Romera *et al.*, 2004) was used to simulate 21 haying policies (combination of area cut, AREA and cutting herbage mass, MASS), across a range of target cattle numbers (SR). The model is climatically driven, and the sub-models representing the farm's biophysical dynamics were taken from the literature. A set of decision rules was developed to represent (on a 100 ha farm) the management applied in Reserva 6, an experimental cow-calf farm located at the INTA-Balcarce Experimental Station. A simplified profit indicator based on current prices (P) was calculated. Sale price per kg of liveweight was 1.0 for calves and 0.8 for cows. The cost for haying was calculated as a function of the hay yield in kg DM/ha ($y = 0.012+0.094e^{(-5E-4x)}$). A factorial experiment was simulated, with different levels of AREA (0 to 60 ha), MASS (3 to 6 t DM/ha) and SR (170 to 350, cows plus heifers), and 20 replicates. Each replicate consisted of 50 years of random weather sampled from the real weather sequence at Balcarce, 1970-2000. The results were analyzed using the risk-efficient frontier method (Cacho *et al.*, 1999).

Results The risk-efficient frontier (Figure 1) shows the best possible combinations of expected P and risk (standard deviation of P, SD). Any policy below (or to the right of) the frontier is inefficient, as higher P is possible from the same level of risk (moving vertically) or the same P is possible with a lower risk (moving horizontally). No combination without hay (NoHay) was included in the frontier, but many policies using hay were clearly inferior to NoHay policy. The risk-efficient set included combinations that made hay at medium herbage mass and that harvested up to 40 ha (decreasing with increasing SR) (Table 1).

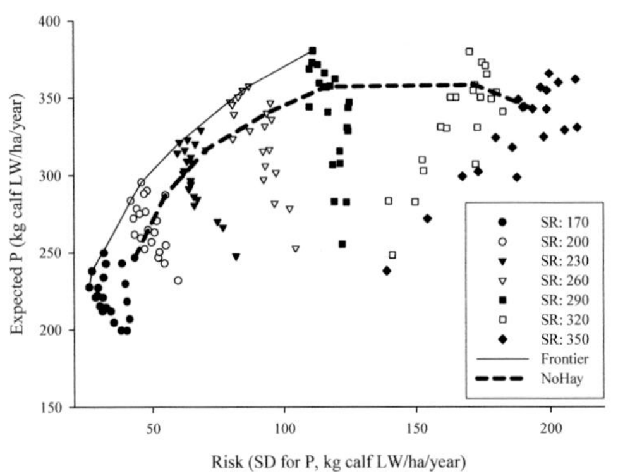

Figure 1 Risk-efficient frontier for the different haying strategies and target cow numbers (SR)

Table 1 Means (±SD) of P for the strategies included in the efficient sets

SR[1]	AREA[2] (ha)	MASS[3] (kg DM/ha)	P[4] (kg LW/ha/year)
170	40	4000	227±26
170	30	4000	238±27
200	20	4000	284±41
200	30	4000	295±46
230	30	4000	321±60
260	20	3000	355±84
260	20	4000	357±86
290	20	3000	381±111

[1]Target cow numbers in a simulated 100 ha farm (emergency sales and deaths can occur so actual stocking rates may be lower). [2]Area allocated to haying. [3]Cutting herbage mass for haying. [4]Expected profit.

Conclusions Making hay is not automatically beneficial for the system but at lower herbage masses it could be more profitable and less risky.

References

Cacho, O. J., A. C. Bywater & J. L. Dillon (1999). Assessment of Production Risk in Grazing Models. *Agricultural Systems,* 60, 87-98.

Romera, A. J., S. T. Morris, J. Hodgson, W. D. Stirling, & S. J. R. Woodward (2004). A model for simulating rule-based management of cow-calf systems. *Computers and Electronics in Agriculture,* 42, 67-86.

Irish dairy farming: effects of introducing a Maize component on grassland management over the next 50 years

A.J. Brereton and N.M. Holden
Department of Biosystems Engineering (Bioresources Modelling Group), University College Dublin, Earlsfort Terrace, Dublin 2, Ireland, Email: abrereton@club-internet.fr

Keywords: grass, maize, dairy system simulation, climate change

Introduction Typical management of Irish dairy units is based on a low-cost spring-calving strategy with 90% of annual feed derived from grass grown on the farm. Almost 70% of feed is from grazed grass managed by rotational grazing, the remainder is conserved forage and concentrates. The objectives of the work were to examine how the management system has to be modified when part of the dairy unit land is allocated to maize silage instead of grass silage production, and to examine how climate change over the next 50 years will impact on grass and maize management within the production system.

Materials and methods A dynamic, mechanistic dairy unit simulator (Fitzgerald *et al.*, 2003) was used to simulate the operation of a hypothetical farm at three locations under current climate and under the climate conditions predicted for the climate period 2055 (2041-2075), and at 0, 10 and 20% allocation of dairy unit area to maize. Unit 1 was located in the warm, dry south-east (Wexford). Unit 2 was located in the cool, humid north (Leitrim) and Unit 3 in the warm, humid south (Cork) (Holden and Brereton, 2004). In each unit/climate/maize scenario the system was simulated for 30 successive years using weather data generated from monthly means by stochastic weather generation (Geng *et al.*, 1986; Richardson, 1985). The same weather data was used to simulate the production of a generic short-season maize type using the Ceres – Maize model (Jones *et al.*, 1984). The grassland management simulated was the same as the blueprint described by O'Donovan (2000).

Results In the current climate the stocking rate (SR) at 0% maize area was 2.4 cows ha^{-1} for units 1 and 2 and 2.7 for unit 3. In the 2055 scenario the rates were 2.6 for unit 1 and 3.1 for units 2 and 3. In all unit/climate scenarios SR was reduced as the maize area increased. At 20% maize area SR was reduced by about 10% in all cases. The number of grass silage paddocks was reduced but the reduction did not compensate fully for the allocation of land to maize. Similarly, the reduction in land area for grazing also created a feed deficit at turnout and late in the season.

Conclusions The conversion of part of a grassland unit to maize silage production would result in a stocking rate reduction over the range of production environments existing or predicted for Ireland. However in all cases, the stocking rate on the grass area of the unit increased, by more than 10% in most cases, as the maize area increased. This suggests that the acquisition of maize silage from outside the unit would enable significant increases in stocking rate within the unit, but the potential increases to approximately 3 cows/ha are environmentally challenging. An alternative interpretation of the results is that the out-sourcing of maize silage would allow a dairy unit to maintain environmentally acceptable SR using less nitrogen fertiliser.

References

Fitzgerald, J.B., A.J. Brereton & N.M. Holden (2003). Dairy farm system simulation for assessment of climate change impacts on dairy production in Ireland. In J. France and L. A. Crompton, Proceedings of the Thirty-fifth Meeting of the Agricultural Research Modellers' Group, *Journal of Agricultural Science, Cambridge 140*, 479-*487*

Geng S., F. W. T. Penning de Vries & I. Supit (1986). A simple method for generating daily rainfall data. *Agricultural and Forest Meteorology 36*, 363-376.

Holden, N. M. & A. J. Brereton (2004). Definition of agroclimatic regions in Ireland using hydro-thermal and crop yield data. *Agricultural and Forest Meteorology 122*, 175-191

Jones, C.A., J. T. Ritchie, J. R. Kiniry, D. C. Godwin & S. I. Otter-Nacke (1984). The CERES Wheat and Maize model. In: *Proceedings International Symposium on minimum datasets for Agrotechnology Transfer*. ICRASET, Pantancheru (India). pages 95-100.

O'Donovan M.A. (2000). The relationship between the performance of dairy cows and grassland management on intensive dairy farms in Ireland, PhD thesis, University College Dublin.

Richardson C.W. (1985). Weather simulation for crop management models. *Tranactions of the American Society of Agricultural Engineering, 28*, 1602-1606.

Lucerne crown and taproot biomass affected early-spring canopy expansion

E.I. Teixeira, D.J. Moot, A.L. Fletcher
Agriculture and Life Sciences Division, Lincoln University, Canterbury, New Zealand,
Email: teixeie2@lincoln.ac.nz

Keywords: alfalfa, modelling, morphology, root reserves

Introduction Leaf area index (LAI) quantifies canopy expansion in crops and is used in lucerne (*Medicago sativa* L.) simulation models to predict daily PAR interception (PAR_i). This then drives yield through radiation use efficiency (RUE) (Gosse *et al.*, 1984). In perennial crops, like lucerne, the level of biomass stored in crown and taproot may affect canopy expansion in subsequent regrowth cycles (Avice *et al.*, 1997). In temperate regions the impact of this is likely to be greatest in early-spring, when low temperatures delay development. The objective of the current research was to identify whether contrasting levels of winter biomass in crown and taproots affected LAI expansion in early-spring regrowth crops.

Materials and methods A two year old fully irrigated 'Kaituna' lucerne crop was subjected to 28 or 42-day grazing rotations from 12 June 02 to 10 June 03 to induce different levels of crown and taproot biomass. Treatments were arranged in a randomized complete block design with four replicates at Lincoln University, NZ (43°38'S and 172°28'E). Taproot biomass (300 mm depth) was 3.0 t/ha in the 42-day crop but ~33% lower for the 28-day crop by 02 June 03 (Teixeira *et al.*, 2005). Shoot yield, radiation interception and main LAI components (stem population, leaves per stem and leaf size) were measured in the following early-spring regrowth cycle (Jun-Oct 03). Bell-shaped functions (Dwyer and Stewart, 1986) were used to describe changes in leaf area with node position.

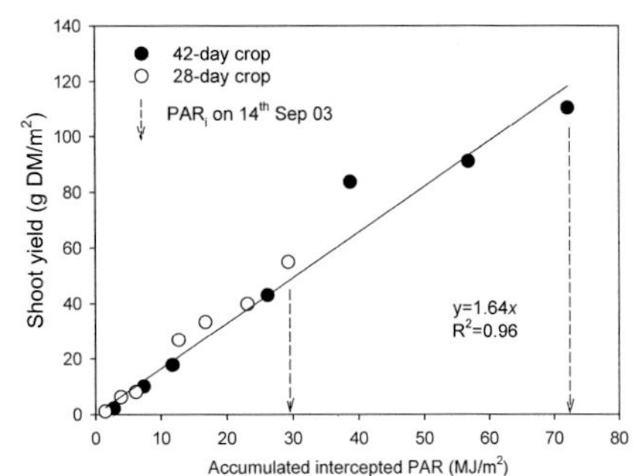

Figure 1 Relationship between intercepted PAR and shoot yield of lucerne crops

Results By 14 September 03, accumulated shoot yield in the 42-day crop was twice ($P<0.05$) that of the 28-day crop (Figure 1). Differences in shoot yield were explained ($R^2=0.96$) by a single linear relationship against PAR_i with an RUE of 1.64 g/MJ (Figure 1). PAR_i was limited in the 28-day crop due to a slower LAI development caused by smaller leaf area/node on primary and axillary nodes. This was expressed as a reduction in the parameter Y_0, which represents the largest leaf area per node position (Figure 2). The position of the largest leaf (X_0) was similar ($P<0.16$) in all treatments, being node 7-8 and 4-5 for primary and axillary leaves, respectively.

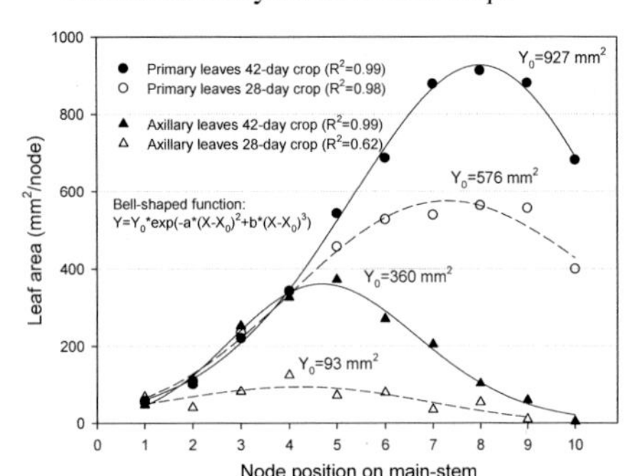

Figure 2 Total leaf area at each main-stem node position of lucerne crops on 1 Oct 03

Conclusions Early-spring shoot yield was reduced in lucerne crops with limited amount of winter crown and taproot biomass. The causal mechanism was a decrease in LAI development through reduced expansion of individual leaves. This limited light interception. Other LAI components and RUE had a minor influence on yield differences. These results indicate that mechanistic approaches are required to quantify the effect of crown and taproot reserves on lucerne shoot yields.

References

Avice, J., G. Lemaire, A. Ourry & J. Boucaud (1997). Effects of the previous shoot removal frequency on subsequent shoot regrowth in two *Medicago sativa* L. cultivars. *Plant and Soil* 188, 189-198.

Dwyer, L. M. & D. W. Stewart (1986). Leaf area development in field-grown maize. *Agronomy Journal* 78, 334-343.

Gosse, G., M. Chartier & G. Lemaire (1984). Predictive model for a lucerne crop. *Comptes Rendus de l'Academie des Sciences, III Sciences de la Vie* 298, 541-544.

Teixeira, E. I., D. J. Moot, H. E. Brown & M. Mickelbart (2005). Seasonal variation of taproot biomass and N content of lucerne crops under contrasting grazing frequencies. In 20th International. Grassland Congress, Dublin, Ireland.

Autumn root reserves of lucerne affected shoot yields during the following spring

D.J. Moot, E.I. Teixeira
Agriculture and Life Sciences Division, Lincoln University, Canterbury, New Zealand,
Email: moot@lincoln.ac.nz

Keywords: alfalfa, modelling, stem, plant density

Introduction Frequent grazing affects shoot yield of lucerne (*Medicago sativa* L.) by limiting radiation interception (Teixeira *et al.*, 2005b) and the accumulation of endogenous reserves (C and N) in perennial storage organs like crowns and taproots (Teixeira *et al.*, 2005a). In temperate regions, the impact of low level of perennial reserves is particularly evident during early-spring, when lucerne regrowth resumes after an overwintering period. The analysis of lucerne yield can be fragmented into its yield components of plant population, shoots per plant and yield per shoot (Volenec *et al.*, 1987). The objective of this research was to quantify the impact of limiting levels of perennial reserves, caused by frequent defoliations, on lucerne early-spring yield and determine the sensitivity of yield components to treatments.

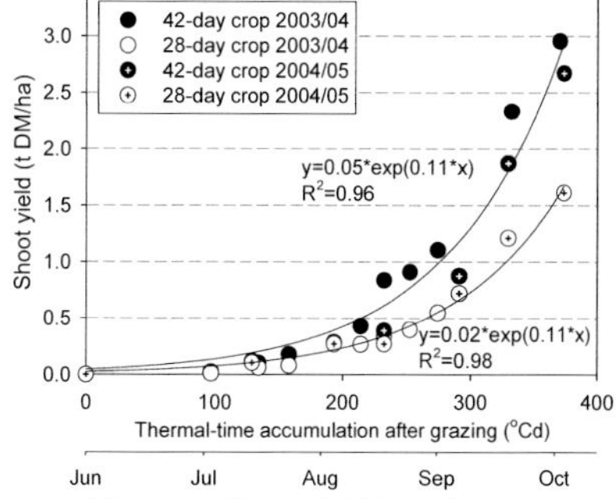

Figure 1 Shoot yield in early spring

Materials and methods A two year old fully irrigated 'Kaituna' lucerne crop was subjected to 28 or 42-day grazing rotations from 12 June 02 to 14 June 04 to induce different levels of crown and taproot biomass. The experiment was conducted at Lincoln University, NZ (43°38'S and 172°28'E) in a randomized complete block design with four replicates. In the autumn of 2003, the total biomass and N content of taproots (300 mm depth) was on average 33% and 20% lower for the 28-day crop than the 42-day crop, respectively (Teixeira *et al.*, 2005a; Teixeira *et al.*, 2005b). During the two early-spring regrowths of the experimental period, shoot yield and yield components were measured weekly.

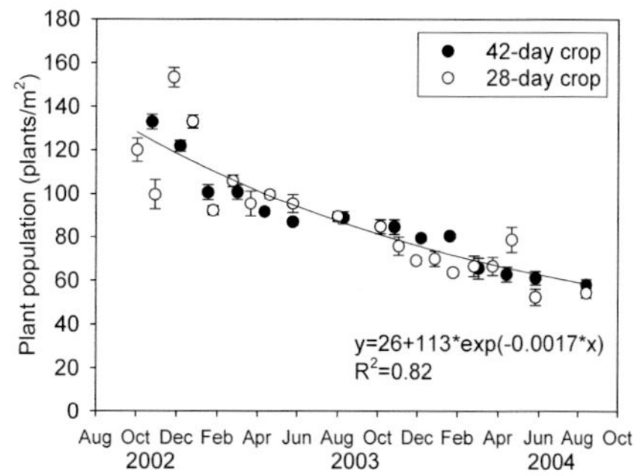

Figure 2 Plant population

Results Shoot-yield in early-spring regrowth was reduced ($P<0.05$) by 45% in the 28-day crop (Figure 1). There was a similar decrease in plant population in both treatments throughout the experimental period from ~130 to 60 plants/m^2 (Figure 2). Maximum shoot numbers were constant (~700 shoots/m^2) due to an increased shoots per plant (Figure 3). Differences in shoot yield were mostly explained (R^2=0.98) by the yield per shoot component.

Conclusions The weight of each individual shoot was the yield component mainly affected by differences in the biomass and N content of lucerne crowns and taproots. Plant population was unaffected by treatments and stem population was maintained through an increase in shoots per plant. The implication is that frequent defoliations reduced the accumulation of C and N in lucerne crown and taproot (Teixeira *et al.*, 2005a). This may then limit canopy expansion and light interception during early-spring regrowth (Teixeira *et al.*, 2005b) reducing lucerne yield through a slower growth of each individual shoot.

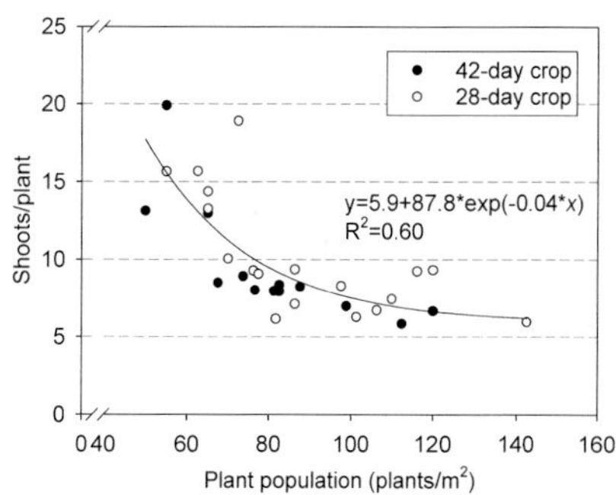

Figure 3 Shoots per plant in relation to plant population

References

Teixeira, E. I., Moot, D. J., Brown, H. E., and Mickelbart, M. (2005a). Seasonal variation of taproot biomass and N content of lucerne crops under contrasting grazing frequencies. Proceedings of XX International Grassland Congress, Dublin, Ireland.

Teixeira, E. I., Moot, D. J., and Fletcher, A. L. (2005b). Lucerne crown and taproot biomass affect early-spring canopy expansion. Proceedings of XX International Grassland Congress, Cork Satellite Workshop, Ireland.

Volenec, J. J., Cherney, J. H., and Johnson, K. D. (1987). Yield components, plant morphology, and forage quality of alfalfa as influenced by plant population. Crop Science 27, 321-326.

Milk production performance based on grazed grassland in Switzerland

P. Thomet and H. Menzi
Swiss College of Agriculture, Laenggasse 85, CH-3052 Zollikofen, Switzerland
Email: peter.Thomet@shl.bfh.ch

Keywords: milk production system, spring-calving, grazing, feed conversion efficiency

Introduction A common feature of profitable dairy systems is the use of large amounts of low cost feed. One approach to improve the competitive ability of the dairy production therefore is to promote grazing (Dillon *et al.*, 1995). A comparison of the actual feeding costs on typical Swiss dairy farms showed that hay and grass silage were four times and concentrates seven times more expensive than grazed grass. A maximum utilisation of grazed grass can be achieved with a seasonal production system, which synchronises the cow's feed requirements with pasture growth. This strategy was implemented and consistently optimised on an experimental farm. The aim was to focus more on the achieving of a high yield per hectare and high feed conversion efficiency rather than high yields per cow.

Materials and methods On a typical site in the Swiss plains the productivity of a dairy production system with maximum proportion of grazing was studied over three years (April 2001 to March 2004). The experimental herd consisted of 14 Red Holstein and 2 Jersey cows with an average live weight of 592 kg/cow. After an average calving date in mid February continuous grazing started at the end of March and lasted until mid November. The experimental land for grazing and winter feed consisted of 6.0 ha grassland, of which 63 % was sown grass legume leys (sown in 2000) and 37 % was permanent grassland (with 33 % *Agrostis stolonifera*). Grass production was measured over the 3-year period, using the method described by Corrall and Fenlon (1978).

Results On average over the three years 14,291 kg ECM/ha were produced which clearly surpassed typical values from conventional Swiss dairy farms. Total grass production was 13.5, 12.0 and 10.9 t DM/ha per year for 2001, 2002 and 2003, respectively (Figure 1). The overall stocking rate on pastures was 2.5 cows/ha during the first two years and 2.0 cow/ha in the extremely dry year 2003. On a dry matter basis, the yearly average ration consisted of 65.7 % grazing, 27.6 % grass silage plus hay and only 6.7 % or 405 kg DM/cow per year concentrates (Table 1).

Table 1 Productivity of the seasonal production system based on grazed grass

Year	2001	2002	2003
overall stocking rate (cows/ha)	2.44	2.48	1.99
grass silage & hay (t DM/ha/yr)	4.4	4.0	3.1
concentrates & potatoes (t DM/cow/yr)	0.42	0.38	0.41
Estimated intake on pasture (t DM/ha/yr)	8.8	9.3	9.2
Performance			
- kg ECM/cow/yr	6'746	6'817	7'826
- kg ECM/ha/yr (incl. conc. & potatoes)	16'461	16'907	15'574
- kg ECM/ha/yr from grassland (excl. conc. & pot.)	14'859	15'451	14'301

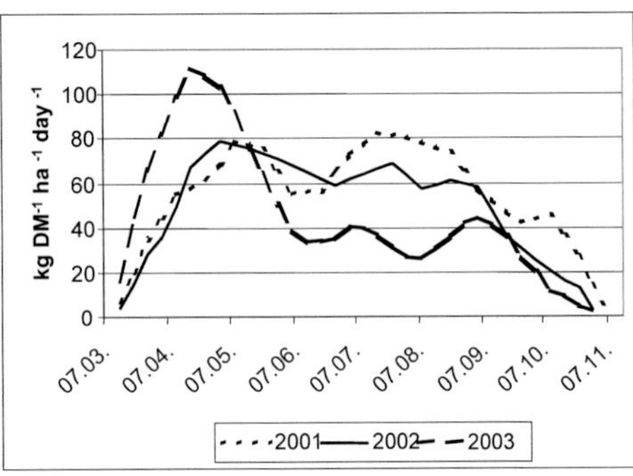

Figure 1 Daily herbage growth rates on the pastures (487 m altitude; 1451, 1289 and 795 mm annual rainfall for 2001, 2002 and 2003, respectively)

Conclusions The full grazing system with seasonal calving end of winter proved to be highly productive under Swiss valley conditions. More than 14'000 kg ECM/ha/yr can be produced on pastures with an annual yield of 12.5 t DM/ha (70 % grazed and 30 % conserved forage).

References

Corrall A.J. & J. S. Fenlon (1978). A comparative method for describing the seasonal distribution of production from grasses. *Journal of Agricultural Science Cambridge,* 91, 61-67.

Dillon P., S. Crosse, G. Stakelum & F. Flynn (1995). The effect of calving date and stocking rate on the performance of spring-calving dairy cows. *Grass and Forage Science,* 50, 286-299.

Extending the grazing season with turnips

P. Thomet and S. Kohler
Swiss College of Agriculture, Laenggasse 85, CH-3052 Zollikofen, Switzerland
Email: peter.thomet@shl.bfh.ch

Keywords: *Brassica*, turnip variety, yield, grazing season, milk production

Introduction A comparison of the actual feeding costs on 86 typical Swiss dairy farms confirmed that the production of hay and grass silage is very expensive with costs from the field to intake in the range of 20-25 Euro cents/kg of dry matter (DM). Options to extend the grazing season for dairy cows in the late autumn and early winter, and hence reduce winter feeding costs, would therefore be welcome (Penrose et al., 1996). The aim of the study was to compare the DM production potential of summer-seeded turnips with other brassicas and Italian ryegrass and to test whether dairy cows can utilize the bulbs of turnips efficiently.

Materials and methods Over three years (1999-2001) the yield potential of ten turnip varieties *(Brassica rapa L. var. rapa (L.) Thell.)* and other autumn crops was studied on three sites in the region of Bern (sowing dates: Aug 4-15; harvest dates: Nov 1-10). A randomized-block design with 3 blocks was used. Plot size was 7.0m x 1.5m. For the statistical analysis of the results the method of linear contrasts according to Clarke and Kempson (1997) was used. Each year farm grazing experiments were conducted on three to six farms (total 14 on farm experiments) to assess the suitability of turnips for grazing. On every field, pre and post grazing DM yield of leaves and bulbs was measured in 6 randomized plots of 1.5m x 1.5m. In the framework of these grazing experiments, blood and milk samples were also analysed to assess potential risks for animal health and milk quality.

Results With an average of 6.5 t DM/ha, the turnip varieties had higher yields than the Italian ryegrass mixtures (2.9 t DM/ha) and were also superior to other Brassica spp. like a chinese cabbage (variety BUKO; Figure 1). In the on-farm grazing experiments grazing losses were 33 %, on average, with a net yield of 4.3 t DM/ha (Table 1). In spite of treading damage no indications of long-term impacts on the soil were found. There was no negative influence on animal health, milk quality and taste. Nevertheless, soiling of the cows increased the risk of contamination of the milk with anaerobic spores, which could be a problem for cheese making with raw milk.

Table 1 DM-yields and losses of grazed turnips in 14 farm experiments (1999-2001)

years	1999	2000	2001
No. of farm experiments	*6*	*5*	*3*
Total yield (t DM/ha)	6.17	6.88	6.27
Intake (t DM/ha) [1]	4.49 [1]	4.07 [1]	4.35 [1]
DM losses (%)	27.5	45.5	30.7
- maximum	49.8	57.2	35.4
- minimum	11.9	24.2	26.0

[1] To compare: the DM-intake on Italian ryegrass plots was 2.29 t DM (mean of 3 trials).

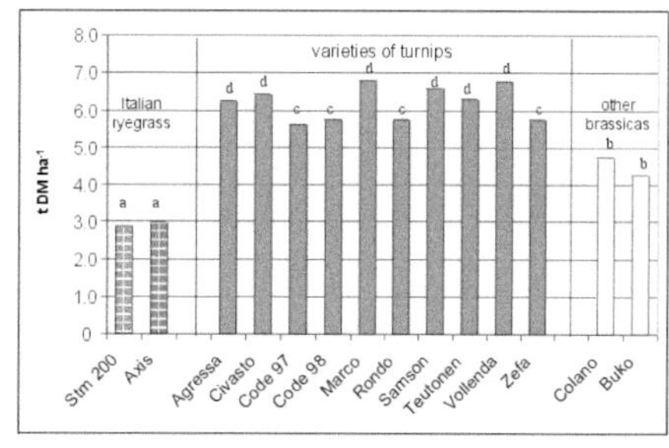

The same letters do not differ significantly at $P < 0.05$.

Figure 1 Yields of turnip varieties in comparison with Italian ryegrass mixtures (SM 200 and AXIS) and other brassica species (varieties COLANO and BUKO without bulbs (means of 9 trials, 1999-2001)

Conclusions Based on the results of the study and a survey on 32 farmers having relevant experience, it is concluded that turnips are a suitable crop to extend the grazing period in autumn and early winter. Summer-seeded turnips produced, on average, twice the yield of Italian ryegrass. With strip grazing, dairy cows can make good use of both tops and bulbs.

References

Clarke G.M. & R. E. Kempson (1997). Introduction to the design and analysis of experiments. Arnold, London.

Penrose. C.D., H. M. Bartholomew, R. M. Schumacher and R. Duff (1996). Performance of brassica cultivars from New Zealand and United States seed sources in Southeast Ohio, Ohio State University, USA. *Proceedings New Zealand Grassland Association,* 57, 111-113.

The effect of inclusion of a range of supplementary feeds on herbage intake, total dry matter intake and substitution rate in grazing dairy cows

S.J. Morrison[1], D.C. Patterson[1], S. Dawson[2] and C.P. Ferris[1]
[1]*Agricultural Research Institute of Northern Ireland BT26 6DR, Email: s.sjm.morrison@talk21.com,* [2]*Biometrics Division, Department of Agriculture and Rural Development, Newforge Lane, Belfast BT9 5PX, United Kingdom*

Keywords: forage maize, grazing, whole crop wheat, dairy cows

Introduction The milk production potential of dairy cows has increased substantially over the past two decades. This development presents new challenges for managing dairy cows during grazing, particularly where the objective is to maximise the proportion of energy in the diet derived from forage. The objective of the current study was to explore supplementation strategies to maintain high total forage intakes from grazed grass supplemented with alternative forage supplements in dairy cows during the grazing season. A second objective of the study was to examine the effect of supplement on substitution rate (SR) and milk yield response.

Materials and methods Twenty-four spring calving dairy cows were used with five periods, each of four weeks duration, and six treatments in a partially balanced, change-over design experiment commencing on 9 May. The treatments were based on a range of supplements including: maize silage (MS); whole crop wheat silage (WS); grass silage (GS); rapidly available energy concentrate (RC); slowly available energy concentrate (SC); and a control which was unsupplemented (U). The components of the SC concentrate were sugar beet pulp, citrus pulp, maize gluten feed, soyabean meal, soya hulls, rape meal, vitamin/minerals, molasses, Megalac and urea included at 150, 190, 125, 175, 242, 25, 40, 30, 20 and 3 g/kg fresh respectively. The RC concentrate included barley, wheat, soyabean meal, sucrose, wheat feed, rape meal, vitamins/minerals and molasses at 200, 255, 200, 125, 100, 50, 40 and 30 g/kg fresh respectively. Forage supplements were offered indoors *ad libitum,* for 2h after the morning milking. Cows on concentrate treatments (RC + SC) received the allocated concentrate in the milking parlour twice daily (4.5 kg fresh/cow/d). The concentrate supplemented and unsupplemented cows returned to grazing immediately, whereas cows receiving forage supplements were retained after morning milking to access the allocated forage supplement using Calan gates. All cows were communally grazed in a rotational system with a target residual sward height of 5-6 cm, with cows offered a new paddock daily after each pm milking. Herbage intake was recorded using the n-alkane technique.

Results Mean herbage intake, supplement intake and total dry matter intake (DMI) data are presented in Table 1. Of all the supplements used in the present study, the MS treatment resulted in a significantly higher ($P<0.001$) supplement DMI than any other treatment, resulting in the highest total DMI of all the forage supplement treatments. The MS supplemented cows however, had a significantly lower ($P<0.001$) herbage intake than any other treatment. The cows offered the concentrate treatments had the greatest herbage intake of all the supplemented cows and achieved the highest total DMI. Unsupplemented cows on the control treatment had a significantly higher herbage intake than all other treatments except SC. Substitution rate, defined as the reduction in herbage DMI (kg) per kg of supplement DMI, was lower with the concentrate supplemented cows compared to all forage supplemented cows. There was no difference in SR between the GS, MS and WS supplemented cows. The RC supplement incurred a numerically higher SR than the SC.

Table 1 Effect of supplement treatments on herbage and supplement intake of grazing dairy cows

		Forage			Concentrate			
	Control	Grass silage	Maize silage	Wheat Silage	Rapid energy	Slow energy	S.e.d.	Sig.
Supplement DMI (kg DM/d)		3.0	6.3	3.6	3.9	3.9	0.37	***
Herbage intake (kg DM/d)	12.9	11.2	8.9	11.0	11.8	12.2	0.50	***
Total DMI (kg DM/d)	12.9	14.2	15.3	14.7	15.7	16.1	0.50	***
Substitution rate		0.56	0.63	0.53	0.28	0.18		

Conclusions The results of the present study suggest that all the forage supplements incurred a similar SR of between 0.5-0.6 kg reduction in herbage DMI for every kg of supplement DMI. The overall level of herbage intake was low, possibly indicating limited herbage availability and therefore influencing the substitution rate. No differences in herbage DMI, total DMI or SR were observed between concentrate types. Of the forage supplements offered, MS is the more favourable to obtain a high total DMI, due to its higher intake potential.

White clover soil fatigue: an establishment problem on large and intensive dairy farms

K. Søegaard[1] and K. Møller[2]
[1]*Danish Institute of Agricultural Sciences, DK-8830 Tjele, Denmark, Email: Karen.Soegaard@agrsci.dk*
[2]*Danish Institute of Agricultural Sciences, DK-4200 Slagelse, Denmark*

Keywords: white clover, establishment, clover cyst nematode

Introduction In recent years a new constraint, clover soil fatigue, has appeared for the establishment of white clover (*Trifolium repens*) in Denmark. Increasing dairy farm size has led to more intensive use of clover in crop rotation schemes in the grazing areas located at convenient distances from stables. It has become common practice to establish new clover/grass in the fields just after ploughing clover/grass swards, and soil fatigue is becoming more common. On fatigued land the clover plants emerge, then become stunted and eventually disappear within the same year. The problem tends to cover the full field area. Obviously, this is a major constraint since the importance of N-inputs derived from fixation is growing in Danish dairy farming.

Materials and methods Soil samples were collected in spring 2004 from grass/clover pastures on 15 dairy farms where pronounced establishing problems were seen in the past few years. A sample from a field, in which clover had not been grown for at least 20 years served as reference. To study the soil fatigue phenomenon, white clover was undersown to spring barley in these soils in pots (27 x 35 cm containers, 31 l) in three replications, and clover and cover crop growth and dry matter yields were studied. In search of possible microbial causes of the problem, samples of clover plants and of top-soil around plants were collected 6-7 weeks after sowing, from farms which reported clover soil fatigue and from selected pots in the pot experiment. For reference, samples collected from soils with no history of soil fatigue were also studied. Plant roots were indexed by size and branching, analyzed for presence of clover cyst nematodes (*Heterodera trifolii*) and stem nematode (*Ditylenchus dipsaci*) by microscopy, and fungi isolated from the surface disinfected roots and identified.

Results In the pot experiment, clover yields from fatigued soils (FS) were considerably lower than from the reference soil (RS) (Figure 1). Pronounced fatigue was found in soils 3, 9 and 15, which yielded only 6% of the RS yield. Five to eight weeks after sowing many of the clover seedlings died, and the production of surviving plants was very low for the rest of the season. In 6 to 7 week old plants, there were no stem nematodes but roots from FS had significantly more clover cyst nematodes per root and root indices were significantly less than roots from RS, clearly indicating that the cyst nematode plays an important part in the soil fatigue. However, the most unambiguous indication for this was the finding that while cysts had formed in a very high proportion of roots from FS, in roots from RS only few nematodes were found (J2 larval stage). Since cyst formation requires about five weeks to complete, this indicates that in FS, an unknown factor may have triggered hatching of cysts and subsequent attacks at the very beginning of plant germination. Table 1 highlights these findings by comparing of roots from soils where management practices were exactly the same, and only the previous crop types differed. Numbers of cysts were comparable in FS and RS soils, hence, high cyst densities do not explain the fatigue. Of 43 fungal taxa obtained from roots, *Fusarium*- and *Cylindrocarpon*-species were the most common in plants from both FS and RS. A complex of some of these species may have aggravated the problem on FS, entering plant roots through wounds made by nematodes, but, so far, its possible role is not clear.

Table 1 Clover cyst nematodes in plants and soil with or without soil fatigue from pot experiment and from two fields in a grass-arable system. Results from 2004

	Pot experiment		Grass-arab. rot.	
Clover soil fatigue	+	-	+	-
Previous crop (2003)	G/C[2]	Barley	G/C	Maize
Days from sowing to sampl.	46	46	50	50
Root index [1]	3.2	4.6	2.9	4.5
Nematode/root (number)	6.6	2.7	7.8	1.2
Roots with cysts (%)	73	0	70	0
Roots with ripen cysts (%)	23	0	60	0
Cysts/kg soil (number)	95	70	380	100
Filled cysts/kg soil (numb.)	0	15	10	5

[1]Root index 1-5 (1 is a thin root <2 cm without branching and 5 is >3 cm with more adventitious roots) [2]grass/clover

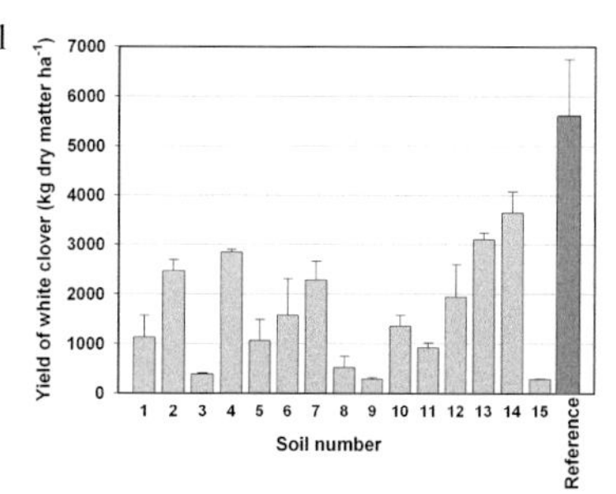

Figure 1 Clover yield in pot experiment after harvest of spring barley. Standard deviation is shown

Conclusions The main reason for white clover soil fatigue was attack of clover cyst nematode.

Effect of farm grass cover at turnout on the grazing management of spring calving dairy cows

M. O'Donovan[1], L. Delaby[2] and P. Defrance[3]
[1]*Teagasc, Moorepark Production Research Centre, Fermoy, Co. Cork, Ireland, Email: maodonovan@moorepark.teagasc.ie,* [2]*INRA, Unité Mixte de Recherche sur la Production du Lait, 35590 Saint-Gilles, France,* [3]*Chambre Régionale d'Agriculture de Bretagne, CS 74223, 35042 Rennes Cedex, France*

Keywords: dairy cows, decision support system, early grazing

Introduction Early spring grazing is an objective for most Irish dairy farmers. If more grass is included in the diet of the cow in early lactation, the profitability of the farm system can be increased. Post turnout, dairy cow feeding management varies with the amount of farm grass cover (FC) available. Experiments on the consequences of different FC at turnout require large resources and all scenarios cannot be accounted for. Consequently, a decision support system, Pâtur'IN (Delaby *et al.*, this volume), was used to describe the effects of various FC at turnout on grazing management in spring.

Materials and methods The model herd was composed of 50 dairy cows (mean calving date 15 February: peak milk yield 35 kg). Cows were allowed access to grass according to the pattern of calving. Concentrate supplementation was pre-planned to decrease from 7 kg/cow per day in February to 0 kg on 10 April. Twenty one-ha paddocks of perennial ryegrass were available for grazing (2.5 cows/ha: 60 kg N/ha per rotation). The grass growth data used with the model corresponded to grass growth measured at Moorepark between 1982-2004. Mean sward density was 250 kg DM/cm per ha. Three different farm grass covers (A - 550, B - 800 and C - 1100 kg DM/ha) were tested at turnout on 15 February. The objectives of all treatments were to finish the first rotation by 10 April, to graze the entire farm in the first rotation and to close 9 ha for silage in the 2nd rotation (4.5 cows/ha). During the first rotation, grass silage was used to fill the grass deficit (Treatment A), while concentrate feeding was reduced when grass supply was in excess.

Results Pre grazing herbage mass and sward height were higher throughout the first rotation for treatment C. With turnout on 15 February, treatment A had a low FC; therefore the herd was supplemented with grass silage (65 kg DM/cow). There was no silage supplemented to treatments B and C. In spite of the reduced level of concentrate fed (115 kg DM/cow), the FC at turnout for treatment C was too high to finish the first rotation on 10 April. Only 16 to 17 ha were required to feed the herd during the 56-day simulation. At the end of the 2nd rotation (2 May), treatment C tended to have the highest FC. A farm cover of 800 kg DM/ha at turnout is in agreement with the level recommended by O'Donovan *et al.* (2000). This scenario makes efficient use of spring grass, avoids silage feeding post calving and does not impose very early pasture closing in autumn.

Table 1 Main characteristics of early grazing with different farm grass covers (15 February to 2 May)

Farm cover at turnout (15 February - kg DM/ha)		A – 550	B – 800	C - 1100
From 15 February to 10 April (56 days)			On average, 35 cows at grazing	
Pre grazing herbage mass (kg DM/ha)		1400	1600	2000
Pre / Post grazing height (cm)		9.6 / 5.0	10.4 / 5.2	12.0 / 5.1
Herbage intake (kg DM/cow)		660	750	820
Concentrate intake (kg DM/cow)		240	240	115
Silage intake (kg DM/cow)		65	0	0
Farm cover (kg DM/ha)	1 March on 20 ha	670	830	1135
	10 April on 11 ha	1085	1060	1195
	2 May on 11 ha	925	975	1065

Conclusions Turnout date in spring should be based on the amount of FC available. On most farms early grazing is a realistic objective, with this simulation showing that medium FC in spring will allow grass silage to be removed completely from the diet of the lactating dairy cow. This can further improve milk production efficiency. However, having too high a FC in spring without a high grass demand, will increase the silage harvesting area and may create grazing management problems in subsequent rotations.

References

Delaby L., J. L. Peyraud & P. Faverdin (2005). Pâtur'IN: Software to assist the dairy cows grazing management. *This volume*.
O'Donovan M. (2000). Grass Measurement. Benefits and Guidelines. 24 pages; Moorepark Publication.

What supplementation type for spring calving dairy cows at grass in autumn?

M. O'Donovan[1], E. Kennedy[1], T. Guinee[2] and J.J. Murphy[1]
[1]*Teagasc, Moorepark Production Research Centre, Fermoy, Co. Cork, Ireland,*
Email: maodonovan@moorepark.teagasc.ie, [2]*Teagasc, Dairy Products Centre, Fermoy, Co. Cork, Ireland*

Keywords: grass, supplementation, milk processability

Introduction In spring-calving herds the requirement for conserved forages for indoor feeding is very limited but these feeds may have a role to play as buffer feeds in the spring and autumn periods when grass supply is less than required or as alternatives to concentrates on pasture for cows in early lactation. The objective of this experiment was to compare alternative forages and concentrates as buffer feeds on pasture with spring-calving cows in the autumn.

Materials and methods Ninety cows were blocked on calving date and milk yield (19.9 ± 1.5 kg/head per day), into groups of six and assigned randomly to the following treatments; (i) 17.5 kg of grass DM allowance (LG), (ii) 24 kg of grass DM allowance (HG), (iii) LG + 4 kg concentrate DM (C), (iv) LG +4 kg maize silage DM (M), (v) LG + 4 kg urea-treated processed whole crop wheat DM (UPWCW) and (vi) LG + 4kg fermented whole crop wheat DM (FWCW). The treatments were in place from 13 September to 7 November 2004 over 2 grazing rotations. The animals grazed as three separate herds. Both LG and HG were grazed separately while the four supplemented treatments were grazed together as a herd of 60 cows. The supplementary forages were group fed from a diet feeder after morning milking. The concentrates were offered individually in the milking parlour in two equal feeds daily. The cows were offered fresh grass after morning milking. Herbage mass (above 50mm) was determined in each grazing paddock by cutting either four or six strips (0.5 × 10m) with an Agria motor scythe. Ten grass height measurements were recorded in each cut strip (pre and post harvesting) to determine the sampled height and calculate the bulk density (kg DM per mm/ha). The sward height before grazing was measured. This sward height multiplied by the mean bulk density from the Agria cuts was used to calculate the herbage mass in the paddocks. Thirty pre and post grazing sward heights were measured daily for each treatment with a rising plate meter. Milk yield was recorded daily. The concentrations of fat, protein and lactose were determined in one successive morning and evening milk sample per week. Composite morning and evening bulk milk samples were taken from each treatment on one day weekly for processability measurements using a Rheometer. Production data were analysed by SAS using covariate analysis.

Results and discussion Mean herbage mass was 2670kg DM/ha (s.d. 372). Mean pre and post grazing sward heights were 16.2, 15.9, 15.7cm (s.d. 1.33) and 6.7, 5.7, 5.9cm (s.d. 0.76) for HG, LG and supplemented herds respectively. Grass disappearance was 18.7, 15.0 and 14.4 kg/head per day (s.d. 2.65 kg) for HG, LG and supplemented herds. Treatment C had a significantly greater milk yield than HG, M and UPWCW, which in turn had a significantly greater yield than LG and FWCW. Solids corrected milk (SCM) yield was significantly greater for C than HG, which was greater than M, UPWCW and FWCW. Treatment LG had the lowest SCM yield. Milk fat, protein and lactose concentrations, body condition score (BCS) or liveweight were not significantly different across treatments. The rennetability of milk tended to be highest in treatments M and FWCW and poorest in C which was largely a reflection of milk protein concentrations in these treatments.

Conclusion There is a large solids-corrected milk production benefit to supplementing grazing cows, on a restricted grass allowance in late lactation, with concentrates. Supplementing with other forages gave smaller responses, with extra herbage allocation being the best. Milk rennetability would appear to be influenced by the type of supplement offered.

Table 1 Effect of fermented (FWCW) and urea-treated processed whole crop wheat (UPWCW), maize silage (M), concentrate (C) and additional grass (LG) on production of grazing dairy cows in autumn

	LG	HG	C	M	UPWCW	FWCW	RSE	Sig
Milk yield (kg/day)	13.2[a]	15.5[b]	18.3[c]	15.0[b]	14.9[b]	14.2[a]	1.56	***
Milk fat (g/kg)	4.31	4.25	4.04	4.21	4.31	4.29	0.37	Ns
Milk protein (g/kg)	3.67	3.72	3.57	3.71	3.63	3.71[a]	0.16	+
Milk lactose (g/kg)	4.26	4.35	4.35	4.26	4.29	4.30	0.13	Ns
SCM yield	12.6[a]	14.9[c]	17.2[d]	14.5[b]	14.3[b]	13.8[b]	1.35	***
Bodyweight (kg)	562	576	568	570	573	575	20.6	Ns
BCS	2.83	2.77	2.81	2.80	2.90	2.87	0.16	Ns

[abc]Means within row with different superscripts differ ($p<0.05$)

Manipulation of grass growth through strategic distribution of nitrogen fertilisation

M. Stettler and P. Thomet
Swiss College of Agriculture, Laenggasse 85, CH-3052 Zollikofen, Switzerland,
Email: matthias.stettler@shl.bfh.ch

Keywords: nitrogen fertilisation, grass growth, grazing, clover

Introduction The objective of this study was to evaluate possibilities and limits of manipulating the grass growth of pastures by different nitrogen (N) application strategies with the aim to better synchronise grass supply and feed demand. In Switzerland, the use of N is strongly restricted by legislation. An efficient and well allocated N fertilisation is therefore important.

Materials and methods Experiments were performed on two sites in the Swiss lowlands, on a sown pasture (sown 2000, with 50% *Lolium perenne*, 20% *Poa pratensis*, 25% *Trifolium repens*) and a permanent pasture (40% *Lolium perenne*, 25% *Poa trivialis*, 20% *Poa pratensis*, 10% *Trifolium repens*). Grass growth was measured with an adapted method of Corrall & Fenlon (1978) for three replicates each of four different N strategies of distributing 150 kg N ha^{-1} as ammonium nitrate fertiliser and a zero N strategy (Figure 1). The trial lasted over two grazing seasons (March 2003 to November 2004). On both sites, $30m^3$ cattle slurry was applied each winter. Analysis of variance was carried out using the general linear models approach.

Results The zero N strategy reached high average annual yields of 12.4 t DM on the sown pasture and 8.9 t DM on the permanent pasture. The annual DM yields of the fertilised plots were higher ($p<0.05$) than of the zero N plots (+ 8.7% on the new pasture, + 14.9% on the permanent pasture). The annual DM yield of the four different N distribution strategies did not vary significantly ($p<0.05$), but differences could be observed in the seasonal growth distribution pattern (Figure 1). The clover content of the zero N strategy was higher ($p<0.05$) (+ 10% on average) than that of fertilised treatments.

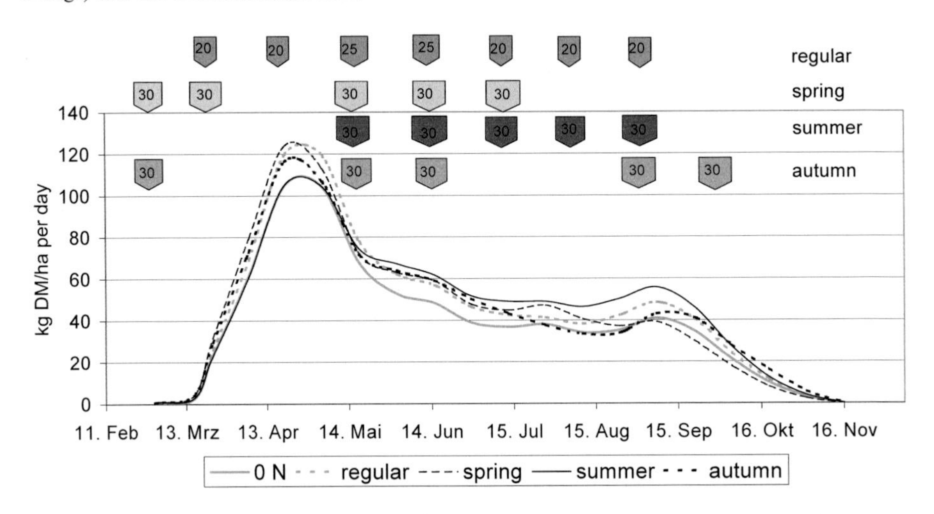

Figure 1 N fertiliser application patterns (kg N/ha) and grass growth curves (mean grass growth over two years) for four N distribution strategies and a zero N strategy.

Conclusions This study demonstrated that the grass growth curve can be manipulated by seasonal N distribution strategies without significantly affecting annual DM yield, especially on pastures with low clover content,. N fertilisation can be an important management tool for farmers with seasonal dairy grazing to better synchronise grass supply and feed demand on pastures.

References

Hennessy D., P. French, M. O'Donovan & A. S. Laidlaw (2004). Manipulation of grass growth by altering nitrogen application. *Grassland Science in Europe*, 9, 705-707.

Corrall A.J. & J. S. Fenlon (1978). A comparative method for describing the seasonal distribution of production from grass. *Journal Agricultural Science Cambridge,* 91, 61-67.

The effect of closing date and type of utilisation in autumn on grass yield in spring

E. Mosimann[1], M. Lobsiger[2], C. Hofer[2], B. Jeangros[1] and A. Lüscher[2]
[1]*Agroscope RAC Changins, CH-1260 Nyon, Switzerland, Email: eric.mosimann@rac.admin.ch, 2Agroscope FAL Reckenholz, CH-8046 Zurich, Switzerland.*

Keywords: grazing, cutting, closing date, herbage yield

Introduction Due to the low cost of grazed grass, most dairy farmers extend the grazing period in autumn. However, delaying the closing date may reduce the grass yield in the following spring (Roche *et al.*, 1996; O'Donovan *et al.*, 2002). The objective of this experiment, conducted in the Swiss lowlands, was to quantify the effects of closing date and type of utilisation in autumn on grass yield in the following spring.

Materials and methods The trial was carried out from autumn 2001 to spring 2004 on two dairy pastures with continuous stocking: Waldhof (wet) and Saint-Livres (dry). It comprised of 6 treatments: two kinds of utilisation (cut or grazed), each with three closing dates (early October, late October or late November) replicated 5 times. Grass yield in the following spring was measured at three different dates (March, April and May).

Results The mean yields in spring averaged over both sites are presented in Table 1. In most cases, the type of utilisation in autumn did not influence DM-yield in spring. There was, however, a highly significant 'closing date' effect, with a reduction in spring yield when the closing date was delayed. When comparing the treatments 'late November' and 'early October' there was a yield reduction of 69% (44 to 87%) in March, 53% (23 to 75%) in April and 19% (9 to 34%) in May. In absolute terms the yield reduction was largest in May (783 kg DM/ha). Regression analysis (herbage mass in November vs. spring yield) demonstrated the importance of an increased standing biomass in autumn and thus leaf area during winter, for a high spring growth. This result is in accordance with observations on white clover in a European multi-site experiment (Wachendorf *et al.*, 2001; Lüscher *et al.*, 2001).

Table 1 Effect of autumn closing date and type of utilisation on herbage dry matter yields in spring (kg DM/ha)

Closing date (CD)	early October		late October		late November		*se*	Significance			
Utilisation (U)	cut	grazed	cut	grazed	cut	grazed		U	CD	U x CD	Site
March 2002	608	754	232	371	148	100	*59*	ns	< 0.001	ns	ns
March 2003	280	281	153	212	99	59	*37*	ns	< 0.001	ns	< 0.001
March 2004	234	598	144	186	130	237	*49*	< 0.001	< 0.001	0.005	< 0.001
April 2002	1355	1435	782	893	618	529	*75*	ns	< 0.001	ns	ns
April 2003	1087	1087	802	941	835	639	*70*	ns	< 0.001	ns	0.05
April 2004	326	362	213	252	123	89	*46*	ns	< 0.001	ns	---
May 2002	3644	3681	3003	2921	2922	2432	*113*	ns	< 0.001	ns	< 0.001
May 2003	4436	4085	3872	4050	3696	3273	*132*	ns	< 0.001	ns	< 0.001
May 2004	4730	5079	4477	4583	4310	4327	*166*	ns	0.005	ns	< 0.001

Conclusions Extending the grazing season or late cutting in autumn reduces the grass yield in the following spring, up until May, thereby, delaying the start of the spring grazing season. To facilitate grazing management in spring, the closing date should be varied between paddocks, so that paddocks closed early are available for grazing early the following spring, whilst those with a late closing date reach the grazing stage later in the spring.

References

Lüscher, A., B. Stäheli, R. Braun & J. Nösberger (2001). Leaf area, competition with grass, and clover cultivar: Key factors to successful overwintering and fast regrowth of white clover (*Trifolium repens* L.) in spring. *Annals of Botany*, 88, 725-735.

O'Donovan, M., P. Dillon, P. Reid, M. Rath & G. Stakelum (2002). A note on the effects of herbage mass at closing and autumn closing date on spring grass supply on commercial dairy farms. *Irish Journal of Agricultural and Food Research*, 41, 265-269.

Roche, J.R., P. Dillon, S. Crosse & M. Rath (1996). The effect of closing date of pasture in autumn and turnout date in spring on sward characteristics, dry matter yield and milk production of spring-calving dairy cows. *Irish Journal of Agricultural and Food Research*, 35, 127-140.

Wachendorf, M., R.P. Collins, A. Elgersma, M. Fothergill, B.E. Frankow-Lindberg, A. Ghesquiere, A. Guckert, M.P. Guinchard, A. Helgadottir, A. Lüscher, T. Nolan, P. Nykänen-Kurki, J. Nösberger, G. Parente, S. Puzio, I. Rhodes, C. Robin, A. Ryan, B. Stäheli, S. Stoffel, F. Taube & J. Connolly (2001). Overwintering of *Trifolium repens* L. and succeeding spring growth: A model approach to plant-environment interactions. *Annals of Botany*, 88, 683-702.

A comparison of three systems of milk production with different land use strategies

L. Shalloo, P. Dillon and J.J. Murphy
Dairy Production Department, Teagasc, Moorepark Production Research Centre, Fermoy, Co. Cork, Ireland, Email: lshalloo@moorepark.teagasc.ie

Keywords: grass, concentrate, maize silage

Introduction Under the Luxemburg agreement FAPRI-Ireland (Breen & Hennessey 2003) projects that milk price will decrease by 5.0 to 5.5 c/l because of reductions in support for butter and skimmed milk powder. These changes mean that many dairy farmers need to reappraise their systems of milk production and consider necessary adjustments that will ensure viability in the longer term. The objective of this study was to model three different systems of milk production in scenarios where quota, cow numbers or land was restricted.

Materials and methods Three systems of milk production were compared 'High Grass' (HG), 'High Concentrate' (HC) and 'High Maize Silage' (HM) in this analysis. The HG and HC systems were managed similarly (Horan *et al*., 2004) with concentrate inputs of 350kg and 1500kg, respectively. The HM system consisted of the addition of maize silage to the diet in early spring and autumn with a response of 0.35 kg milk/kg of maize silage DM. The modeled farm had 29.5 ha, 50 cow-housing places and 323,327l of milk quota, which was representative of Teagasc monitor farms, while housing and milk quota were included at €1,600/cow and €0.153/l respectively. All milk, male calf and cull cow prices and opportunity cost of land were based on projections from FAPRI-Ireland (Binfield *et al*., 2003). The Moorepark Dairy Systems Model (MDSM-Shalloo *et al.,* 2004) was used to simulate the three milk production systems under four different milk quota scenarios. In scenario 1 milk quota was restricted (QR) and therefore an increase in milk yield per cow resulted in reduced cow numbers. In scenario 2 milk yield was increased per cow but cow numbers were restricted (CR) at 49.4. In scenario 3 land available for expansion was limited (LL) to that of the HG in scenario 1 (19.6 ha). In scenario 4 there was unlimited availability of land for expansion (LU) with the same amount of quota purchased as in scenario 3. Similar amounts of milk quota were purchased in scenario 4 as in scenario 3 with the HC and HM.

Results Table 1 shows the key herd parameters for the three systems of milk production under the four different scenarios. In QR farm profit was €2,617 and €1,279 lower in the HC and HM systems, respectively than in HG. In CR farm profit was €1,079 and €503 higher in the HG than in the HC and HM systems, respectively. In LL farm profit was €770 lower and €137 higher in the HC and HM systems than in the HG system. In LU farm profit was €1,701 and €794 higher in the HG system than in the HC and HM systems when compared to LL.

Table 1 Key herd output parameters for the three systems of milk production under four scenarios

	QR			CR		LL		LU
	HG	HC	HM	HC	HM	HC	HM	HG
Hectares used (Ha)	19.6	15.3	15.4	18.1	16.4	19.6	19.6	24.9
Quota purchase (kg)	-	-	-	53,562	19,257	82,724	81,251	81,308
Cows calving	49.4	41.9	46.4	49.4	49.4	53.5	59.0	62.8
Stocking rate (LU/ha)	2.37	2.57	2.82	2.57	2.82	2.57	2.82	2.37
Labour costs (€)	-	-	-	4,165	1,497	6,432	6,295	6,346
Farm profit (€)	30,582	27,965	29,303	29,503	30,079	29,812	30,719	31,513

Conclusions The results indicate that the most profitable system of milk production is where grazed grass is maximised in the diet. Expansion through increasing output per cow reduces farm profitability. Where land area is limiting there may be an advantage in going to high input systems but it is very much dependent on the supplement to milk price ratio and the ability of the farmer to operate higher input systems effectively.

References

Binfield J., T. Donnellan, K. Hanrahan & P. Westhoff (2003). The Luxembourg CAP Reform Agreement: Analysis of the Impact on EU and Irish Agriculture, Teagasc, Dublin. Page 1-69.

Breen J. & T. Hennessy (2003). The impact of the Luxembourg Agreement on Irish Farms. The Luxembourg CAP Reform Agreement: Analysis of the Impact on EU and Irish Agriculture, Teagasc, Page 70-78.

Horan B., P. Dillon, P. Faverdin, L. Delaby, F. Buckley & M. Rath (2004). The interaction of strain of Holstein-Friesian and pasture based feed systems on milk yield, body weight and body condition score *Journal of Dairy Science (accepted).*

Shalloo L., P. Dillon, M. Rath & M. Wallace (2004). Description and validation of the Moorepark Dairy Systems model. *Journal of Dairy Science,* 87, 1945-1959.

Project Opti-Milk: optimisation and comparison of high yield and low input milk production strategies on pilot farms in the lowlands of Switzerland

H. Menzi, T. Blaettler, P. Thomet, B. Durgiai, S. Kohler, R. Staehli, R. Mueller and P. Kunz
Swiss College of Agriculture, CH-3052 Zollikofen, Switzerland, Email: harald.menzi@shl.bfh.ch

Keywords: dairy production systems, strategy optimisation, high yield strategy, full grazing strategy

Introduction Compared to other European countries milk production costs in Switzerland are high. Therefore Swiss milk producers must drastically reduce their costs. The high yield strategy (HY) and the full grazing (FG) or low-cost strategy appear most promising, at least for the lowland regions. Although the basic knowledge is available for both strategies, they have been applied very little in Switzerland in a consistent and optimised form.

Materials and methods The HY strategy and the FG strategy with seasonal calving in spring, were each consistently implemented and optimised on nine pilot farms. The farms initially differed little from a group of reference Swiss dairy farms. In a participatory approach the strategy was individually optimised together with the farmer and a controlling concept was defined. To guarantee an intensive exchange of experience, the pilot farmers were organised into two learning groups,. The farms were studied in detail for three years (2001 – 2003) and economic model calculations were performed up to 2010.

Results No major problems were encountered with the implementation of the new strategies and all 18 pioneer farms were satisfied with the improvements achieved. The initial average milk yield was approximately 6000 kg for the FG farms and 8200 kg for the HY farms. During the project it increased by about 900 kg for the HY farms whereas a slight decrease due to the change to seasonal calving in early spring was observed on the FG farms. The milk yield of the HY farms may appear low compared to yields achieved in other countries. This can manly be attributed to the price of concentrate being about three times as high in Switzerland as in the EU. The average concentrate use per cow per year was approximately 1300 kg and 350 kg for the HY and the FG herds, respectively. The area required for roughage production was 0.56 ha and 0.51 ha per cattle unit (1 cow = 1 cattle unit) for the FG and the HY farms, respectively. While grazing contributed, on average, 55 % on FG farms it contributed only 17% on HY farms. Consequently the proportion of conserved feed made up 41% and 70% on the FG and the HY farms, respectively. For both strategies, no abnormal or drastic decrease in body condition score (BCS) was observed during the first 100 days of lactation. No serious fluctuations were detected and most animals reached their initial BCS towards the end of the lactation at the latest. With two exceptions, the FG farms had calving intervals well below 400 days (mean 388 days, s.d. 25 days) in the second year after the introduction of the new strategy. The mean annual veterinary costs per cow were low compared to Swiss average values and were, on average, 40% lower on FG than on HY farms. They ranged between 2.1 and 3.0 Swiss cents /kg of milk for both strategies. Fertility problems and mammary infections were the major causes for the preliminary elimination of cows in both strategies.

On average, the milk produced per labour unit could be improved by about 15% and 10% on the HY and FG farms respectively. The amount of milk produced per ha of forage remained constant at about 14,600 kg/ha on HY farms and increased by 13% to about 8000 kg/ha on FG farms from 1999 to 2002. The effect of the new strategies on the economic results was not large over the duration of the project. For the HY farms economies of scales were impeded by high growth costs and neutralised by declining milk prices. For the FG farms the cost reduction was retarded, because costs of existing infrastructure had to be depreciated. Nevertheless, a considerable decrease in labour demand per unit could be observed in both strategies. The model calculations for 2010 showed that both strategies have a considerable cost reduction potential of around 40% per unit. The high Swiss ecological production standards can be met in spite of the increasing specialisation. A socially sustainable implementation seems also possible, provided that the labour reduction potential is consistently used and the interests of the family are consciously considered. The learning groups played an important role for the quick and consistent implementation and optimisation of the new strategies. They demonstrated that the sound specialist knowledge of researchers and the practical experience of farmers can complement each other very well within the framework of a structured guided processes.

Conclusions Both the HY and FG production strategies seem to be realisable in the Swiss lowlands. If they are consistently implemented both strategies have a cost reduction potential of around 40% over the 10 years from 2000. The consistent but individual implementation of the FG strategy can improve the liquidity and the work load quickly. From the economic point of view, the HY strategy, which is based on strong and, under present Swiss conditions, expensive growth must be considered ambitious and risky for the first decade of its implementation. In the long term, the potential for a good competitiveness, even on the European level, is good for both strategies.

Assessment of grass production and efficiency of utilisation on three Northern Ireland dairy farms

A.J. Dale, P.D. Barrett and C.S. Mayne
Agricultural Research Institute of Northern Ireland, Hillsborough, Co Down BT26 6DR, United Kingdom, Email: andrewdale2000@yahoo.com

Keywords: utilisation, calibration, efficiency

Introduction Recent research has shown that grazed grass can be an expensive forage for milk production, particularly if herbage production is low or utilisation is inefficient. There is very limited data on the level of herbage grown and utilised on commercial farms. The objective of this project was to quantify grass production and efficiency of utilisation on farm to substantiate the potential of grazed grass for profitable milk production.

Materials and methods Three dairy farms (Farm 1, 2, and 3) with 145, 68 and 135 dairy cows, respectively, were selected for detailed on-farm assessment during 2004. Farm 1 had predominantly medium loam soils and a grazing area of 30 ha, with Holstein Friesian cows managed under a winter/spring calving system. Farm 2 included 28 ha of heavy clay soils in the grazing area and a winter/spring calving herd of Holstein and crossbred animals (Norwegian x Holstein). Farm 3 incorporated 50 ha of light loam soils with crossbred dairy animals (Jersey x Holstein) managed within a spring calving system. Concentrates were fed throughout the summer on Farms 1 and 2, with part time feeding on Farm 3, and average quantities of concentrates offered during the project were 3.7, 3.6, and 1.5 kg/cow/day for Farms 1, 2 and 3 respectively. Cows were fully grazed for 157, 160, and 256 days on Farms 1, 2 and 3, respectively, with this extending to 183, 219, 299 days respectively, including days of partial grazing. The average grazing stocking rates (cows/ha) for the grazing season were: Farm 1 (4.5); Farm 2 (3.4); and Farm 3 (2.9). Nitrogen fertiliser application totalled 304, 332, and 262 kg N/ha for Farms 1, 2 and 3, respectively. Sward height (cm) (b) was converted to herbage mass (Y) (kg DM/ha) using the linear equation (Y=316(b)+330). Herbage mass was then calibrated using a rolling adjustment based on detailed cutting measurements undertaken periodically during the season (Barrett *et al.,* 2005). Utilised Metabolisable Energy (UME) was calculated using the equations of FIM (2004) from known quantity and quality of milk and supplements, with assumptions made for liveweight. Four grazing paddocks per farm were intensively measured throughout the grazing season with pre- and post-grazing herbage mass recorded using a rising plate meter. All data were recorded during the period March to November.

Table 1 Summary of sward and animal measurements recorded on farm

		Farm 1		Farm 2		Farm 3	
			s.d		s.d		s.d
Pre-grazing	Sward height (*cm*)	10.7	2.51	11.9	2.70	11.5	2.62
	Herbage mass (*kg DM/ha*)	3239	578.0	3119	805.9	3491	704.7
Post-grazing	Sward height (*cm*)	6.1	1.09	5.5	0.70	5.5	1.11
	Herbage mass (*kg DM/ha*)	1967	295.1	1585	341.7	1842	290.6
Average farm cover (*kg DM/ha*)		2480	332.2	2515	343.8	2589	307.5
Average milk yield during grazing (kg/cow/day)		24.0	4.03	19.6	2.82	21.6	3.55
Total grass accumulation (*kg DM/ha*)		8936		11736		10786	
Utilised herbage (*>1500kg DM/ha*) (%)		75		78		83	
Utilised Metabolisable Energy (*GJ/ha*)		85		64		119	

Results The average pre- and post-grazing sward information for the three farms is shown in Table 1, with the average farm cover and milk yield during the grazing season also shown. Average utilisation and total grass accumulation for the three farms are shown in Table 1. The pre- and post-grazing sward data and the average herbage mass presented for the three farms are within recognised targets for grazing management, with the UME measured for Farms 1 and 3 indicative of good grazing management.

Conclusions All three farms demonstrated excellent grassland management throughout the summer, reflected in pre- and post-grazing herbage mass and average farm cover. These farms demonstrate that grazed grass can be an effective forage for milk production with high levels of growth and utilisation being achieved.

References

Barrett, P.D, & A. J. Dale (2005). Assessment of rising plate meter calibration on Northern Ireland dairy farms. *Irish Grassland and Animal Production Association, (In press)*

Feed Into Milk. (2004). A new applied feeding system for dairy cows. Ed. Thomas, C. Nott. University Press

The effect of grassland management on bovine nitrogen efficiency

N.J. Hoekstra[1,2], R.P.O. Schulte[1], E.A. Lantinga[2] and P.C. Struik[2]
[1]*Teagasc, Johnstown Castle, Wexford, Ireland, Email: nhoekstra@johnstown.teagasc.ie,* [2]*Department of Plant Sciences, Wageningen University, Droevendaalsesteeg 1, 6708 PB Wageningen, The Netherlands,*

Keywords: nitrogen, herbage, nitrogen partitioning, high sugar grass

Introduction Nitrogen (N) losses through grazing bovines are at the heart of the current debate on environment and agriculture. N utilisation of grazing bovines is predominantly determined by the form and amount of energy and protein in their diet, which in Ireland consists mainly of grazed grass. The two main problems of grazed grass with respect to animal N utilisation are 1) the imbalance between total N content and energy content, and 2) the lack of synchronisation between the release of N and carbohydrates in the rumen. It was hypothesised that both the balance and synchronisation of N and energy in herbage could be improved through grassland management. The objective of this study was to study the effect of grassland management on herbage carbohydrate and protein fractionation.

Materials and methods The effects of perennial ryegrass variety (high sugar (HS), Aberdart; low sugar control (LS), Respect), N-application rate (0, 90 and 390 kg N/ha per year), rotation length (T) (2, 4 and 6 weeks), and cutting height (8 and 12 cm) on N and carbohydrate fractions in lamina and sheath material were examined, in a cutting plot experiment at Johnstown Castle Research Centre over 4 periods in 2002 and 2003. The data were subjected to multiple regression analyses and the results of the summer period in 2003 are presented here.

Results The water soluble carbohydrate (WSC) content was increased by (i) the high sugar variety, (ii) increasing T and (iii) decreasing N application rate (Table 1). In the lamina, the total N content decreased with lower N application rates, and with longer T. For sheaths, the total N-content decreased with longer T only at higher N application rates. Lower N application rates generally lowered the estimated degradation rate of N (increasing NDIN, decreasing NPN), whereas this rate increased with longer T.

Table 1 Regression equations of carbohydrate and N fractions

		Regression equation[1]	R^2	p
Lamina	WSC[2]	10.4+7.76V+0.12T-0.01N-0.0003VTN	0.88	<0.0001
	N-total	2.12+0.0048N–0.00007TN	0.90	<0.0001
	NDIN (%N)			ns
	NPN (%N)	17.7+0.16T+0.02N	0.52	<0.0001
Sheath	WSC	26.83+5.5V+0.26T-0.032N+0.0007TN	0.88	<0.0001
	N-total	0.73+0.005T+0.0024N-0.0007TN	0.89	<0.0001
	NDIN (%N)	38.1-0.16T-0.026N	0.59	<0.0001
	NPN (%N)	25.1+0.14T+0.03N	0.71	<0.0001

[1]V=Variety (LS=0, HS=1), T=regrowth length (days), N=N application rate (kg N/ha/yr)
[2]WSC=water soluble carbohydrates, NDIN=Neutral Detergent Insoluble Nitrogen (slowly degradable cell wall N), NPN=Non protein N (very fast degradation in rumen)

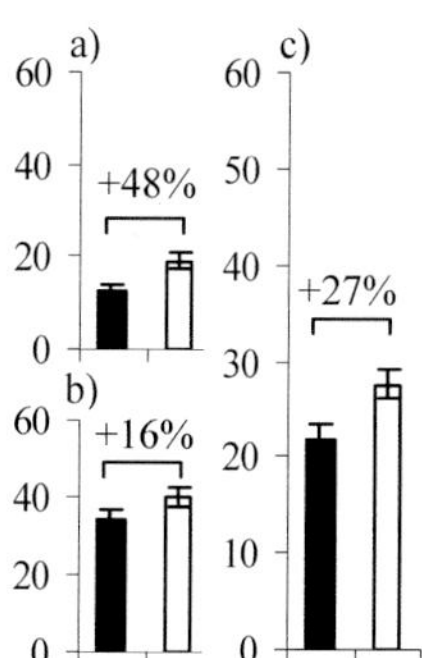

Figure 1 The WSC (%DM) content in lamina (a), sheath (b) and whole plant (c) material for the LS (solid bar) and HS (open bar) variety. Error bars represent 2xSE.

The reported increase in sugar content for the HS variety was more distinct in the lamina material compared to the sheath material, especially on a relative basis (Fig. 1). As the lamina material forms the bulk of the intake during grazing, this might explain why HS varieties have shown more promising results when managed for grazing (Lee *et al.,* 2001) compared to silage (Conaghan *et al.,* 2002).

Conclusions The balance and synchronisation of N and energy in the rumen can be improved through decreasing N application, increasing T and using HS varieties which appear especially effective under grazing.

References

Conaghan, P., P. O'Kiely, F.P. O'Mara & P.J. Caffrey (2002). Yield and chemical composition of lines of Lolium perenne L. selected for high water-soluble carbohydrate concentration. In: Proceedings of the Agricultural Research Forum, March 2002, Tullamore, pp. 39.

Lee, M.R.F., E.L. Jones, J.M. Moorby, M.O. Humphreys, M.K. Theodorou, J.C. Macrae & N.D. Scollan (2001). Production responses from lambs grazing on *Lolium perenne* selected for elevated water-soluble carbohydrate concentration. *Animal Research*, 50, 441-449.

The effect of stocking rate and initial grass height on herbage production and utilization, and milk production per unit area under set stocking by lactating dairy cows

H. Nakatsuji, T. Endo, S. Bawm, T. Mitani, M. Takahashi, K. Ueda and S. Kondo
Graduate School of Agriculture, Hokkaido University, Sapporo 060-8589, Japan,
Email: nakahiro@anim.agr.hokudai.ac.jp

Keywords: set stocking, stocking rate, herbage utilization, milk production, dairy cows

Introduction In our previous study with lactating dairy cows (Nakatsuji *et al.*, 2004), annual herbage production and utilization, and milk production per unit area under set stocking was not always lower than under rotational grazing at the same stocking rate (6 cows/ha). Furthermore, there was a possibility that set stocking could increase pasture utilization and milk production over rotational grazing when cows were grazed at adequate stocking rate and at the appropriate initial date of grazing. The purpose of the present study was to evaluate the effect of stocking rate and initial grass height on herbage production and utilization, and milk production per unit area under set stocking by lactating dairy cows.

Materials and methods Three experimental pastures (0.66 ha each) differing in stocking rate and initial grass height were prepared: HL (6.1 cows/ha; 15 cm), HH (7.6 cows/ha; 15 cm) and LL (6.1 cows/ha; 8 cm). Four, five and four lactating Holstein cows were grazed for 5 hours per day on HL, HH and LL, respectively. In addition to grazing herbage, maize and grass silage were fed *ad libitum* to all cows in the barn. Concentrates were supplemented with roughages at 25% of daily milk yield. Herbage production and utilization were measured using protective cages set on each pasture. Milk production from pasture was calculated as follows: Total milk production x (herbage TDN intake / Total TDN intake). Total digestible nutrients (TDN) content of grazing herbage and silage were estimated by chemical compositions (Deguchi *et al.,* 1997).

Results The results of pasture utilization and milk production throughout the grazing period are shown in Table 1. Mean grass height was lower ($P<0.05$) in HH and LL than in HL. However, TDN content of herbage did not differ across groups. Total herbage production did not differ across groups although stocking rate was higher in HH and grazing days were greater in LL than HL. Herbage utilization was higher in HH and LL than in HL because of a decrease in dung pats in HH and LL. Total TDN intake tended to be higher in HH than in HL and LL. Fat-corrected milk (FCM) yield from pasture in LL was similar to that in HL because low total FCM yield was compensated for by a high ratio of herbage to total TDN intake in LL. The FCM yield from pasture in HH was also similar to HL, in spite of high stocking rate in HH, because total FCM yield and the ratio of herbage to total TDN in HH were nearly equal to HL.

Table 1 Sward characteristics, herbage production and utilization and milk production from pasture throughout the grazing period

	HL	HH	LL
Initial grazing date	6 May	6 May	26 April
Total grazing days	162	162	172
Grass height, cm	15.7[a]	11.9[b]	10.8[b]
Herbage mass, tDM/ha	3.3[a]	2.6[b]	2.2[c]
TDN content of herbage, %DM	64.5	65.0	64.8
Herbage production, tDM/ha			
Growth before grazing	1.5	1.8	1.3
Regrowth	9.6	10.0	10.3
Total	11.1	11.7	11.6
Herbage utilization, tDM/ha	8.4	9.8	9.9
TDN intake, t			
Herbage (1)	3.6	4.2	4.3
Total (2)	10.1	11.8	10.3
(1) / (2), %	40.2	41.1	47.1
Total FCM yield, t	17.7	17.8	14.9
FCM yield from pasture, t/ha	9.6	9.7	9.5

[abc]Means within rows with different superscripts differ ($p<0.05$)

Conclusions Under set stocking as in the present study, the modification of initial grass height relative to stocking rate may have a positive effect on pasture utilization and milk production per unit area.

References

Deguchi, K., M. Amari, S. Masaki & A. Abe (1997). Estimation of TDN content pooled several species of temperate grasses and differences in precision of estimation between the species. *Grassland Science,* 43 (Supplement.), 290-291.

Nakatsuji, H., T. Endo, M. Kurata, T. Mitani, M. Takahashi, K. Ueda & S. Kondo (2004). Herbage production and utilization, and milk production per unit area under set stocking and rotational grazing by lactating dairy cows. *Proceedings of the 11th Asia-Australasian Animal Production (AAAP) Congress,* Volume III, 517-519.

Section 6

Other aspects of grassland/grazing

Better dairy farm management increases the economic return from phosphorus

J.D. Morton
AgResearch, Invermay Agricultural Centre, PB 50 034, Mosgiel, New Zealand,
Email: jeff.morton@agresearch.co.nz

Keywords: dairy farm, management, milksolids, pasture utilisation, phosphorus, soil Olsen phosphorus

Introduction Some 60% of New Zealand dairy farms on allophanic and sedimentary soils have soil Olsen phosphorus (P) levels to 75 mm depth above the target range for near-maximum pasture production of 20-30 µg/ml (Roberts & Morton 1999). For an economic response in milksolids (MS) production from high Olsen P, the pasture needs to be capable of high production, and the extra pasture grown be converted to milk. This paper seeks to justify this contention and outline some of the farm management practices required to achieve it.

Pasture production responses to Olsen P Results from several mowing trials at different sites and three grazing trials at one site have been used to establish the relationship between soil Olsen P and relative pasture production for allophanic soils (Figure 1). Each data point represents a P treatment from one trial site for one year. From the Flexi-fitted curve, on average, 97% of maximum pasture production occurred at an Olsen P level of 22µg/ml. There was a large degree of variability about the average curve so that for some sites, near-maximum pasture production was achieved at Olsen P less than 20 µg/ml, whereas at other sites there were still small pasture production responses above 30 µg/ml. Examination of the soil and pasture properties at each site could not clearly identify the factors that determined the differences in responsiveness of pasture production to Olsen P. However, Sinclair et al. (1997) reported that trial sites with high absolute pasture production gave lower relative production for a specific Olsen P than sites with low production. High pasture production (17.5 t DM/ha/yr) in the P responsive 1996-2000 grazing trial was achieved from a high ryegrass/clover content (67%), nitrogen (total soil N uptake 550 kg/ha) inputs and water (>192 mm rainfall/season) and adequate drainage.

Utilisation of the extra pasture grown The three P grazing trials carried out in South Taranaki demonstrated that the key farm management practices required to achieve 90% pasture utilisation and convert the 5-15% increases in pasture production above an Olsen P of 22 µg/ml to more milk were high stocking rate associated with >300 kg MS/cow, identification and conservation of pasture surpluses in spring as silage and the feeding of high quality silage to extend the lactation. A comparison between one of the trial farmlets and 65 commercial dairy farms with similar soils, Olsen P, pasture production and cow live weight is shown in Table 1. This data indicated that the average dairy farm in South Taranaki was not gaining the full financial benefit from high Olsen P because stocking rate was too low, all surplus pasture was not conserved and the lactation was too short.

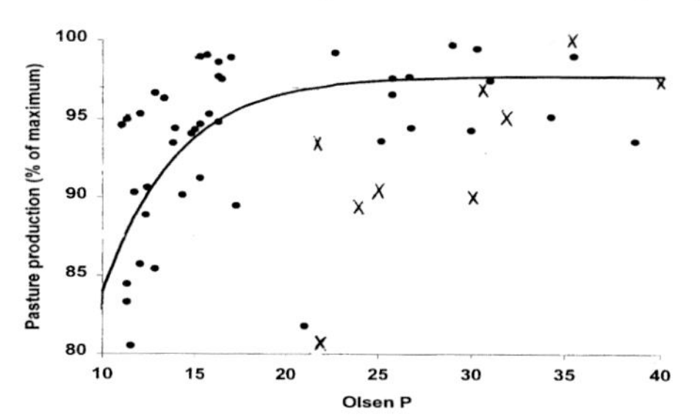

Figure 2 Effect of Olsen P on relative pasture production-average for allophanic soils (- = mowing, x = grazing).

Table 1 Comparison of trial farmlet and average for dairy farms on allophanic soils in South Taranaki (1994/95)

Parameter	Farmlet	Dairy farm
Olsen P (µg/ml)	35	37
Annual pasture (t DM/ha)	17.0	17.0
Cows/ha	3.8	3.0
Pasture intake (t DM/ha)	15.0	13.2
Conserved feed (t DM/ha)	0.83	0.48
% utilisation	90	80
Kg MS/cow	308	302
Lactation (days)	240	228
Kg MS/ha	1100	915
Economic surplus ($/ha)	2280	1700

Conclusions Management practices to help ensure profitable returns from soil P levels above the recommended target ranges include pastures with potential for high production together with sufficient cows/ha, complete conservation of surpluses and a long lactation length to utilise the extra pasture grown and convert it to milk.

References

Roberts, A.H.C.; Morton, J.D 1999. Fertiliser use on New Zealand dairy farms. DRC/Dexcel/AgResearch booklet. 37 pp.

Sinclair, A.G.; Johnstone, P.D.; Smith, L.C.; Roberts, A.H.C.; O'Connor, M.B.; Morton, J.D. 1997. Relationship between dry matter yield and soil Olsen P from a series of long term trials. *New Zealand Journal of Agricultural Research 40*: 559-567.

The effect of two magnesium fertilisers, kieserite and MgO, on herbage Mg content

M.B. O'Connor[1], A.H.C. Roberts[2] and R. Haerdter[3]
[1]*AgResearch, Ruakura Research Centre, PB 3123, Hamilton, New Zealand,*
Email: mike.oconnor@agresearch.co.nz, [2]*Ravensdown Fertiliser, P.O.Box 608, Pukekohe, New Zealand* [3]*K+S Kali,Bertha-von-Suttner-Str.7, 34131 Kassel, Germany*

Keywords: magnesium fertilisers, kieserite, MgO, pasture Mg

Introduction Supplementing Mg to dairy cows is widely practised in New Zealand. Various methods are used including drenching, pasture dusting, water trough treatment and adding to hay, silage and other feedstuffs (Young *et al.*, 1979). Fertiliser Mg (calcined magnesite, MgO) is widely used to maintain soil Mg status but research has shown that using fertiliser Mg to achieve good soil, pasture and animal Mg status requires large inputs of Mg (120 kg/ha) and maintaining blood serum Mg status in dairy cows tends to be short-lived without further animal supplementation (O'Connor *et al.*, 1987). The objective of these experiments was to test whether a more soluble Mg product like kieserite ($MgSO_4$), when applied to pastures, could achieve an immediate but short-term boost in Mg status when applied at critical times of the year.

Materials and methods Field trials were established in Northland in 2002 and Rotorua in 2003 in the North Island of New Zealand.. Each trial consisted of 2 products, MgO and kieserite, 3 rates of application (25, 50 and 100 kg Mg/ha) and 2 times of application (spring and autumn). Herbage samples were taken 5 to 6 times at 3-4 weekly intervals following either spring or autumn application and analysed for % Mg content.

Results Mg content in pasture was significantly higher from kieserite than MgO in the first sampling after application at both the Northland and Rotorua sites (Table 1). Both sites showed a marked rate effect to kieserite relative to MgO indicating that kieserite is a more soluble, quicker-acting material (Table 2). At subsequent samplings there was no difference between products at the Northland site but still differences at the third sampling at the Rotorua site before differences disappeared (Table 1). Other research has also indicated that long-term there will be very little difference between Mg fertilisers (Hogg & Karlovsky, 1968). Results suggest kieserite could be applied at critical times of the year to boost pasture Mg content and animal Mg requirements.

Table 1 Percentage Mg in pasture for control (no Mg) kieserite and MgO (mean of rates and times of application, samples taken at approximately monthly intervals post application)

Sample	Northland				Rotorua			
*	Control	Kierserite	MgO	SED	Control	Kiererite	MgO	SED
1	0.22	0.24	0.22	0.006	0.23	0.29	0.26	0.007
2	0.20	0.23	0.21	0.008	0.23	0.26	0.25	0.011
3	0.20	0.22	0.21	0.007	0.25	0.29	0.27	0.009
4	0.20	0.21	0.20	0.007	0.24	0.29	0.28	0.013
5	0.21	0.23	0.23	0.007	0.24	0.27	0.27	0.009
6					0.23	0.26	0.26	0.009

Table 2 Percentage Mg in pasture one month after application (mean of two sites)

Rate of Mg (kg/ha)	0	25	50	100
Kieserite	0.23	0.25	0.26	0.29
MgO		0.23	0.25	0.25

Conclusions The results from these trials demonstrate that kieserite is a much quicker acting fertiliser than MgO and could be used to provide a significant lift in Mg content of pasture when applied at critical times of the year. Although short-lived this effect could be important in terms of animal requirements.

References

Hogg, D.E., & J. Karlovsky (1968). The relative effectiveness of various magnesium fertilisers on a magnesium deficient pasture. *New Zealand Journal of Agricultural research*, 11,171-183.

O'Connor, M.B., M.G. Pearce,I.M. Gravett & N.R. Towers (1987). Fertilising with magnesium to prevent hypomagnesaemia (grass staggers) in dairy cows. *Proceedings of the Ruakura Farmers' Conference*, 39, 47-49.

Young, P.W., M.B. O'Connor & C. Feyter (1979). The importance of magnesium in dairy production. *Proceedings of the Ruakura Farmers' Conference*, 31,110-120.

Effect of different phosphorous sources and levels on the productive behaviour of a *Lotus pedunculatus* cv. Grasslands Maku oversown pasture

R.E. Bermúdez and W. Ayala
INIA National Institute of Agricultural Research, Uruguay, R8 Km 281, Treinta y Tres, Uruguay, PC 33000, Email: rbermudez@tyt.inia.org.uy

Keywords: Lotus, phosphorous sources, phosphorous response, oversown pastures, rhizomes

Introduction The organic meat production protocol of Uruguay (INAC, 2003) requires that the animals graze pastures that receive no chemical fertilisers. The oversown legume pastures in Uruguay used to be fertilised with soluble phosphorous (P) sources that are not accepted by the protocol. The relative efficiency of different P sources would be useful data for farmers. This information is not available for the acid soils of the eastern region of Uruguay. *Lotus pedunculatus* cv. Grasslands Maku is one of the most adapted legumes to be included in this type of pasture, as the sown area has increased in the last few years. The objective of this experiment was to evaluate the P levels response and the relative efficiency of different P fertilisers for forage production.

Material and methods The experiment was established on a farm (32° 20′S) with the following chemical soil characteristics pH5.3, OM 4.0%, P Ac. Citric4.5 µg/g, K 0.29 meq/100g and Al 0.41 meq/100g. The treatments were four levels (0, 13, 26 and 39 kg of P/ha) of natural ground phosphate rock (NPR) with a relation P soluble/P total (Pt/Ps) of 4.4/12.2, one level (26 kg of P/ha) of granulated partially acidulate reactive phosphate rock (H) (Ps/Pt) of 6.1/11.8 and one level (26 kg of P/ha) of granulated superphosphate (S) (Ps/Pt) of 9.2/10.0.
The design was a randomised block with two replicates and three blocks; plot size was 2*5 m. Three cuts were taken on 24/9/02, 29/10/02 and 10/4/03 at 4 cm height. The cutting area per plot was 6.27 m^2, a sub sample of 1 kg was taken to estimate dry mater (DM) content and lotus content by manual separation. The measurements were total dry matter (TDM), lotus DM (LDM) (kg/ha). Rhizome length (m/m^2), diameter (mm) and dry weight mass (g/m^2) were estimated in autumn 2003, taking four cores of8 cm of diameter and 5 cm depth per treatment.

Results The TDM production was not affected (P>0.05) by P levels but the LDM yield response was 65.4 kg/ha/kg P (P<0.01) applied as NRF, which added between 0 and 39 kg of P/ha (Table 1). There were no differences (P>0.05) between P sources in TDM production but NRF increased LDM production 24 % (P<0.05), with the H source giving an intermediate response. Rhizome lengths were not affected (P>0.05) either by P level or by P sources.

Table 1 Effect on total DM (TDM), lotus DM (LDM) production and rhizome characteristics

				Rhizome		
	P (kg /ha)	TDM (kg/ha)	LDM (kg/ha)	Length (m/m^2)	Diameter (mm)	Dry weight (g/m^2)
	0	7745	1647	56	2.2	4,0
NRF	13	8308	2498	85	2.6	6,1
	26	7443	3348	68	1.8	3,9
	39	7288	4198	109	2.1	5,4
	SEM	463 (6)	1222 (6)	22.4 (4)	0.94 (4)	1.6 (4)
NRF	26	7443	3348	68	1.8	3,9
H	26	4199	3682	67	2.0	3,0
S	26	4719	2980	129	2.1	7,1
	SEM	220 (6)	350 (6)	21.9 (4)	0.73 (4)	1,3 (4)

SEM standard error of the mean; (n) observations for each treatment mean

Conclusions TDM and rhizomes were not significantly affected (P>0.05) by P levels of NRF and by the different P sources evaluated. There were a significant LDM P response (P<0.01) of 65.4 kg/ha per P applied as NRF. NRF was significantly more efficient (P<0.05) than S producing LDM. NRF could be efficiently used to produce organic meat on this type of soil.

Reference

INAC, 2003 Protocolo de producción de carne natural. http://www.naturalmeaturuguay.com/sis-prod.shtml

Field testing of a turnip growing protocol on New Zealand dairy farms

J.P.J. Eerens[1] and P.M.S. Lane[2]
[1]*AgResearch Ruakura, Private Bag 3123, Hamilton, New Zealand* [2]*Wrightson Seeds, PO Box 390, Cambridge, New Zealand, Email: han.eerens@agresearch.co.nz*

Keywords: management practice, dairy farming, herbicide, insecticide, turnip

Introduction Summer droughts are a regular occurrence in central North Island districts of New Zealand, which causes pastures to wilt, lose their nutritive value and stop growing. The resulting summer feed gap depresses farm productivity (Clark *et al.,* 1996). Turnips optimally sown mid-late October (more often sown in November or even December) are grown to fill this feed gap. Recorded average yields of 7.4 t dry matter (DM)/ha are below the economic breakeven point of 8-10 t DM/ha (Clark *et al.,* 1996). A turnip growing protocol was developed from published data (Eerens & Lane 2004) and tested on commercial dairy farms.

Material and methods Dairy farmers who intended to grow a turnip crop were provided with a copy of the protocol in the 2002/03 and 2003/04 growing seasons. The protocol covered pre sowing paddock and seedbed preparation, target early sowing date (middle of October), timing and frequency of herbicide and insecticide application, target soil pH levels and fertiliser inputs. One hundred and eighty eight paddocks in the central North Island were assessed. Data were obtained on a variety of crop management and environmental factors. Thirty-five dairy farmers (19%) returned complete information. Turnip yields were assessed 90 days after sowing (from late January onwards) on a single representative 6 m^2 area. In the field, leaf and bulb material were separated and weighed, leaf material was cut off at around 1 cm above the neck of the bulb. The weighed samples were oven dried for 24 hours at 85°C to determine DM yields. Most of the crops were of the turnip variety Barkant, with some of the cultivar Green Globe.

Results The data reported here were not generated from controlled experiments and replication was unbalanced. The crops were grown as part of commercial operations and business considerations determined the need for action. For many factors all farmers did the same, resulting in a lack of contrast between treatments for those factors and an inability to comment on their impact. Variation was observed in post-emergence herbicide and insecticide application timing, ranging from within 14 (pre-emptive) to up to 60 (reactive) days later than what the protocol prescribed. This was the only factor on which the protocol could be tested. Total turnip (leaf and bulb) and leaf yields are given in Table 1. In one instance the yield was less than 10 t DM/ha when the protocol was followed but in more than 60% of the cases yields exceeded 12 t DM/ha. Where the protocol was not followed, the yield was under 10 t DM/ha in 33% of the cases and in excess of 12 t DM/ha only in 22% of the cases.

Table 1 Total dry matter (DM) and leaf DM yields achieved when applying post-emergence herbicide and insecticide either in accordance (n=13) with the protocol or not (n=22)

	Followed protocol		SED
	Yes	No	
Total yield (t DM/ha)	12.5	10.9	0.513
Leaf yield (t DM/ha)	8.1	6.8	0.403
Leaf:Bulb ratio	1.87	1.71	0.145

Conclusions Pre-emptive post-emergence herbicide and insecticide application, appears to lead to increased crop yields. However, similar high yields were achieved in the absence of either spray which points to variation in farmer cropping experience and variation in local weed and pest loadings. Crops are generally used to break disease cycles and improve poor and generally weedy paddocks, implying a need for weed and pest control.

References

Clark, D.A., S. W. Howse, R. J. Johnson, A. Pearson, J. W. Penno & N. A. Thomson (1996). Turnips for summer milk production. Proceedings of the New Zealand Grassland Association 57, 145-150.

Eerens, J.P.J. & P. M. S. Lane (2004). Best management practice turnip production; how to target a 14 tonne/ha crop on a central North Island dairy farm. Proceedings of the New Zealand Grassland Association 66, 203-206.

Labour input associated with grassland management on Irish dairy farms

B. O'Brien[1], K. O'Donovan[1,2], J. Kinsella[2], D. Ruane[2] and D. Gleeson[1]
[1]*Teagasc, Dairy Production Department, Moorepark Research Centre, Fermoy, Co. Cork, Ireland*
Email: bobrien@moorepark.teagasc.ie, [2]*Department. of Agribusiness, Extension and Rural Development, University College Dublin, Ireland*

Keywords: dairy farm, grassland management, labour input

Introduction The issues of labour and work organisation (working hours, working conditions) must be seriously addressed on Irish dairy farms if dairy farming is to have a viable future. The objective of this study was to quantify the annual labour input per cow on Irish dairy farms, with a specific focus on the task of grassland management, and to establish monthly patterns of labour utilisation over a two-year period for a range of herd sizes.

Materials and methods Labour input was recorded for 29 dairy farm tasks during the period February 2000 to January 2002. The study incorporated 98 and 73 spring calving herds in years 1 and 2, respectively. Herds were categorised according to cow number as small (< 50 cows), medium (50-80 cows) and large (> 80 cows). All farm operators recorded the duration of the different farm tasks conducted throughout the day. Records were made on consecutive 3 or 5-day periods on one occasion per month, using both timesheets and data loggers.

Results Grassland management accounted for approximately 12 % of total dairy labour input. Total annual dairy labour input to all dairy tasks, and the task of grassland management, on farms of three different herd-size groups, expressed on a per cow basis, is presented in Table 1. Both of these parameters decreased ($P<0.001$) with increasing herd-size group. Average monthly grassland labour input per cow was at a maximum in June at 1.14 h, 0.82 h and 0.52 h and at a minimum in February at 0.17 h, 0.12 h and 0.08 h on small, medium and large farms, respectively (Figure 1). Relationships between a number of factors and grassland labour input per cow were tested for significance. Using a multivariate model, an R^2 of 0.49 was measured for herd-size group, frequency of allocation of fresh grass and the use of back fencing, with significant ($p<0.05$) positive, positive and negative correlations, respectively, with respect to grassland labour input per cow per year.

Table 1 Annual total dairy labour input per cow for all dairy tasks and the task of grassland management

	Herd-size group			s.e.m.	Sig.
	Small (n=51)	Medium (n=78)	Large (n=42)		
Total dairy labour input (h/cow/year)	49.7 [a]	42.2 [b]	29.4 [c]	1.64	***
Dairy labour input to grassland management (h/cow/year)	6.3 [a]	5.1 [b]	3.2 [c]	0.31	***

n=number of farms
[abc] means on the same line without a common superscript are significantly different
*** = p<0.001

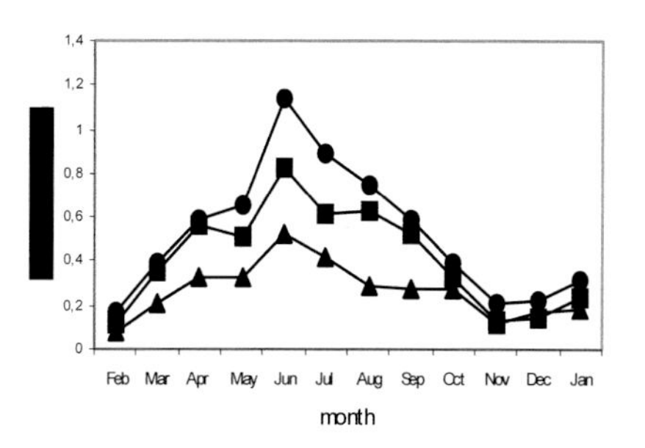

Figure 1 Average labour input per cow per month (h) over 12 months associated with grassland management, showing small (●), medium (■) and large (▲) herd-size groups

Conclusions These data indicate where labour supply and technology should be directed in terms of time of year and scale of enterprise, in order to maximise labour efficiency associated with grassland management on dairy farms. Specifically, allocation of fresh grass at a frequency of > 24 h would reduce labour demand and not have an adverse effect on cow production characteristics (Dalley *et al.*, 2001).

References

Dalley, D.E, J.R. Roche, P.J. Moate & C. Grainger (2001). More frequent allocation of herbage does no improve the milk production of dairy cows in early lactation. *Australian Journal of Experimental Agriculture*, 41, 593-599.

Factors affecting Italian ryegrass (*Lolium multiflorum* L.) seed distribution

R.D. Williams and P.W. Bartholomew
USDA-ARS-GRL, Research and Extension, Langston University, P.O. Box 1730, Langston, OK 73050, USA
Email: rdwms@luresext.edu

Keywords: seed dispersal, dispersal distance, seed movement, seed shadow

Introduction Italian ryegrass (*Lolium multiflorum* L.) can be a productive and high-quality cool-season forage in the Southern Great Plains of the U.S.A, if it is managed to produce sufficient seed for effective re-establishment without compromising forage yield. Before the re-seeding dynamics of Italian ryegrass can be modeled an understanding of seed production, seed-shed, and seed dispersal is necessary. Here two factors affecting Italian ryegrass seed dispersal and distribution are examined – wind and cultivation practice (mowing and raking).

Materials and methods Italian ryegrass was no-till overseeded into dormant unimproved warm-season pasture in the fall of 2002. Before heading the established ryegrass was mowed leaving four 1-m^2 uniform, grass blocks. Seed traps (15-cm dia.) were placed at intervals of 0, 0.3, 0.6, 0.9, 1.2 and 1.8 m from the edge of the stand in the eight cardinal directions. Trapped seed were counted every 7 to 10 days until the ryegrass was harvested on July 28, 2003. Immediately before harvesting, the seed traps were removed and a 0.09 m^2 area was vacuumed along the same cardinal directions and at the same spacing. After harvest additional vacuum samples were collected at each edge of the plot, as well as in the direction the forage was raked and piled for removal.

Results Since prevailing wind was from the south, the majority of the trapped seed was found north of the plots (Figure 1). Mean wind speeds recorded at the site ranged from calm to 6 m/sec, while gusts ranged from calm to 12 m/sec. Using a simple ballistic equation and mean wind speeds travel distances were estimated at 0.7 to 3.0 m, while at maximum wind speeds the estimated travel distance exceeded 6.0 m. Although some seed (8%) were trapped at 183 cm, 80% of the seed was found at 1 m or less from the edge of the plot. Forage removal by cutting and raking increased the seed deposition up to 61 cm from the plot's edge, as compared to the pre-harvest (trapped) seed estimates (Figure 2). Less seed than anticipated was recovered after harvest, and examination of the baled forage showed significant material still attached to the seed heads.

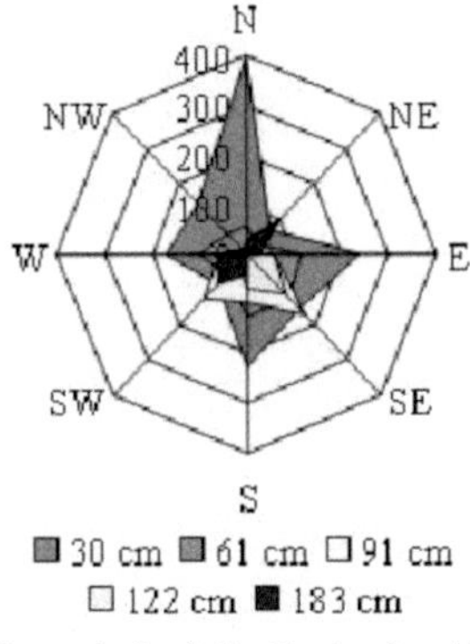

Figure 1 Seed distribution (seed/m^2).

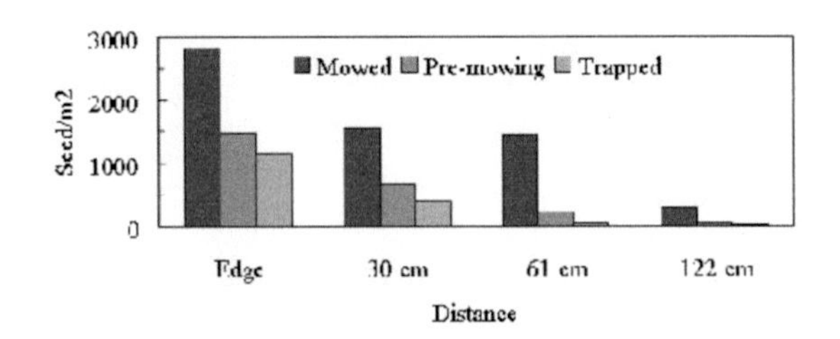

Figure 2 Seed distribution after forage harvest and removal.

Conclusions Some Italian ryegrass seed was wind dispersed, however most seed were found well within the estimated 3 to 6 m travel distance. Numerous seed were deposited at the edge of the plot, and the majority of the seed was within 1 meter of the plant stand. Mowing increased the seed deposited at the edge of the plot and further seed was deposited along the direction the forage was raked for removal. Passive seed distribution provides only limited dispersal of Italian ryegrass seed and is unlikely to allow uniform re-establishment by self-seeding.

Practical application of a one-parameter approach to assess the accuracy of two different estimates of diet composition in sheep

C. Elwert and M. Rodehutscord
Institut für Ernährungswissenschaften, Martin-Luther-Universität Halle-Wittenberg, 06099 Halle /S., Germany
Email: christian_elwert@web.de

Keywords: diet composition, estimation, accuracy

Introduction The composition of ingested herbage mixtures can be estimated using the alkane technique (Dove & Moore, 1995). Until now, the accuracy of the estimates is assessed by linear regression of estimated and actual proportions of the dietary components. The authors presented an approach to compare actual and estimated diet compositions using only one parameter named *Distance* (*D*; Elwert & Rodehutscord, 2005), thus enabling a statement regarding the similarity of estimated and known diet composition. In a feeding trial with sheep, diet composition was estimated using two different levels of information. The accuracy of the two estimates was assessed and compared using *Distance*.

Materials and methods In a balance trial with adult wethers (liveweight ~ 70 kg) selective intake of three roughages (perennial ryegrass [P], meadow fescue [M], red clover [R]) and barley was simulated. Diets A to G contained (P:M:R g air-dry matter): A 750:0:0; B 0:750:0; C 0:0:750; D 250:250:250; E 375:75:300; F 525:187.5:37.5; G 75:525:150. Diets A+ to G+ contained further 100 g air-dry matter of beeswax-labelled barley. Diets were fed once daily to a total of 4 animals per diet. Alkane concentrations in feed and faeces were determined according to Elwert (2004). Faecal alkane recovery rates (FARR) were calculated for each animal separately. Diet composition was estimated using EatWhat (Dove & Moore, 1995), and it was assumed, that all animals had potential access to all dietary components. Methods of estimation differed by the level of information included: Method 1 was based upon experimental means of FARR and alkane patterns of the components; method 2 was based upon dietary means. *Distance* was calculated as the square root of the sum of the squared differences of estimated from known proportions of each dietary component in the diet according to Elwert & Rodehutscord (2005).

Results *Distance* between estimated and actual diet compositions averaged 42 and 32 g/kg for methods 1 and 2, respectively. This difference was significant. The inclusion of barley in the diets had no effect on *Distance*. The accuracy of method 1 - which represents current situation in grazing experiments - was satisfactory. The inclusion of more specific information increased the accuracy of the estimate of diet composition: Out of 52 observations, 10 and 3 observations (methods 1 and 2, respectively) showed a *Distance* of more than 70 g/kg, which was chosen as the threshold to declare estimated and known diet compositions as dissimilar. It could be noticed, that those diets that had a high *Distance* contained large amounts of M. This could be attributed to a high degree of heterogeneity within that component throughout the experiment.

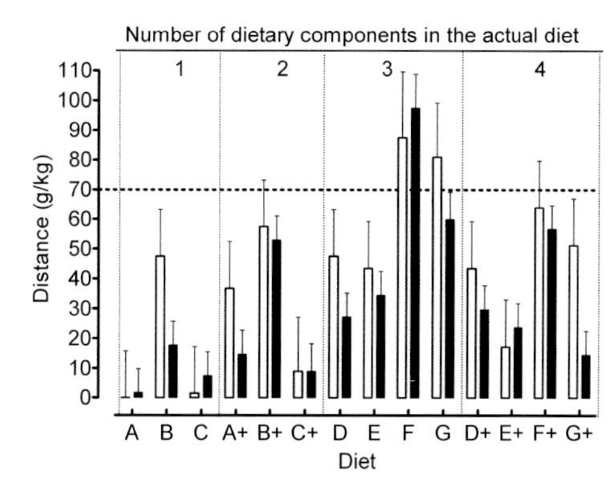

Figure 1 Accuracy of estimated diet composition in 14 diets based upon experimental (blank bars) or dietary (black bars) means of FARR and dietary component alkane concentrations (means and s.e.)

Conclusions Using the parameter *Distance*, the accuracy of an estimate of diet composition can be assessed. It does not allow for the assessment of the accuracy of individual dietary components.

References

Dove, H. & A.D. Moore (1995) Using a least-squares optimization procedure to estimate botanical composition based on the alkanes of plant cuticular wax. *Australian Journal of Agricultural Research*, 46, 1535-1544.

Elwert, C. (2004) Studies on the use of alkanes to estimate diet composition, intake, and digestibility in sheep. PhD thesis, Martin-Luther Universität Halle-Wittenberg, Germany.

Elwert, C. & M. Rodehutscord (2005) Theoretical considerations on an one-parameter approach to compare actual and estimated compositions of multi-component diets. *Procedings of the XX International Grassland Congress*. Dublin, Ireland.

Use of alkanes to estimate dry matter intake of beef steers grazing high quality pastures

G. Scaglia, H.T. Boland, I. Lopez-Guerrero, R.K. Shanklin and J.P. Fontenot
Department of Animal and Poultry Sciences, Virginia Polytechnic Institute and State University, Blacksburg, VA 24061-0306, USA, Email: billgs@vt.edu

Keywords: beef steers, intake, forages, alkanes

Introduction Pastures remain the most important source of nutrients for ruminant livestock and nutrition is critical to optimize animal production. The daily quantity of dry matter that is consumed by an animal is a critical measurement to make nutritional inferences about feed and subsequent animal response. Researchers are facing the dilemma that, while estimates of individual animal performance are readily obtained, it is still difficult to estimate the herbage intake of individual animals. The objectives of this experiment were to estimate forage intake in beef steers grazing tall fescue (*Festuca arundinacea Schreb.*) and alfalfa (*Medicago sativa*)/tall fescue pastures and to measure the recovery rate of artificial alkanes from a controlled release device under these conditions.

Materials and methods The experiment was conducted on two pastures: a pure stand of tall fescue and a mixture of alfalfa and tall fescue. Six steers (330±11 kg) were allotted to each pasture (0.61 ha). Steers were previously trained to the use of a harness and a faecal collection bag, so that the expected negative effect on animal behaviour could be minimized. A controlled-released capsule (Nufarm, Auckland, NZ) containing alkanes (C_{32} and C_{36}; release rate: 400 mg/d) was administered on d 0 to all 12 steers. During the experimental period (d 8 to 14) faecal collection bags were fitted to all steers in the fescue pasture and to three steers in the fescue/alfalfa pastures. Collection bags were emptied twice daily (at 0830 and 1630). Faeces were weighed, mixed, and an aliquot of 0.5 kg frozen. Rectal grab samples were obtained twice daily from each animal at the time of switching bags. Forage mass in each pasture was estimated from two 3-m strips obtained on d 8 using a push mower. From d 6 to 12 forage samples for alkane determination were obtained by walking the pasture in an "X" and clipping every 20 steps with a set of hand-held clippers at a height of 2.54 cm from the ground. Faecal and forage samples were freeze-dried and ground through a 0.5 mm screen. Alkane determination and dry matter intake estimation (using C_{31} as odd-chain alkane) followed the method described by Mayes *et al.* (1986). The effect of pasture on herbage intake was estimated by analyses of variance. A paired *t*-test was used to compare herbage intake estimated from grab and bag samples. All statistical analyses were conducted using SAS (SAS Institute, Cary, NC).

Results Forage availability (3078 vs. 3088 kg DM) and quality (CP: 19 vs. 17.5%; NDF: 50 vs. 55%; ADF: 31 vs. 29%) were similar for fescue/alfalfa and fescue pastures, respectively. Similar concentrations of C_{32} (mg/kg of faeces) were found in grab and bag samples, with an average ratio (C_{32} grabs/ C_{32} bags) of 0.99 and 0.92 for alfalfa/fescue and fescue pastures, respectively. The recovery rate of C_{32} in faeces was higher than previously reported elsewhere (Mayes *et al.*, 1986; Dove and Mayes, 1991) with an average of 0.985 and 0.991 for fescue/alfalfa and fescue pastures, respectively. Using information obtained from faecal samples collected from bags there was no difference ($p>0.05$; SEM=0.48) in the average daily dry matter intake of steers grazing fescue/alfalfa (10.8 kg DM) and fescue pastures (10.3 kg DM). Similarly, no differences ($p>0.05$) were detected in average daily dry matter intake when using C_{32} from grab samples (9.9 vs. 9.6 kg DM for steers grazing fescue/alfalfa and fescue pastures, respectively; SEM=0.39). When sampling method was compared (grabs vs. bags) there was no difference ($p>0.05$) in the estimation of dry matter intake for steers grazing fescue or fescue/alfalfa pastures.

Conclusions The results of the present study demonstrate that knowing the recovery rate of the marker used (in this case C_{32}) faecal grab samples can be used to estimate dry matter intake in beef steers under grazing conditions. However, the recovery rate for C_{32} observed in this experiment was 5 – 15% higher than those found in the literature. Therefore, further research is needed to study the effectiveness of the controlled-release capsules under different conditions and factors that might be affecting the release rate of the marker.

References

Dove, H. & R. W. Mayes (1991). The use of plant wax alkanes as marker substances in studies of the nutrition of herbivores: A review. *Australian Journal of Agricultural Research*, 42, 913-952.

Mayes, R. W., C. S. Lamb & P. M. Colgrove (1986). The use of dosed and herbage n-alkanes as markers for the determination of herbage intake. *Journal of Agricultural Science, Cambridge*, 107, 161-170.

Effects of rumen fill on intake and milk production in dairy cows fed perennial ryegrass

A.V. Chaves and A. Boudon
UMRPL-INRA 35590 St-Gilles, France, Email: Alexandre.CHAVES@rennes.inra.fr

Keyword: intake, performance, rumen fill

Introduction Physical limitation often limits dry matter intake (DMI) of high producing cows or cows fed high forage diets. The extent to which DMI is regulated by distention in the rumen depends upon the cow's energy requirement and filling effects of the diet offered (Allen, 2000). The objective here was to challenge middle lactation dairy cows with rumen fill (rumen inert bulk – RIB) feeding ryegrass fresh cut (indoors) or grazed to determine whether RIB affects intake and milk production.

Material and methods Four high and four medium (34 and 24 kg milk/day, respectively) producing fistulated Holstein cows in mid lactation were blocked by days in milk, milk yield and body weight. The experiment was a cross-over 4 x 4 Latin square design for milk production level consisting of four 14-d periods with a factorial arrangement of treatments: indoors versus grazing, RIB versus control (without RIB). Herbage height post-cut was aimed to be identical to herbage height post-grazing. The RIB consisted of indigestible coconut fibre (NDF=92%, lignin=53% and DM degradation rate 0.0% over 360h) placed in a weighted lingerie bags. The mean volume of added RIB was 18 L of rumen displacement per cow (mean wet weight 14 kg/cow). Calculations of DMI and feeding management are described by Chaves *et al.* (2005). Wet weight of rumen digesta was measured by rumen emptying on day 13 during the measurement period at 0830 h. A weight sample of total rumen contents (5% w/w) was taken for DM determination (80°C for 72 h).

Results Ryegrass heights, biomass and density were similar for cows fed indoors and grazing either at pre-cut (pre-grazed) or at post-cut (post-grazed). Ryegrass was of high quality with average values of protein concentration of 19.2% of DM and in vitro organic matter digestibility of 0.83. Herbage intake, milk yield and milksolids were lower at grazing than for cows fed indoors (Table 1). Rumen wet and DM digesta were higher for cows fed indoors compared to grazing cows. Ryegrass DM intake for cows challenged with RIB was on average 2.3 kg/d lower compared to treatments without RIB ($P<0.01$). This effect did not differ between grazing or indoors feeding and level of production Addition of RIB decreased both milk production and milksolids ($P<0.02$). The interaction RIB x ryegrass fed indoors or grazing was not significant for any variable tested. Rumen wet digesta decreased in both treatments when RIB was added and also affected both medium and high producing cows. Weight of rumen digesta DM also decreased in both treatments when RIB was added.

Table 1 Ryegrass intake (kg/d), milk yield (kg/d) and rumen mass (kg) for 4 cows fed indoors or 4 cows grazing without or with added rumen-inert bulk (RIB)

	Treatments					Main effects (*P*<)		
	Indoors	Indoors + RIB	Grazing	Grazing + RIB	SE	RIB	Indoors/ grazing	Production level
DM intake estimated by Yb marker	19.8	17.5	17.6	15.3	0.8	0.02	0.02	0.02
Milk	26.3	24.9	24.7	22.4	0.7	0.02	0.02	0.02
Milksolids (milk fat+milk protein)	1.78	1.68	1.68	1.54	0.04	0.02	0.02	0.08
Rumen wet digesta	109	87	91	73	3.6	0.001	0.001	0.012
Rumen wet digesta plus RIB	109	101	91	87	3.5	0.11	0.001	0.011
DM rumen digesta	13.2	10.0	9.9	8.7	0.7	0.007	0.006	0.004

SE = standard error; DM = dry matter.

Conclusions Addition of rumen inert bulk (RIB) decreased ryegrass intake and milk production in both medium and high producing dairy cows. This proves the hypothesis that ruminal distention influences forage intake in ruminants even when high quality forage is fed.

References

Allen, M. S. (2000). Effects of diet on short-term regulation of feed intake by lactating dairy cattle. *Journal of Dairy Science*, 83, 1598-1624.

Chaves, A.V., A. Boudon, J.L. Peyraud & R. Delagarde (2005). Does the feeding behaviour of dairy cows fed ryegrass indoors or grazing differ? *In these proceedings.*

Both grass development stage and grazing management influence milk terpene content

G. Tornambé[1,4], A. Cornu[1], N. Kondjoyan[2], P. Pradel[3], M. Petit[1] and B. Martin[1]

[1]*Unité de Recherches sur les Herbivores, INRA, Theix, 63 122 Saint Genès Champanelle, France, Email: tornambe@clermont.inra.fr, 2Unité de Recherche sur la Viande, INRA, Theix, 63 122 Saint Genès Champanelle, France,* [3]*Domaine de la Borie, INRA, 15 330 Marcenat, France,* [4]*Università degli studi di Palermo, Dipartimento S.En.Fi.Mi.Zo., viale delle Scienze, 90128 Palermo, Italy*

Keywords: milk terpene, pasture plants, grazing management, plant maturity

Introduction Terpenes are a wide group of molecules originating from plants' secondary metabolism. Forage terpenes vary according to the botanical composition and in particular to the proportion of plants such as Apiaceae, Lamiaceae or Asteraceae. These molecules are considered effective milk markers for the presence of diversified forages in dairy cow diets. The variation in terpene content in the milk of grazing cows would depend on the period of development of terpene-rich plants and on the grazing management, whereby cows do or do not have the opportunity to choose and to modify the botanical composition of the ingested grass. The aim of this trial was to quantify the respective effects of grass development stage and grazing management on milk terpene content.

Material and methods A diversified mountain grassland (1100 m elevation) located in the Cantal (France), was exploited from 31 May to 1 July (first growth) and again from 1 to 6 October 2003 (re-growth). The portion was divided into two parts each grazed by 6 cows, either by strip grazing (SG), the strip limits moving ahead every 2 days or by paddock grazing (PG), the paddock being changed on the 13 and 24 of June 2003. Bulk milk was sampled twice a week for analysis. Volatile compounds were extracted from the milk fat by Purge and Trap Dynamic Head Space, separated by gas chromatography and terpenes were selectively detected by mass spectrometry using their characteristic fragments at 93, 136, 161 and 204 ions. Monoterpenes and sesquiterpenes were semi-quantified by integrating the 93 and 161 ion chromatograms respectively.

Results In SG milk obtained in June, monoterpenes increased linearly with time (R^2=0.91) up to 8 times the initial value. Sesquiterpenes also increased linearly until 20 June and then decreased. This strong increase was linked with the DM proportion of dicotyledons in the pasture (R^2 = 0.51 and 0.35 for mono and sesquiterpene respectivily) and strip-grazing cows had limited selectivity of ingested grass.

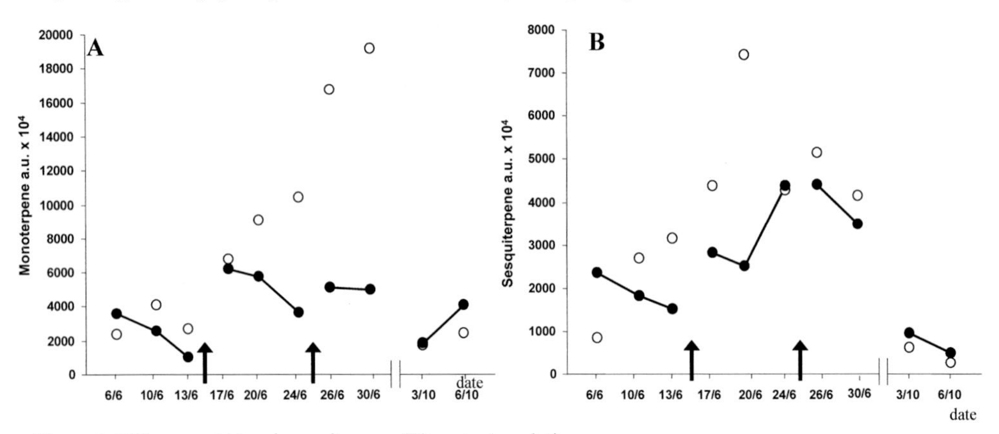

Figure 1 Milk mono-(A) and sesquiterpene (B) content evolution
•: paddock grazing (PG); o: strip grazing (SG); ↑: paddock changing for PG cows

Milk terpene content in PG at the entrance of cows in to each paddock increased less than in SG milk. PG milk monoterpene content was much lower at the entrance of the third paddock. Milk from re-grown grass was very low in terpenes probably because this grass was almost entirely vegetative, including regrowths of dicotyledons that still account for 20% of the biomass.

Conclusions Milk from the SG cows who had limited selectivity of the plants they ate was much higher in terpenes. Our results show that the milk terpenes are fingerprints of both the botanical composition and the grazing strategy.

Supplementation under intensive grazing, silage- or grain-based diets for beef production on steer performance and meat fatty acid composition

J. Martínez Ferrer[1], E. Ustarroz[1], C.G. Ferrayoli[2], A.R. Castillo[3] and D. Alomar[4]

[1]*Instituto Nacional de Tecnología Agropecuaria, Estación Experimental Agropecuaria Manfredi, Ruta Nacional N°9, km 636 (5988) Manfredi (Córdoba), Argentina, Email: martinezferrer@correo.inta.gov.ar,* [2]*Agencia Córdoba Ciencia S.E, Unidad CEPROCOR, Alvarez de Arenales 230, X5004AAP, Córdoba, Argentina*
[3]*University of California, Davis, 2145 Wardrobe Ave, 95340 Merced, California, USA,* [4]*Universidad Austral de Chile, Instituto de Producción Animal, Casilla 567, Valdivia, Chile*

Keywords: grazing, alfalfa, dietary regimen, meat, fatty acid composition

Introduction Alfalfa (*Medicago sativa L.*) is the main cultivated pasture in Argentina. In beef production enhanced productivity and profit depend on high stocking rates and pasture utilisation, with grain supplementation necessary to maintain high individual live weight gains (LWG) and to increase production per ha (Ustarroz, 1999). Substitution of grazed grass by concentrate can affect meat fatty acid (FA) composition (French *et al.*, 2000). The objective of this study was to evaluate the effects of intensifying an alfalfa-based grazing system and two confinement dietary regimens for beef steer finishing on animal performance and meat FA composition.

Materials and methods The study had 7 treatments (T) with 2 replicates in a complete randomised design. Experimental units (pens or paddocks) had 5 Angus steers (7-8 months and 165.9±7.0kgLW). Grazing T comprised 2 forage allowances (A, g/kg LW: low (LA=15) or medium (MA=30) combined with 2 supplementation levels (S, g/kg LW: low (LS=7.5) or high (HS=15). The last grazing T received solely a high A (HA=40-60). Supplement consisted of maize (0.7) and rye (0.3) grains. Steers grazed first rye (winter) and then alfalfa (spring-autumn). Confined T were corn grain (CG, 73:14:12:1% CG: alfalfa hay: soybean meal: urea) or sorghum silage-based (SS, 65:17:11:6:1% whole plant BMR SS: corn grain: soybean meal: rye grain: urea) diets offered *ad libitum*. Steers were slaughtered at 10mm of back fat (BF). *Longissimus* muscle (11st-13^{rd}rib) FA composition was analysed by gas chromatography.

Results Forage consumption was higher for HA ($p<0.01$). Total intake (pasture + grain or confined diets) was lower ($p<0.05$) while pasture utilisation was higher in LA-LS ($p<0.01$). CG reached target BF earlier due to higher LWG ($p<0.01$, Table 1). Carcass weight, BF and *longissimus* muscle area were consistently higher for MA-HS ($p<0.1$). Secondary production (kg LWG/ha) was higher in SS and LA-HS ($p<0.01$). HA had higher $_{c9,t11}$CLA, SFA and *n*3-PUFA, and lower MUFA and *n*6/*n*3 ratio, while the opposite occurred in confined T (Table 1).

Table 1 Weight gain (g/d) and meat fatty acid composition (g/100gFAME) from steers on different diets

	CG	SS	LA-HS	LA-LS	MA-HS	MA-LS	HA	s.e.m.
LWG (g/d)	1148 ^{a}	967 ^{b}	805 ^{bcd}	704 ^{d}	863 ^{bc}	755 ^{cd}	858 ^{bc}	39
$_{c9,t11}$CLA	0.44 ^{bc}	0.34 ^{c}	0.70 ^{a}	0.70 ^{a}	0.62 ^{ab}	0.69 ^{a}	0.77 ^{a}	0.043
SFA	41.0 ^{c}	43.2 ^{bc}	42.8 ^{bc}	43.4 ^{bc}	43.9 ^{b}	44.5 ^{b}	48.1 ^{a}	0.607
MUFA	47.3 ^{a}	47.2 ^{a}	45.1 ^{ab}	41.1 ^{c}	43.5 ^{bc}	41.6 ^{c}	40.3 ^{c}	0.800
PUFA	10.4 ^{cd}	9.6 ^{d}	12.1 ^{bc}	15.5 ^{a}	12.5 ^{bc}	13.9 ^{ab}	11.6 ^{bcd}	0.556
*n*3-PUFA	1.17 ^{d}	1.24 ^{d}	3.28 ^{c}	4.83 ^{ab}	4.15 ^{bc}	4.47 ^{ab}	5.33 ^{a}	0.441
PUFA/SFA	0.25 ^{cd}	0.22 ^{d}	0.29 ^{bc}	0.36 ^{a}	0.29 ^{bc}	0.31 ^{ab}	0.24 ^{cd}	0.013
*n*6/*n*3	7.98 ^{a}	6.67 ^{b}	2.46 ^{c}	2.11 ^{cd}	1.91 ^{d}	1.99 ^{cd}	1.04 ^{e}	0.697

Different superscripts indicate differences between treatments, LSD (α=0.05).

Conclusions Confinement and grazing dietary regimens affect beef production and produce meat with a different fat profile. Under intensive grazing systems grain supplementation can affect not only animal performance but also meat fatty acid composition and as a consequence meat quality.

References

French, P., C. Stanton, F. Lawless, E.G. O'Riordan, F.J. Monahan, P.J. Caffrey & A.P. Moloney (2000). Fatty acid composition, including conjugated linoleic acid, of intramuscular fat from steers offered grazed grass, grass silage, or concentrate-based diets. *Journal of Animal Science,* 78, 2849-2855.

Ustarroz, E. (1999). Utilización de alfalfa en pastoreo. *Revista Argentina de Producción Animal,* 19, 57-70.

Section 7

Open debate - What should grassland/grazing research focus on now?

Research into the types of cows and systems required to utilise grazed pastures sustainably in 100 years from now

C. Holmes
Institute of Veterinary Animal and Biomedical Sciences, Palmerston North, New Zealand

Dairy farming has undergone rapid intensification in many countries over the past 60 years, as the result of technological developments, trade policies, and financial incentives. For example, the use of soluble fertilisers, irrigation and concentrated feeds have increased enormously, and antibiotics and hormone treatments did not even exist in 1940. However, the dramatic increases in milk yields per cow (e.g. in North America) and per hectare (e.g. in New Zealand) and in cows per person (in most countries) have been associated with growing concerns about the health and fertility of cows, and their metabolic stress and welfare and about the adverse effects of high stocking rates, plus related inputs, on soil water quality.

Pastoral dairy systems incorporate three key biological "subsystems", the soils, the pastures and the cows, which must be kept in balance, with each other and with the wider external environment.

Systems research, modelling, and intensive monitoring of farms are urgently required to ensure that these balances can be maintained within sustainable, profitable and acceptable limits. Two key aspects of the cows and the systems are outlined below.

Use of genetic information about the cows for the design and management of pastoral systems

Milk is produced by a wide variety of dairy systems in the world, with an enormous range in milk yields per cow (and presumably also in feed intakes per cow). It is now clear that there is no one type of cow that is best for all dairy systems.

In seasonal, pastoral dairy systems maximum daily feed intake capacity by grazing is lower (by 10 to 20%) than on other more concentrated, non-grazing diets, and the interval between successive calvings must average 365 days. Energy demand and fertility of the cow must therefore be compatible with these two essential requirements, and the cows' Genetic Values can be used for these purposes.

The cows' Genetic Feed Demand (GFD) can be calculated from their Genetic Values for liveweight and yields of milk, fat and protein, and used to estimate the optimum stocking rate for a particular type of cow and feed supply. In future, the GFD will be calibrated against the maximum daily intake per cow achievable from grazed pasture, in order to identify those cows that are able to eat enough by grazing to prevent unsustainable negative energy balances.

The cows' Genetic Values for Fertility can be used to identify those cows that are able to meet the dates of conception and calving required in the seasonal system; for example, for more than 90% of the herd to conceive in a 10 week mating period.

Logical use of Genetic Values for key traits in these and other ways in future will result in soundly based use of cows for particular systems.

Research into systems for sustainable, resource-efficient pastoral dairying This is essential when seen against the background of the recent, rapid intensification in dairying. Two New Zealand examples are described briefly below.

The current Resource Effluent Dairying study, with 6 separate farmlet systems at Dexcel, Hamilton, is a bold step to meet some of these needs. It will provide important information for systems operated at widely different stocking rates and levels of supplementary feeds, and will guide decisions about future research.

The philosophy behind organic methods, to improve the biological activity of soils, and to prevent stresses on cows and adverse effects on the environment, is generally compatible with the objectives of sustainable systems. But some organic "rules", for example prohibition of soluble fertilisers, antibiotics and some other conventional animal health remedies seem to run counter to pragmatic sustainability. Nevertheless, the present comparison between an organic and a conventional pastoral system, at Massey University, will help to reemphasize the benefits of "good biological husbandry" of soils, pastures and animals as effective methods to prevent problems and to reduce environmental and financial costs.

Conclusions

Recent research in dairying on pasture has explored the limits of technical possibilities, which may have enabled production to outstrip the biological limits within which these systems must operate. It must now refocus within these limits, to ensure mutual compatibility between the cows, the systems and the environment, and their ongoing sustainability.

Educational programmes in dairy production systems, and exchange programmes for postgraduate students between countries with pasture-based dairy industries, would make important contributions to increased international understanding of the main problems and their solutions.

Farming for fun & profit

M. Murphy
Longstrand, Castlefreek, Clonakilty, Co.Cork, Ireland

We live in a competitive capitalist economy. Farming is rapidly moving from a highly subsidised sheltered existence into a more competitive future. Milk is a commodity. In commodity production the low cost producer survives and prospers. High cost farmers go out of business. So farmers, researchers and advisors need to maintain a relentless focus on low cost. We need durable, simple systems that exploit low feed cost based on grazed grass where temperate climate allows e.g. Ireland and New Zealand.

We strongly need to avoid the false gods of the production driven trap – which so frequently ensnare the dairy sector. "Profit is sanity, production is vanity". Correct measurement (metrics) is essential to capture the real key success factors. I suggest that growth in net worth, return on capital, return on equity, free cash, free time, profit/hectare (outside a quota system); and profit/quota unit (within a quota system) are among the relevant metrics. Note well that production/cow emphatically does not feature as a key metric in measuring a farm business.

To properly exploit grassland a focus on systems must be the priority. Researchers must be aware that "component research" which isn't compatible with an efficient "system" – is doomed to a quick death. Systems of grass dairy farming should seek the following benefits:

Lower feed costs
Higher labour productivity (but need cows suitable for system)
Lower capital costs (except for land in Ireland and New Zealand)
Lower veterinary and replacement costs (again with "right" cows)
Lower miscellaneous costs from cows that walk to grass, harvest it themselves, and spread their own slurry.

Other often-overlooked benefits of long grazing seasons are:

Higher milk protein
Higher milk CLA a future potential major competitive advantage.
Welfare benefits to cows outside on well managed grassland for most of the year. We should start to label confinement dairying as battery cow farming.
Farmers need to think and use their brains when pastoral farming; confinement dairying is so numbingly boring it quickly leads to "brain-dead" farmers.

Researchers in pastoral counties, should bear in mind that the combination of once a day milking and very long grazing seasons with suitably selected cows is quite likely to lead to huge breakthroughs in labour productivity. I envisage one person managing up to 400 cows with some seasonal help in the near future.

Efficiency of the system and scaling up when regulations/quota allow will be a strong feature of Irish dairying over the next 10 years. Scale alone, without efficiency won't work. Cost efficiency must be excellent before scale, and scaling-up will work.

In pastoral economies such as Ireland and New Zealand where strong economies and full employment looks likely to be the norm for the foreseeable future we won't attract young people into farming unless dairy systems are profitable, enjoyable and people friendly. Efficient grass based dairy farming hammers "battery cow farming" systems on all these metrics by a country mile.

Grassland research: goals for the future

J.R. Roche
Dexcel Ltd., Hamilton, New Zealand

Our goal as scientists must be to provide sufficient food for the world's population, while returning sufficient income for effort to the food producer. It has been established elsewhere in the conference that pasture-based systems can return high milk production per hectare at low input cost, but are limited to highly fertile, temperate regions with evenly dispersed rainfall patterns and moderate maximum and minimum temperatures. There should be two goals of grazing research in the future; to sustain or improve the profitability of current grazing production systems, and to develop plant and animal varieties that allow grazing in currently less suitable environments. Both of these goals can, and will be achieved through improvements in biotechnology.

Traditional breeding shuffles whole genomes, making it difficult to predict the transfer of desired traits and the loss of others. Bioengineering differs in that only one or two specifically identified trait-inducing genes are typically introduced into the background of tens of thousands of genes (Conko & Prakash, 2002). Moreover the introduced gene(s) could be isolated from the host plant, mimicking traditional breeding but in a more precise and predictable approach, so called Cisgenics®. For example, the breeding of current maize varieties took thousands of years, but only involved a change in five or so of its own genes resulting in a loss of plant hardiness, a common problem in plant breeding. Similar improvements could be achieved in 10 to 15 years using current bioengineering techniques without any loss in plant hardiness (K. Elborough, personal communication). Future grassland research must identify production and nutritional limitations in currently used herbage species, and focus on molecular improvements. According to the NRC (1989), "we are in a better, if not perfect, position to predict the characteristics" of organisms modified by molecular methods.

Pasture-based cows have traditionally been low yielding when compared with cows offered total mixed rations. Kolver *et al.* (2002) reported milk production differences of 2,500 kg between cows of similar genetic merit for milk production, either grazing or offered a total mixed ration in confinement. Low yielding cows partition a greater proportion of energy eaten into maintenance and activity and are therefore less efficient converters of energy into saleable product. This will be a significant limitation to profitable grass-based farming in the future and must be the focus of any future research. Improvements in the ability of forage species to withstand temperature (cold or hot) and moisture stress, and/or improvements in the nutritional quality of grazed forage (lower protein, higher sugar, less indigestible fibre) will increase milk production per hectare and per cow with very little additional cost.

Biotechnology will also provide land (and potentially animals) that would previously have been unsuitable for grazing systems. Advances in biotechnology are being made – acid-, saline- and drought-tolerant plants will allow forage production in some of the world's traditionally less suitable places. Animals that are more suited to grazing systems or are more efficient at converting energy into milk are also not beyond the realms of imagination.

The challenge for grassland scientists is to integrate this new technology into sustainable profitable farming systems.

References

Conko, G. and Prakash, C.S., 2002. The attack on plant biotechnology. Chapter 7 in Global warming and other eco-myths: how the environmental movement uses false science to scare us to death. R. Bailey, ed. pp 179-217.

Kolver, E.S., Roche, J.R., De Veth, M.J., Thorne, P.L. and A.R. Napper. 2002 Total mixed rations versus pasture diets: evidence for a genotype x diet interaction in dairy cow performance. Proceedings of New Zealand Society of Animal Production. 62:246-251.

National Research Council, 1989. Field testing genetically modified organisms: Framework for decisions, Washington, DC: National Academy Press. Pg 13.

What research is required for economically and environmentally sustainable farming?

W. Taylor
Glastry Farm, 43 Manse Road, Kircubbin, BT22 1DR, Northern Ireland, Email: glastryfarm@btconnect.com

This Congress is being held on an island that is a Nitrate Vulnerable Zone.No other land area in the western world has achieved such a status!

With this designation come completely new parameters for agriculture in general and grassland production in particular. Alongside this change in emphasis for the grass based industry is the implementation of the Common Agricultural Policy reform. For farming within the European Union it is not completely about maximising production, about "growing two blades of grass where one grew before". It's also about creating a diverse landscape, about less pollution about greater recreational opportunities, about sustainability about protection of flora and fauna. For the first time in our history, we as grassland farmers and livestock producers will be paid for our multifunctionality in 2005.

That change in emphasis brings immense challenges for me as an "intensive" dairy farmer. How can I react to a 170 kgs/ hectare N limit on my enterprises? How do I manage a system that produces vast quantities of waste? How do I reduce phosphorous inputs within my farming system? How do I maintain a competitive farm business within an increasingly global marketplace that is sustainable economically and environmentally? Can I adapt my grass-based farming systems to be both nutrients efficient, input efficient and profitable? Can I cope with the legislation coming down the "track" on air emissions in a business where ammonia and methane are produced in large quantities?

A survey of research projects at a number of centres throughout the British Isles would suggest to me that much of the previous and current research is serving yesterday's industry and there are few answers to my list of concerns about how I shape my business to meet current legislation or future legislation on water and air quality. Much of the current research has three main strands:

- The first seeks to maintain or increase production whilst constraining or reducing costs.
- The second seeks to develop land management approaches to reduce the impact of farming on the landscape.
- The third seeks to create added value away from the farm gate, using farm outputs as global commodities.

These strands are far from integrated, certainly do not enhance sustainability, and in most cases do no more than quantify the problems rather than identify the solutions.

The areas of food chain connections that maximise the human health attributes of grass-based livestock products, the role of technology transfer and the public perception of science are the challenges ahead.

Keyword index

acoustic monitoring 154
agro-climatic zones 159
alfalfa 202, 226, 227, 253
alkanes 250
alternatives 217
animal intake 190
animal performance 119, 190
animal production 19
animal welfare 119
beef 62
beef production 192
beef steers 168, 250
biodiversity 162
body condition score 177
body weight 177
botanical composition 172
Brassica 229
Brittany 214
calculator 219
calibration 238
cash cost 131
cattle 62, 151, 166, 217
Caucasian clover 167
chew-bites 154
climate 223
climate change 225
closing date 235
clover 234
clover cyst nematode 231
clustering 223
computer model 224
concentrate 236
confinement TMR system 131
conjugated linoleic acid 160
cow size 177
cow strain 182
cow-calf systems 224
crop model 193
cultivar evaluation 216
cutting 235
Dactylis glomerata 163, 201
dairy 62, 186, 213
dairy cows 89, 152, 165, 184, 185, 198, 207, 215, 230, 232, 240
dairy farm 243, 247
dairy farming 246
dairy goats 191
dairy production 183, 194, 237
dairy system simulation 225
decision support 79, 105, 207, 211, 215, 232
decision-making 157
dietary regimen 253
digestibility 201
diploid 161
dispersal distance 248
diversity 164
DSS 217
early grazing 232
eating time 184
efficiency 238
electronic calculator 212
endangered species 209
environment 49, 119, 223
environmental response 159
establishment 167, 231
evaluation 89
ewes 180, 181
facial eczema 161
farm cover 189
farmer knowledge 157
fatty acid composition 160, 253
feed budget 212
feed budgeting 157, 208
feed system 182
fertility 176
fescue toxicosis 168
fine wool 179
forage 193
forage maize 230
forage mixtures 162, 164
forage quality 171
forages 250
forecast 218
forecasting 79
full grazing strategy 237
G×E 62
generative 203
genetic strain 175
genetics 62
goats 153
grass 184, 203, 225, 233, 236
grass growth 214, 234
grass silage 189
grass varieties 165
grass/clover 218
GrassGro 105
grassland 19
grassland budgeting 79
grassland management 155, 247
grazing 37, 49, 62, 89, 131, 152, 153, 169, 176, 185, 202, 212, 217, 218, 229, 230, 234, 235, 253
grazing behaviour 168, 198
grazing frequency 156, 172
grazing management 207, 209, 214, 215, 252
grazing systems 105, 178, 179, 190, 210
greenhouse gases 119
growth 213
growth rate 218
haying policy 224
heading date 166
Hedysarum coronarium 156
heifers 180, 181
herbage 158, 160, 213, 235, 239
herbage allowance 37, 185
herbage intake 180
herbage mass 37, 156

herbage utilization 240
herbicide 246
high sugar grass 239
high yield strategy 237
Holstein-Friesian 175, 182, 183
insecticide 246
intake 37, 154, 164, 191, 250, 251
intake rate 151
Internet 210
irrigation 163
jaw movements 154
kieserite 244
Kura clover 167, 171
labour input 247
lamb finishing 195, 196
lambs 178
leaf and stem yield 169
linear programming 192
Lolium perenne 201
long term simulation 224
longevity 186
Lotus 245
lucerne 202, 211
magnesium fertilisers 244
maize 225
maize silage 236
management 189, 213, 243, 246
mastitis 176
mathematical model 192
meal criterion 198
meat 253
MgO 244
micro-budgeting 208
milk 176
milk acetone 177
milk processability 233
milk production 164, 185, 229, 240
milk terpene 252
milksolids 243
mixed farming 211
mixed pasture 169
modelling 89, 105, 197, 198, 202, 203, 213, 226, 227
morphology 226
native swards 179
New Zealand 186, 223
nitrogen 163, 194, 199, 200, 234, 239
nutritive value 201, 216
orchardgrass 163
oversown pastures 245
Paspalum notatum 193
pasture 119, 162, 164, 175, 177
pasture based-system 131, 195, 196
pasture feeding 194
pasture intake 89, 183
pasture Mg 244
pasture plants 252
pasture quality 49, 208
pasture renovation 171
pasture utilisation 181, 243
perennial ryegrass 151, 158, 161, 166
performance 153, 175, 251
persistence 156, 202
Phleum pratense 201
phosphorus 243, 245
photosynthetically active radiation 199, 200
Pithomyces chartarum 161
planning 157
plant density 227
plant maturity 252
plant testing 159
ploidy 166
prairie grass 168
problem solving 210
protein 176
radiation use efficiency 199, 200
rangeland health 209
registration 159
replacement series 181
reproductive performance 182
rhizomes 245
root reserves 226
rotational stocking 151
rumen degradability 166
rumen fill 251
ruminating time 184
ryegrass 199, 200
seed dispersal 248
seed mixture 219
seed movement 248
seed shadow 248
seeding rate 219
set stocking 240
sheep 208, 212
shrubland 191
simulation 79, 190, 203, 210, 216, 217
simulation modelling 195, 196
sod seeding 171
soil Olsen phosphorus 243
stem 227
stocking rate 152, 165, 178, 179, 180, 240
strategy optimisation 237
subterranean clover 211
suckler cow 155
supplementary feeds 194
supplementation 233
survival analysis 186
sward height 37, 178
systems 19, 192
tactical management 157, 208
tall fescue 168
temperate pasture 49
tetraploid 161
tissue N concentration 193
tissue turnover 197
Trifolium ambiguum 167
tropical pasture 172
tropical perennial grass 193
turnip 229, 246
utilisation 238

varieties 151, 158
vegetation dynamics 190
vegetative 203
virtual plants 202
visual models 202
volatile fatty acids 160
voluntary intake 185
water quality 209
water-soluble carbohydrates 160
wheat 211
white clover 167, 231
whole crop wheat 230
winter 197
wool industry 210
yield 163, 229

Author index

Abaye, A.O. 168
Abrahamse, P.A. 198
Agnew, R.E. 158
Alomar, D. 253
Améndola-Massiotti, R.D. 180, 181
Anderson, B.E. 170
Angus, J.F. 211
Ayala, W. 245
Baram, H. 154
Barrett, P.D. 79, 213, 238
Barry, S.J. 209
Bartholomew, P.W. 248
Baumont, R. 190
Bawm, S. 240
Ben-Moshe, E. 154
Berg, B.P. 219
Bermúdez, R.E. 245
Berntsen, J. 218
Berretta, E.J. 179
Berry, D.P. 186
Beukes, P.C. 189
Black, A.D. 167
Blaettler, T. 237
Blum, J.W. 177
Boland, H.T. 168, 250
Boote, K.J. 193
Boudon, A. 184, 251
Brenner, S. 154
Brereton, A.J. 225
Brookes, I.M. 157, 194, 195, 196, 208
Bryant, J.R. 223
Brzezinski, N. 162
Buckley, F. 61
Cabiddu, A. 191
Camesasca, M. 178
Carson, A.F. 19
Castillo, A.R. 253
Champion, R.A. 151
Chardon, X. 189
Chaves, A.V. 184, 251
Clark, D.A. 119
Cohen, R.D.H. 105
Cook, J.E. 151
Cornu, A. 252
Crosse, S. 19
Crosson, P. 192
Curnow, M. 212
Daily, H.G. 210
Dale, A.J. 238
Dawson, S. 230
De Barbieri, I. 179
Decandia, M. 191
Defrance, P. 214, 215, 232
Delaby, L. 152, 165, 185, 207, 214, 215, 232
Delagarde, R. 89, 184, 185
Dighiero, A. 179
Dijkstra, J. 198
Dillon, P. 131, 182, 236
Donnelly, J.R. 105, 216
Doran, M. 209
Dove, H. 216
Doyle, P.T. 37
Drennan, M.J. 19, 155, 192
Durgiai, B. 237
Eerens, J.P.J. 161, 246
Elgersma, A. 160
Ellen, G. 160
Elwert, C. 249
Endo, T. 240
Faverdin, P. 207
Fearon, A.M. 158
Ferrayoli, C.G. 253
Ferris, C.P. 230
Flanagan, J. 15
Fletcher, A.L. 226
Fontenot, J.P. 168, 250
French, P. 197
Galbraith, J.M. 217
Genizi, A. 154
Gilliland, T.J. 158, 159
Glassey, C.B. 175
Gleeson, D. 247
Gomez, M. 201
González-Montagna, S.J.C. 180, 181
Goslee, S. 162
Gray, D.I. 157, 208
Groover, G.E. 217
Guenther, K. 209
Guinee, T. 233
Haerdter, R. 244
Harris, B.L. 186
Hayes, G. 209
Hennessy, D. 197
Hodgson, J. 224
Hoekstra, N.J. 239
Hofer, C. 235
Hofstetter, P. 177
Holden, N.M. 225
Holmes, C.W. 61, 175, 183, 223, 257
Horan, B. 131, 182
Horne, D.J. 157, 194
Hurley, G. 165
Hutton, G. 219
Hyder, M. 212
Jamieson, P.D. 163
Jeangros, B. 235
Jiménez-Guillen, R. 169, 172
Jones, J.W. 193
Jouven, M. 190
Keane, M.G. 61
Kemp, P.D. 156, 157
Kennedy, E. 152, 233
Kenyon, P.R. 157, 195, 196
Kinsella, J. 247
Kohler, S. 177, 229, 237
Kondjoyan, N. 252

Kondo, S. 240
Krishna, H. 156
Kunz, P. 177, 237
Kyne, S. 155
Laberge, G. 171
Laidlaw, A.S. 79, 197, 213
Lambert, R. 199, 200, 201
Lancaster, J.A.S. 175
Lane, P.M.S. 246
Lantinga, E.A. 239
Larson, R. 209
Levy, G. 189
Liu, ZL. 105
Lobsiger, M. 235
Lopez-Guerrero, I. 250
López-Villalobos, N. 223
Lucas, R.J. 163, 167
Luginbuhl, J-M. 153
Lüscher, A. 235
Macdonald, K.A. 175, 183
Martin, B. 252
Martínez Ferrer, J. 253
Martínez-Hernández, A. 169, 172
Martínez-Hernández, P.A. 180, 181
Matthew, C. 157
Mayne, C.S. 238
McGee, M. 155
McKenzie, B.A. 163
Mederos, A. 179
Mee, J.F. 182
Mellsop, J.M. 161
Menzi, H. 228, 237
Mills, A. 163
Mitani, T. 240
Molle, G. 191
Møller, K. 231
Montgomerie, W. 186
Montossi, F. 178, 179
Moore, A.D. 211
Moot, D.J. 163, 167, 226, 227
Morel, P.C.H. 195, 196
Morris, S.T. 157, 224
Morrison, S.J. 230
Morton, J.D. 243
Mosimann, E. 235
Mueller, J.P. 153
Mueller, R. 237
Muendermann, L. 202
Muller, L.D. 162, 164
Mulligan, F.J. 166
Munger, A. 177
Murphy, J.J. 166, 233, 236
Murphy, J.P. 152
Murphy, M. 259
Nader, G. 209
Nakatsuji, H. 240
Nichol, W.W. 161
Nielsen, K.A. 218
Nolla, M. 178
Norman, H.D. 176
Norriss, M.G. 161
O'Brien, B. 247
O'Connor, M.B. 244
O'Connor, P. 182
O'Donovan, K. 247
O'Donovan, M. 89, 152, 165, 166, 197, 215, 232, 233
O'Kiely, P. 192
O'Mara, F.P. 152, 166, 192
O'Neill, B. 155
Olivares-Pérez, J. 169, 172
Olsson, V. 166
Orr, R.J. 151
Palliser, C.C. 189
Patterson, D.C. 230
Peeters, A. 199, 200, 201
Pérez-Pérez, J. 169
Petermann, R. 177
Peterson, P.R. 171
Petit, M. 252
Peyraud, J.L. 184, 185, 207
Pinna, G. 191
Powell, R.L. 176
Pradel, P. 252
Preve, F. 178
Pryce, J.E. 175, 223
Purchas, R.W. 195, 196
Ramaswami, S. 195, 196
Rath, M. 182
Ravid, N. 154
Rearte, D.H. 49
Reid, J.I. 157
Reid, J.M. 210
Reynaud, D. 198
Ribeiro Filho, H.M.N. 185
Roberts, A.H.C. 244
Roche, J.R. 131, 260
Rodehutscord, M. 249
Rojas-Hernández, S. 169, 172
Romera, A.J. 224
Rook, A.J. 151
Rossi, J.L. 183
Ruane, D. 247
Rymph, S.J. 193
Salamanca, M.E. 201
Salmon, L. 105, 211
San Julián, R. 178
Sanderson, M.A. 162, 164
Scaglia, G. 168, 250
Schacht, W.H. 170
Schulte, R.P.O. 203, 239
Scott, J.M 210
Seguin, P. 171
Seuret, J.M. 214, 215
Shalloo, L. 131, 236
Shanklin, R.K. 250
Sheaffer, C.C. 171
Singh, A. 202
Skinner, H. 162
Smit, H.J. 160

Smith Jr., S.R. 202
Smith, S.R. 168, 217
Soder, K.J. 162, 164
Søegaard, K. 218, 231
Stack, J.L. 164
Staehli, R. 237
Stakelum, G. 165
Steiger Burgos, M. 177
Stettler, M. 234
Stirling, W.D. 224
Stockdale, C.R. 37
Stone, N.D. 217
Struik, P.C. 239
Takahashi, M. 240
Tamminga, S. 160, 198
Taube, F. 162
Taylor, W. 261
Teixeira, E.I. 226, 227
Thomet, P. 177, 228, 229, 234, 237
Thorrold, B.S. 175, 183, 189
Thysen, I. 218
Tornambé, G. 252
Trolove, M.R. 161
Ueda, K. 240
Ungar, E.D. 154
Ustarroz, E. 253
Valentine, I. 157
van Wijk, A.J.P. 159
Venuto Jr, J.J 217
Verkerk, G.A. 175
Wachendorf, M. 162
Wales, W.J. 37
Wallace, M. 192
Waller, J. 161
Wastney, M.E. 189
Wildbore, B. 195, 196
Williams, R.D. 248
Wilson, F.E.A. 158
Winkelman, A.M. 186
Woodward, S.J.R. 224
Wright, J.R. 176
Xin, XP. 105
Yonatan, R. 154
Young, K.L. 151
Zada, T. 154